Digital System Test and Testable Design

Zainalabedin Navabi

Digital System Test
and Testable Design

Using HDL Models and Architectures

 Springer

Zainalabedin Navabi
Worcester Polytechnic Institute
Department of Electrical & Computer
Engineering
Worcester, MA
USA
navabi@ece.wpi.edu

ISBN 978-1-4899-7927-8   ISBN 978-1-4419-7548-5   (eBook)
DOI 10.1007/978-1-4419-7548-5
Springer New York Dordrecht Heidelberg London

Printed on acid-free paper

Springer is part of Springer Science+Business Media (www.springer.com)

This book is dedicated to my wife, Irma, and sons Aarash and Arvand.

Preface

This is a book on test and testability of digital circuits in which test is spoken in the language of design. In this book, the concepts of testing and testability are treated together with digital design practices and methodologies. We show how testing digital circuits designing testable circuits can take advantage of some of the well-established RT-level design and verification methodologies and tools. The book uses Verilog models and testbenches for implementing and explaining fault simulation and test generation algorithms. In the testability part, it describes various scan and BIST methods in Verilog and uses Verilog testbenches as virtual testers to examine and evaluate these testability methods. In designing testable circuits, we use Verilog testbenches to evaluate, and thus improve testability of a design.

The first part of the book develops Verilog test environments that can perform gate-level fault simulation and test generation. This part uses Verilog PLI along with Verilog's powerful testbench development facilities for modeling hardware and programing test environments. The second part of the book uses Verilog as a hardware design tool for describing DFT and BIST hardware. In this part, Verilog is used as a hardware description language describing synthesizable testable hardware. Throughout the book, Verilog simulation helps developing and evaluating test methods and testability hardware constructs.

This book professes a new approach to teaching test. Use of Verilog and Verilog PLI for test applications is what distinguishes this book from other test and testability books. As HDLs were used in late 1970s for teaching computer architectures, today, HDLs can be used to illustrate test methodologies and testability architectures that are otherwise illustrated informally by flow charts, graphs, and block diagrams. Verilog eliminates ambiguities in test algorithms and BIST and DFT hardware architectures, and it clearly describes the architecture of the testability hardware and its test sessions. Describing on-chip test hardware in Verilog helps evaluating the related algorithms in terms of hardware overhead and timing and thus feasibility of using them on SoC chips. Further support for this approach comes in use of testbenches. Using PLI in developing testbenches and virtual testers gives us a powerful programing tool interfaced with hardware described in Verilog. This mixed hardware/software environment facilitates the description of complex test programs and test strategies.

Acknowledgments

When I first thought of using a hardware description language for test purposes, I started using VHDL models for test purposes in my course on digital system testing at the University of Tehran. After several years of teaching this course, we switched to Verilog and a set of library components that facilitated this usage of Verilog was developed. The groups of students who developed the software and helped me in the formation of the materials are important contributors to this work. The student, who took the responsibility for the development of the software package was Nastaran Nemati. She managed the development of the complete library by the time of her graduation in 2010. Her efforts contributed significantly to this work. I thank my students at Worcester Polytechnic Institute in Massachusetts, USA, and the University of Tehran for sitting at my presentations or watching them online and making useful suggestions.

When the actual development of the book started, my graduate student, Fatemeh (Negin) Javaheri, became the key person with whom I discussed my ideas. She was always available for consulting with me and her ideas helped significantly in shaping the structure of the book. She later took responsibility for developing the material for the chapter on test compression. Negin continues to work with me on my research, and she is looking forward to the next book that I want to write. Another important contributor, also a graduate student at the University of Tehran, is Somayeh Sadeghi Kohan. Somayeh developed the materials for the chapter on boundary scan, and in the final stages of this work she was very helpful reviewing chapters and suggesting changes. The feedbacks she provided and changes she suggested were most helpful. Nastaran Nemati helped developing the HDL chapter, and Parisa Kabiri and Atie Lotfi also contributed to some of the chapters and helped reviewing the materials.

As always, and as it is with all my books, Fatemeh Asgari, who has been my assistant for the past 20 years, became responsible for managing the project. She managed the group of students who did research, developed software, collected materials, and prepared the final manuscript with its text and artwork. Fatemeh's support and management of my writing and research projects have always been key to the successful completion of these projects. I cannot thank her enough for the work she has done for me throughout the years.

My work habits and time I spend away from my family working on my research and writing projects have been particularly difficult for them. However, I have always had their support, understanding, and encouragement for all my projects. My wife, Irma, has always been a great help for me providing an environment that I can spend hours and hours on my writing projects. I thank Irma, and my sons Aarash and Arvand.

August 2010

Zainalabedin Navabi
navabi@ece.wpi.edu

Contents

Introduction

The main focus of the book is on digital systems test and design testability. The book uses Verilog for design, test analysis, and testability of digital systems. In the first chapter, we discuss the basics of test and testable design while discussing the aspects of test that hardware description languages can be used for. This part discusses the entire digital system testing in terms of test methods, testability methods, and testing methods, and it discusses the use of HDLs in each aspect. After the introductory parts, we have included a chapter on the basics of Verilog and using this language for design and test. The body of the book assumes this basic Verilog-based RT-level design knowledge and builds test and simulation concepts upon that.

Starting in Chap. 3, the focus of the book turns to test issues, such as fault collapsing, fault simulation, and test generation. This part that is regarded as covering "test methods" has four chapters that start with the presentation of fault models, followed by fault simulation methods, and then two chapters on test generation, discussing random HDL-based test generation and deterministic test. For such applications, we use Verilog gate models and PLI-based testbenches that are capable of injecting faults and performing fault simulation and test generation. In this part, Verilog testbenches act as test programs for managing the structural model of a circuit for performing fault simulation, calculating fault coverage, and test pattern generation.

The testability part of the book begins in Chap. 7. There are four chapters in this part (Chaps. 7–10) in which various testability methods, built-in self-test architectures, and test compression methods are discussed. We use Verilog coding for describing the hardware of various testability methods and BIST architectures. Verilog testbenches, in this part, act as virtual testers examining testability and test hardware embedded in a design. Together, PLI-based fault simulation and test generation Verilog environments of Chaps. 3–6, and Verilog coding of testability hardware provide a complete environment for testability and BIST evaluation. Design and refinement of test hardware can be achieved after such evaluations. This part also discusses IEEE standards for boundary scan and core testing, and uses Verilog for describing these standards and using them in designs. The last chapter is on memory testing with a focus on MBIST that we describe in Verilog.

Chapters

Basics of Test and Role of HDLs

Basics of test are covered in this chapter. We talk about the importance of digital system testing and define various test terminologies. Economy of test is discussed and reducing test time by means of better test methods, more testable designs, and more efficient testing is discussed. Relation between design and test are discussed in this chapter.

Verilog HDL for Design and Test

This chapter talks about the Verilog hardware description language for the description of digital systems and the corresponding testbenches. We discuss combinational and sequential circuit modeling and present several examples. Only the key language constructs that is needed for understanding models and architectures in the rest of the book are presented here. This chapter shows the use of Verilog for developing good design testbenches. Several templates for testbench development are discussed. In doing so, the use of PLI and developing PLI functions are presented. The testbench part is extended in the chapters that follow.

Fault and Defect Modeling

Transistor and gate-level faults are described first. Verilog simulations show the correspondence between lower-level transistor faults and upper level gate faults. We discuss functional and structural faults and the distinction between them. Structural gate-level faults are discussed and the justification of this model is illustrated by the use of simulations. We elaborate on the stuck-at fault model and show PLI functions for fault injection. After this, the chapter discusses fault equivalence and several fault collapsing techniques. We develop Verilog and PLI functions and testbenches for generating fault lists and fault collapsing. Several benchmark circuits are tested with these testbenches.

Fault Simulation Applications and Methods

The chapter begins discussing the use of fault simulation and its applications in design and test. We then discuss various fault simulation techniques, including serial, parallel, concurrent, deductive, differential, and critical path tracing fault simulation. For several of these methods, we develop a testbench that injects stuck-at faults and performs simulations. Verilog PLI-based testbenches for partial implementation of other fault simulation techniques are also discussed. Fault dictionaries are discussed and created using Verilog and PLI testbenches. Using these utilities, we also discuss and implement test coverage, fault dropping, and other fault simulation-related concepts. The format of fault lists is taken from the previous chapter in which Verilog and PLI testbenches generated such lists. Several complete Verilog testbenches with PLI are developed and utilized in this chapter.

Test Pattern Generation Methods

This chapter begins with the presentation of various testability techniques, including probability based, structural, and SCOPE parameter calculation. PLI functions in Verilog testbenches are developed for calculating controllability and observability parameters of internal nodes of gate-level circuits. Detectability and its role in the determination of random tests are also discussed. After this first part, we discuss various random test generation methods and take advantage of testability measures of the first part. This chapter uses Verilog testbenches for generating random tests and evaluating them by Verilog-based fault simulation.

Deterministic Test Generation

We started the presentation of test generation in Chap. 5 with presenting random test generation. This chapter discusses deterministic test generation, which we consider Phase 2 of the test generation process. We discuss algorithms like, the D-algorithm, PODEM, CPT, and some of the simplified and derivatives of these algorithms. Verilog testbenches using Verilog PLI functions for deciding when to stop random test generation and when to start deterministic TG are developed. Test compaction can be regarded as the next phase of test generation. A part of this chapter is dedicated to this topic and several test compaction methods and their Verilog implementations are discussed.

Design for Test by Means of Scan

In this and the chapters that follow, in addition to using Verilog for developing test environments, Verilog is also used for describing actual hardware constructs. In this chapter on DFT, we show synthesizable Verilog codes for the DFT architectures that we present. The chapter begins with the presentation of several ad hoc design-for-test techniques. We then show full-scan and various partial-scan architectures, and for the purpose of unambiguous description of such hardware structures, their corresponding Verilog codes are shown. Test methods and testbenches that we developed in the previous chapters of this book are utilized here for scan design evaluations and refinements. We show how a testbench can be used for helping us configure a scan design and generate tests for it. We also show how a Verilog PLI testbench can be used for the application and testing of a scan-based design. This latter application of testbenches is what we refer to as a virtual tester.

Standard IEEE Test Access Methods

This chapter discusses IEEE 1149.1 test standards. Hardware structures corresponding to these standards are discussed in Verilog. Virtual testers that operate board and core testing hardware are described as Verilog testbenches. Through the use of Verilog, we are able to show the architecture and utilization of these standards. Interfacing between various components of IEEE Std.1149.1 and how the standard interacts with the circuit under test, on one side, and the test equipment, on the other side, are clarified here by the use of Verilog hardware descriptions and testbenches.

Logic Built-in Self-test

This chapter starts with the methods of designing on-chip test data generation and output analysis. We then incorporate these components in built-in self-test architectures for on-chip testing. In this chapter, we show synthesizable Verilog codes for all BIST architectures that we present. We show classical BIST architectures, such as RTS, BILBO, and BEST; furthermore, on-line BIST, concurrent BIST, and BISTs for special architectures are discussed. We use Verilog for unambiguous description of such hardware structures. And, we use Verilog and PLI testbenches, which we developed for fault injection and fault simulation, for BIST evaluation and configuration. Determination of BIST test sessions for a better fault coverage and the determination of the corresponding signatures are done by the use of Verilog simulations.

Test Data Compression

This chapter discusses test compression techniques, their usage in design-for-test, and their corresponding hardware implementations. We discuss Huffman, Run-length, Golomb, and other coding techniques used in test compression. In addition, scan compression techniques and their corresponding on-chip scan structures are discussed. Compression algorithms are discussed and hardware for decompression hardware as it is placed on a chip is described in Verilog.

Memory Testing by Means of Memory BIST

This chapter begins with a presentation of memory structures and corresponding fault models. Various March test techniques are described and an analysis justifying various test algorithms is given. We then discuss several memory BIST architectures and show their corresponding Verilog descriptions. Operating an MBIST is demonstrated by means of a Verilog testbench.

Appendixes

A. Using HDLs for Protocol Aware ATEs
B. Gate Components for PLI Test Applications
C. PLI Test Utilities
D. IEEE Std.1149.1 Boundary Scan Verilog Description
E. Boundary Scan IEEE Std.1149.1 Virtual Tester
F. Generating Netlist by RTL Synthesis (*NetlistGen*)

Software and Course Materials

The material for this book has been developed while teaching courses on test and testability at several universities. We are making these materials available.

We have developed a set of Verilog PLI functions for test development that have been used in several courses, and have reached an acceptable maturity. For test applications using original RT-level designs, a software program for a netlist generation has been developed that uses Xilinx ISE to synthesize and convert the output to a netlist that our PLI functions can use. Many of the netlists of the examples used in this book have been generated by this software. Applications that can be performed with the set of our PLI functions and netlist include fault collapsing, random test generation, fault simulation, and testability measurements. Also, virtual testers discussed in this book use the netlist and PLI functions for simulation.

Presentation materials for all the chapters are also available (PowerPoint slides). In addition, when teaching this course, I have noticed that students sometimes need a review of Verilog or logic design concepts. I have developed short videos that students can use to get ready for the material presented in the book. Videos and manuals for the use of software are also available and can be obtained from the author.

A list of materials that are available at the time of publication of this book is shown below. Other materials and software programs are being developed. Interested readers and users can contact the author for obtaining the updates.

- PowerPoint presentation slides for the chapters (PowerPoint files).
- The complete Verilog PLI Test package (Software).
- Netlist generator from RTL descriptions (Software).
- Design and testbench files used the book (Verilog).
- Verilog tutorials (Video).
- Logic design tutorials (Video).
- Problem sets, exams, and projects (PDF files).

Author's email: navabi@ece.wpi.edu; zain@navabi.com

Chapter 1
Basics of Test and Role of HDLs

As the first chapter of a book on digital system test, this chapter tries to cover some of the basics. Also, as the first chapter of a book that emphasizes the use of HDLs in test, we try to cover HDL-based design flow and discuss how various test applications fit in this process, and we show various places where HDLs can help the test process of digital systems. In this chapter we try to answer some of the important questions about digital system testing. The primary questions that need to be answered are what it is that we are testing in digital system test, and why we are testing it. The answer to these questions are: A manufactured chip or device, which will be elaborated in the sections that follows, and we are testing the device for physical defects. Other questions that remain to be answered are regarding the methods that we use for testing, ways of making a chip or a manufactured device more testable, how HDLs can help the test process, and finally what constraints we have in digital system testing.

After this chapter, it will be easy to justify why we dedicate at least three chapters to test methods, three chapters to making circuits testable, and in all the chapters we continuously talk about reducing test data volume and test time. Furthermore, the use of HDLs in the chapters of this book is justified by the HDL discussions of this chapter.

The next section discusses HDL-based design and where testing fits in this process. The discussion of test concerns and test methods and testability methods to address such concerns come next in four subsections. Section 1.3 discusses the role that HDLs can have in the design and test process. The roles described here are exercised thought this book in the later chapters. Section 1.4 of this chapter discusses test equipment that plays a very important role in shaping test and testability techniques.

1.1 Design and Test

Producing a digital system begins with a designer specifying his or her design in a high-level design language, and ends with manufacturing and shipping parts to the customer. This process involves many simulations, synthesis, and test phases that are described here.

1.1.1 RTL Design Process

In a register transfer level (RTL) design process, the designer first writes his or her design specification in an RT level language such as Verilog. As shown in Fig. 1.1, this description uses Verilog high-level constructs such as an **always** statement. Using standard HDL (Hardware Description Language) descriptions and testbenches in the same HDL, this description will be simulated and tested for design errors [1, 13–15].

Z. Navabi, *Digital System Test and Testable Design: Using HDL Models and Architectures*, DOI 10.1007/978-1-4419-7548-5_1, © Springer Science+Business Media, LLC 2011

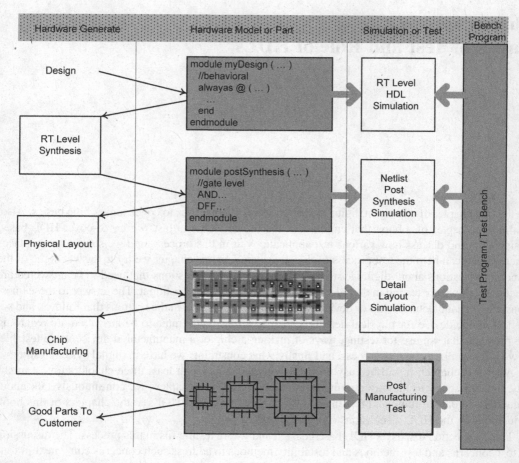

Fig. 1.1 Design and test

1.1.1.1 RTL Simulation

The HDL model input of the RT level simulator of Fig. 1.1 is taken from the original problem description, and consists of synthesizable module interconnections in a top level Verilog model. The testbench shown is originally developed for checking the functionality of the design. The testing of the design here is primarily functional that is extracted from the original specification of the circuit being designed. Detailed timing checks and physical flaws are not addressed at this level of simulation.

For analyzing the behavior of the design, the testbench can inject design errors to predict the behavior of the design under unanticipated circumstances. Furthermore, the testbench can be made to issue warnings when and if it detects that the design's behavior contradicts the expected functionality. Verification and assertion-based verification methods are useful in analyzing the design and checking its performance against the specifications. Various verification methods are either part of an HDL simulator, or they are used as standalone programs.

After a satisfactory simulation, and when the designer is reasonably sure that his or her description of the design meets the design specifications, the next step, that is RT level synthesis, is taken.

1.1.1.2 RT Level Synthesis

As shown in Fig. 1.1, RT level synthesis takes the behavioral description of a design (*myDesign*) as input, and produces a netlist of the design. The netlist, shown here by *postSynthesis* module, specifies interconnection of low-level basic logic components such as AND and OR gates and D-type flip-flops. The exact set of gates used in this level of description depends on the target library, which is the library of components provided by the chip manufacturers. The format for the netlist can be specified to be the same HDL as the original design.

Before going to the next step, this netlist must be tested. The testing here is done by simulating it with an HDL simulation tool. This simulation phase is referred to as postsynthesis simulation. With this simulation we are checking for delay issues, races, clock speed, and errors caused by misinterpretation of the RT level design by the synthesis tool [1, 14].

In general, the same testbench used for testing the RT level design can be used for testing the postsynthesis netlist. The majority of test vectors developed for the former can be migrated for testing the netlist. In some instances, new test vectors may be needed for checking corner cases where timing issues become important. For making sure that the postsynthesis netlist conforms to the presynthesis HDL description, a testbench in the same HDL can instantiate both descriptions and simulate them simultaneously with the same test data. Looking at the right-hand side of Fig. 1.1, it is important to note that, ideally, the same platform should be used for testing outcomes of various design steps.

An alternative to simulation is the formal method of verification. Equivalence checkers are of this category. An equivalence checker, without using any test data (thus, static), generates pre- and postsynthesis models of the circuit and uses formal methods to prove that they are the same. After completion of postsynthesis verification (by simulation, or formal, or a mixture of both), the next step that takes the design one step closer to a final product is performing physical layout.

1.1.1.3 Physical Layout

As shown in Fig. 1.1 (left-hand side column), the verified postsynthesis netlist of the design is used by a layout and placement tool for generating the design's layout and routing of cells. As in the previous stages, the output of this phase also needs to be tested and verified for correctness.

The simulation here verifies wire lengths, wire widths, and transistor sizes, and detects layout and placement flaws that can be introduced from a netlist to its layout. After a successful completion of simulation, the layout will be ready for manufacturing.

1.1.1.4 Chip Manufacturing

After obtaining a working (simulated and tested) layout, it is used for the final step of manufacturing. As shown in Fig. 1.1, the manufactured chips must be individually tested and verified against flaws before they are shipped to the customers. The testing phase for testing the actual, postmanufactured, chips is what is referred to as *test* or *testing* in digital system design industry [2]. Unlike the other steps that testing is done on the model of the design (i.e., simulation), testing in the last step is done on the physical part.

In this testing, which is the focus of this book, we are testing the circuit under test (CUT) against manufacturing defects. The defects are broken wires, shorts, open resistive wires, transistor defects, and other physical problems that affect the functionality of a manufactured part.

In the simulations shown in Fig. 1.1 (third column from the left) unsuccessful test results require modifications in the design or changes in the synthesis or hardware generation process (first column from the left). For example, if postsynthesis simulation results do not conform to the original design specification, the hardware generation phases above it, i.e., design phase and RT level synthesis, must be modified and rerun. This is different if the last test (postmanufacturing test) fails. In most cases, the device that has failed the test is discarded and testing continues. In very rare cases, when all manufactured devices fail, the hard decision of going to the first design step is taken.

As opposed to the other three forms of testing in Fig. 1.1 that a software program (a simulator) performs the testing of the model, in manufacturing test, a physical device, which is a hardware component or a test equipment, performs the testing. However, the expected response and the testing procedure are fully or partially based on the lessons learned from the testings that have been performed in the earlier stages of design. Ideally, the same test platform used in the three boxes above the postmanufacturing box in Fig. 1.1 should be translated to a test program that runs on a test equipment for testing the finished part.

1.1.2 Postmanufacturing Test

The four boxes in Fig. 1.1 that designate testing and simulation have similar properties. However, the box at the bottom that represents postmanufacturing test has certain characteristics that makes it conceptually different than the other three, and thus needs special attention. In this section we discuss characteristics of this testing that puts it apart from other forms of testing (simulation) that we discussed.

In digital systems, testing is referred to the exercise of checking a part or a model to see if it behaves differently than its specification. What distinguishes between various testings are what is being tested, how is test data obtained, what it is being tested with, what procedure we use for testing it, and what we do with the test results. These are some of the issues that will be discussed next.

1.1.2.1 Device and Its Test Data

In digital system testing, device being tested can be a system, a board, a packaged chip, a chip, a die on a wafer, a core on a die, or a section of a core. Regardless of what it is that is being tested, it is treated as a closed box that can only be controlled and observed from the outside. Figure 1.2 shows what can be regarded as a chip or a core that is the part that is being tested.

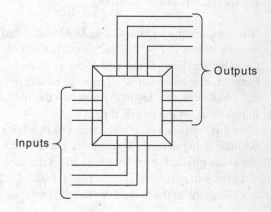

Fig. 1.2 Part to be tested

The key here is that once the part that is being tested is configured for a certain set of inputs and outputs (actual or virtual), we have no more access to the inside of the part, neither to control, nor to observe.

The device configured as such is referred to as device under test (DUT), circuit under test (CUT), or other similar acronyms. An input that is applied to a DUT is referred to as *test vector* or *test pattern*, the set of all inputs for testing a DUT is *test set*. *Stored response* refers to expected response from a test set that is stored on a disk or memory prior to test taking place.

A test set is prepared ahead of time by the *test-generation* process. Test generation is done using a model of DUT (upper three boxes in model column of Fig. 1.1). Functional test generation, where test vectors are made to examine various functions of a DUT, uses the RT level or behavioral model of the DUT (first box in model column in Fig. 1.1). Structural test generation, where test vectors are made to examine interconnections within a DUT, uses the netlist model of the DUT (second box in model column in Fig. 1.1).

The expected response for a test set that can be saved for stored response testing is prepared by simulating a working model of DUT (first two boxes in model column in Fig. 1.1). The model from which the expected response of a circuit is obtained is called *good circuit model* or *golden model*.

1.1.2.2 Testers

Now that we know the nature of device being tested, and data that is used for testing it, the next topic in postmanufacturing test puzzle is the device or equipment that is used for testing our DUT. Just as a simulation program on a computer is used for running simulations and testing a model, a test program running on a special test computer can be used for testing a DUT. However, for testing there are other options that we will discuss.

Whichever option we take, a tester is a device or equipment that applies test vectors to a DUT, collects DUT's responses, and makes comparisons with the expected data. Figure 1.3 shows a tester testing a DUT. The tester only uses inputs and outputs of the DUT.

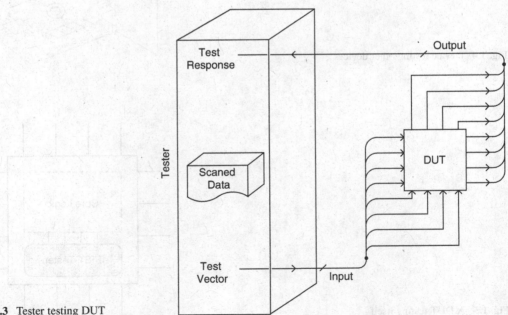

Fig. 1.3 Tester testing DUT

A DUT can be tested by a device especially made and designed for testing it. An example of this type of tester is a test board on a system that takes responsibility for testing other system boards and chips mounted on the boards.

Figure 1.4 shows an example of a tester board that communicates through a test bus with other boards in the system for testing them. The test board consists of storage, memory, processors, and communication busses. A device testing another device can also be a chip testing another chip on the same board. Obviously in this case, test data and response that can be used to test the DUT are limited.

A DUT can test itself. Without requiring an external chip or device, it can be tested by a built-in hardware that has been primarily designed for testing the rest of the hardware of the DUT. This kind of testing is depicted in Fig. 1.5, and is called Built-in Self-test (BIST).

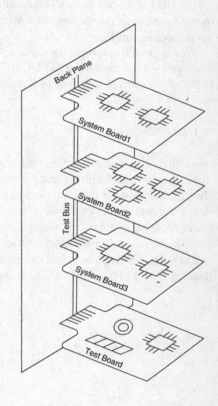

Fig. 1.4 Device testing other devices

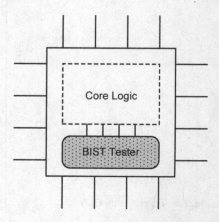

Fig. 1.5 A DUT testing itself

Fig. 1.6 UltraFLEX tester
docked to wafer prober (courtesy
of Teradyne)

Another implementation of Fig. 1.3 is using an automatic test equipment (ATE). An ATE is an equipment that consists of processors, storage, and fixtures for mounting devices. ATEs are used for die testing, wafer testing, or testing a packaged product. An ATE runs a test program that implements the specific procedure designed for testing the DUT.

Most ATEs have *multisite testing* capability that can test multiple devices in parallel. Also *concurrent testing* in an ATE reduces test time by testing various parts of a DUT concurrently. Figure 1.6 shows Teradyne UltraFLEX ATE (in the back) that is docked to a wafer prober that fetches wafers from a stack of wafers for testing. Section 1.4 in this chapter discusses ATEs and their capabilities in more detail.

1.1.2.3 Using Test Results

As discussed, the purpose of test is to find the manufacturing defects and assumes that the design is sound. The manufacturing defects can either cause catastrophic failure or parametric failure. In either case, the test results can impact the product itself, the design process, or the implementation of the design.

Devices having catastrophic failures are always discarded. However, devices with parametric failures that cannot meet the ideal performance specifications can be degraded according to the specifications that they can meet, and sold as cheaper products. For example, a processor chip that does not pass the original frequency test, but passes the 10% reduced frequency test can still be sold at a lower price in a market that does not require the 100% performance. Test results can also be used to identify functionalities whose failures do not necessarily make a device unusable. For example, if the arithmetic accelerator of a processor fails, the processor can still be used by using the software for arithmetic operations. In this case, the result of the test is used for the elimination of certain functionality and modifying the specification of the device.

Test results may also impact the design and implementation process. For example, the cause of catastrophic or parametric failures of an unusual number of devices may be due to problems in layout and mask preparations. In such cases, diagnostic must be done to find the source of failure and correct the implementation process accordingly. Failures in test can also be causes of modifications in design for a better reliability or performance.

1.1.2.4 Types of Tests

Depending on the device being tested, the equipment testing it, and the purpose of test, various types of testing are performed. Below, some commonly used terminologies are discussed.

External Testing. A device is tested by an external device that can be a chip, a board, or a computer or test equipment.

Internal Testing. The tester for a device is in the same packaging as the device. Often, in case of BIST, the tester hardware is integrated with, and on the same chip as the device.

Online Testing. Testing is done while a device is performing its normal functions.

Offline Testing. Device being tested must cease its normal operation, and then be tested. Offline testing can be done by internal or external test hardware.

Concurrent Testing (*Online*). In online testing, concurrent testing is when the normal data the device is using in performing its normal functions are used for testing it.

Concurrent Testing (*ATE*). In ATE terminology, concurrent testing is when a tester is testing various parts of a chip concurrently. For example, the analog, memory, and the logic parts tested at the same time, while the device is on the tester head.

At-Speed Testing. Device being tested at its normal speed of operation [3]. This is also called *AC Testing*.

DC Testing. Device is tested at much slower speed than its operation frequency. This allows all the events to propagate before the outputs are sampled.

In-Circuit Testing. Device being tested is not removed from its mounting place for testing.

Guided Probe Testing. In a process of probing backwards from outputs towards inputs, testing is done to find the source of an error that has appeared on the circuit's outputs.

Diagnostic. Diagnostic is when testing is done to find the cause of failure.

There are many other terminologies used for various types of testing that can be told by the context they are used in. As the above terminologies are not standard, they may be used in the literature for slightly different meanings than what we have presented.

1.2 Test Concerns

The main concern of testing a digital system or a device is to test it as thoroughly as possible, and in as little time as possible. With the number of transistors doubling every 24 months (Moore's law 1965), we already have chips with billions of transistors, and this number continues to grow. Testing devices with this number of transistors quickly and thoroughly is a very challenging task and requires proper strategies and systematic test approaches for generating test, making devices testable, and using testers. This section discusses *test methods*, *testability methods*, and *testing methods* that try to reduce the complexity of electronic testing.

1.2.1 Test Methods

Test methods are algorithms and methodologies that lead to generating tests that can quickly and accurately identify defective parts. To understand why such a requirement that seems so basic presents such a big test challenge, consider the circuit in Fig. 1.7 that is being tested.

Suppose that the CUT shown here is a combinational circuit with 64 inputs, 64 outputs, and 12 ns internal delay. Also let us say that we are testing this circuit with a tester running at 1 GHz clock frequency, and it takes four clock cycles (4 ns) to fetch a new test vector and apply it to the circuit. Also let us say that at this tester frequency, it takes four clock cycles to lookup the output and compare it with the expected result.

Considering that we have 64 inputs, the round of testing described above must be repeated 2^{64} times for all input combinations. Adding the test times with the circuit's internal delay, and multiplying it by the number of input combinations results in the total test time for this circuit as below:

$$\text{Test time}: 2^{64} \times (12 + 4 + 4) \times 10^{-9}\,\text{s}$$

According to this, the CUT in Fig. 1.7 can be completely tested in 11,698 years. The situation gets worse if we are dealing with sequential circuits in which internal states of the circuit must be considered for exercising all circuit structures. Since this way of testing is not possible, we have to look for test methods for simplifying this situation.

In digital system testing, a set of algorithms and methods help reduce the number of test vectors by selecting them more wisely than just trying every combination. The following list of things to do work in this direction:

- Simplify faults that can occur.
- Use a reduced number of faults.
- Find mechanisms for evaluating test vectors.
- Find parts of circuit that are harder to test.
- Generate tests that target hard to test areas.
- Evaluate test vectors and keep more efficient ones.
- Compact test vectors.

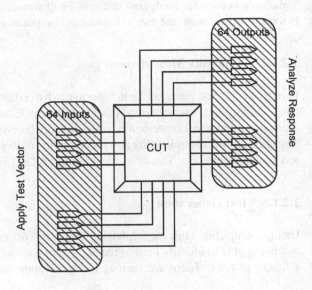

Fig. 1.7 Circuit under test

We try to achieve the above goals by developing test methods discussed in the following subsections. All the test methods are based on gate level circuit descriptions.

Before getting into the methods that are specifically devised for digital system testing, it must be pointed out that a very important tool or method in design and test of digital systems is simulation, and it is used extensively in test. As one of its applications that comes ahead of everything that is discussed below is good circuit response calculation. Using a gate-level or an RT-level simulator, the behavior of a circuit that has no defects is calculated. The response obtained as such will be used by other test methods that are discussed below.

1.2.1.1 Fault Model

To simplify defects that we are trying to test for, we develop a fault model. The fault model that we use is only a model to simplify analysis of a circuit for finding better test vectors and evaluating them. Based on this model we define fault coverage as the percentage of faults that have been detected. For a given test vector, *fault coverage* determines its efficiency in the number of faults it can detect.

1.2.1.2 Fault Reduction

To reduce our efforts for finding test vectors and analyzing a circuit for its faults, we try to reduce the number of faults that we deal with by eliminating redundant ones, and, perhaps, ignoring some that do not occur often. Reducing number of faults can be incorporated in our fault model, i.e., using a simple model consisting of very few fault types. In addition, reducing can be done by eliminating faults that have the same output effect; this process is referred to as *fault collapsing*.

1.2.1.3 Fault Simulation

A test vector or a test set is graded by the number of faults it detects. To find this number, a fault is introduced in the circuit we are testing, and tests are applied to see what faults it detects. The circuit in which a fault is introduced is called a faulty model, and simulating this circuit is referred to as fault simulation. We use fault simulation to decide if a test vector is worth keeping. In addition, fault simulation is used for analyzing test sets. Fault simulation is used for fault coverage calculations, it is the most important test method, and it is computationally very complex.

1.2.1.4 Testability Measurement

In order to find best paths to take in order to reach a certain point in a circuit for testing it, we might need controllability and observability of other circuit areas. Controllability is defined as the ease of controlling a line in a circuit, and observability is the ease of observing the effect of its value on a primary output. Test methods for controllability and observability, or in general for testability, analysis are used in the test generation process. Testability measurement refers to calculation of these parameters.

1.2.1.5 Test Generation

Using a simplified fault model, having tools and methods for evaluating a test vector or a test set, and being able to identify hard to reach areas of a circuit are some of what we use for generating an efficient test set. There are various test generation methods and algorithms, ranging from pure random to deterministic. Repeated use of fault simulation in test generation and its np-complete algorithms make this test method computationally complex.

1.2.1.6 Test Compaction

Another test method that comes into play after test generation has been done is test compaction. Test compaction tries to reduce the number of test vectors without significantly affecting the fault coverage of a test set. A more compact test set tests a CUT quicker.

Chapters 3–6 of this book present various test methods. While the above methods help generation and evaluation of test, there are other methods that help making circuits more testable. We will discuss these in the following subsection.

1.2.2 Testability Methods

Testability methods are hardware methods that are added to a CUT for making its testing easier, faster, and more effective in terms of number of faults that are detected. In most cases, testability methods involve hardware modifications or insertion of new hardware in the original CUT. For making better choices, testability methods use algorithms and methods discussed in Sect. 1.2.1.

To get a better picture as to what a testability method can do for making testing easier, consider the counter shown in Fig. 1.8. The circuit shown is a 64-bit counter with a clock and a reset input. Suppose this circuit is operating at the clock rate of 1 GHz. To test this circuit, let us assume that we can clock it to generate a new count, and it will take the equivalent of five clock cycles to read its output and check it against its expected value. At this rate, the test time for the counter becomes:

$$\text{Test time}: 2^{64} \times (1+5) \times 10^{-9}\,\text{s}$$

According to this, the CUT of Fig. 1.8 takes 3,509 years to test. Obviously, for testing this circuit, a method or technique other than just applying clock inputs and reading circuit outputs is needed. The methods for reducing test time by reducing the number of test vectors, as we did for the example of Fig. 1.7, do not work here, because we are not applying any test vectors. The example of Fig. 1.8 has no data input ports, and, thus, reducing test data inputs to reduce its test time is not relevant. What is needed here is to make hardware modifications to the circuit to make it test better, or test faster. Such methods are referred to as testability methods.

One such method is partitioning hardware into smaller segments each of which can more easily be tested. Consider for example, breaking up the counter of Fig. 1.8 into four segments and gaining independent access to each 16-bit segment by use of multiplexers, as shown in Fig. 1.9. This circuit has a new input that is labeled *NbarT* (*N* for Normal, *T* for Test). When the circuit is to count in the normal mode, the *NbarT* input is 0, and carry out from a lower order counter segment is used for

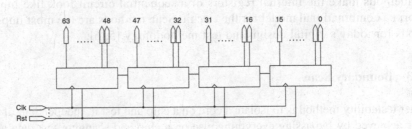

Fig. 1.8 A counter to be tested

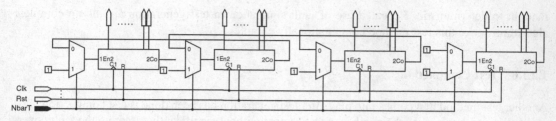

Fig. 1.9 Partitioning into smaller counters

enabling its next immediate upper order counter. Considering the counter notation that *2Co* (carry out) is enabled by the *En2* (enable) input, the structure shown behaves as a 64-bit counter when *NbarT* is 0.

To test the counter, we set *NbarT* to 1 (test mode). With the multiplexers feeding the enable inputs of the 16-bit counters, *NbarT* of 1 enables all counters at the same time. Therefore, to test the circuit in this mode we only need to apply 2^{16} clock pulses, and the test time becomes:

$$\text{Testable circuit test time}: 2^{16} \times (1 + 5) \times 10^{-9}\,\text{s}$$

The test time for the revised testable circuit is only 0.4 ms. This simple modification has caused a circuit that was otherwise impossible to test, testable in a reasonable time.

As with most other testability methods that are outlined below, the testability method we used for the counter improved the controllability and observability of the CUT by adding hardware to the circuit. Testability methods have hardware overhead that must be kept to a minimum. In addition, testability methods may add delays that can slow down the normal operation of a testable CUT.

1.2.2.1 Ad Hoc Testability

Some of what we can immediately think of in terms of making a circuit testable are things like adding extra output pins to observe internal nodes of a CUT, or adding a jumper to make certain parts of a CUT more controllable and/or observable. Partitioning, input/output pins, and multiplexing inputs and outputs are considered as ad hoc testability methods [4].

1.2.2.2 Scan Insertion

In presenting test methods we will show that, although such methods work reasonably well for combinational circuits and certain types of sequential circuits, they are not as effective in making testing easier for all types of sequential circuits. A method of turning a sequential circuit into a combinational circuit for making it accept combinational test methods is the use of scan. Scan methods make the internal registers of a sequential circuit look like inputs and outputs, and form a combinational model for the circuit. Scan methods are the most important testability methods for today's digital design and test methodology [5, 6].

1.2.2.3 Boundary Scan

Another testability method is to isolate a core on a chip and test it, independent of its surroundings. This is achieved by bypassing everything else on a chip and scanning test data all the way to the boundaries of the CUT. This testability method, that is referred to as boundary scan, was originally used for board level testing, and has now been extended to chip and core level testing.

1.2.2.4 Built-in Self-test

Another testability method for a CUT is to let it test itself. BIST is a testability method that involves adding a semi-processing unit to a CUT, with the sole responsibility of testing various parts of the CUT that it shares chip area with. BISTs save test time by not having to pull a component out of its mounting for testing. Furthermore, BIST saves time by interlacing testing with normal operation of a CUT. BIST and scan methods combine to test a complete SOC by each taking responsibility for testing various parts of the SOC.

Starting in Chap. 7 of this book we get into testability methods. In today's electronic design and test, testability methods play a crucial role. Computer aided design (CAD) tools for digital design have ways of automatically inserting testability hardware in a synthesized design.

1.2.3 Testing Methods

Another aspect of testing that can also improve and speed up the testing process is by using smarter and faster testers (ATE). Unfortunately, the ATE industry has been regarded as a different industry from electronic design for many years, and many of the technological advances such as use of HDLs have not penetrated the ATE industry. However, in the last few years ATE industry is looking at ways of incorporating the use of HDL in their ATEs for improving test speed. This will result in a better understanding of test engineering issues by the designers; and at the same time, it makes test engineers get more involved in the design process and early testing of designs. This subject is treated in Appendix A of this book that discusses work done by Teradyne in this respect.

1.2.4 Cost of Test

Another important aspect of electronic testing is the issue of cost. Keeping cost to a level to make production profitable, and still be able to deliver right parts for the right markets imposes a proper test strategy. Test strategy is defined as incorporation of testing in the design development cycle so that a device gets the right test, at the right time, and with the right level of completeness. This section intends to discuss some of the cost-related issues in test.

1.2.4.1 Rule of 10

An important rule to consider in electronic testing is that the cost of testing increases by a factor of 10 when going from one level to the next higher level [7]. This means that we should try to detect faults as early as possible in the design process. Figure 1.10 shows a core on top of the cost

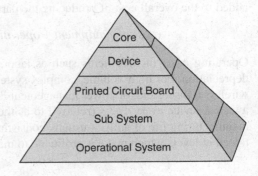

Fig. 1.10 Cost pyramid

pyramid where the pyramid area it occupies represents its testing cost. When designing a core, it is important to develop proper test strategy, apply proper test methods to the core, incorporate test-ability in it, and test it as thoroughly as possible. Faults that go undetected in the core will cost 10 times as much if they are to be detected when incorporated in a device that consists of intercon-nection of many cores.

A device is built by integration of already tested cores. When cores are put together on a chip, their interconnections, their interfaces, and random logic on the chip must be thoroughly tested. Test and testability methods discussed earlier in this section, in particular boundary scan and core testing testability insertion schemes, are particularly important at the core integration level.

The cost as shown in Fig. 1.10 increases as we go down in the cost pyramid. The cost of detecting a fault in an operational system is 10,000 times that of detecting it at the core level. The section below covers some chip cost-related issues.

1.2.4.2 Chip Testing

In addition to the test and testability methods, issues related to the test equipment also play an important role in the cost of testing a chip or a wafer.

In any manufacturing process, the end product is tested "just enough" so that a certain level of performance is verified at the lowest possible cost. This cost/performance trade-off is different depending on the end-use for the device. For low-end mobile phone components, where a failure in the end use application – the "cheap" phone not working – is not terribly consequential, the cost/quality trade-off is much different than it would be for a anti-lock braking system (ABS) for a car – where a device failure in real life is a much more serious matter.

For integrated circuits, there are usually at least two test steps. The first is a test of the bare die while it is still a part of the silicon wafer. Since most of the cost of an integrated circuit is actually in the packaging, the idea is to identify bad devices so that they can be discarded before they are packaged. The second step is to test the device after it has been packaged, mostly to make sure that there are no bonding issues and to verify that the silicon was not damaged during the assembly process. Most manufacturers try to perform enough wafer-level test to attain good device yields of 90% or more at the package level in order to minimize the cost of packages that need to be discarded.

For very cost-sensitive devices, where designs and fabrication processes are well controlled and silicon yield is high, the wafer test step can be omitted. For devices which use very small geometry fabrication processes or are very mission critical, several testing steps may be added at both wafer and package to test the full functionality of the device and to test at different temperatures.

High-volume test is performed using ATE, along with equipment to transport either a wafer or a device package to the ATE (Fig. 1.6). The capital cost for a combination of tester and material han-dler (referred to as a test cell) can amount to well over a million US Dollars. The cost of test (COT) added to the overall cost of producing the part is generally calculated as:

$$(cost\ of\ equipment + operating\ costs/number\ of\ devices\ tested)$$

Operating costs include items such as labor, and facilities cost such as electricity, cooling, and depreciation. For high-volume, complex system on chip (SOC) devices such as mobile processors, wireless transceivers, or power management devices, it is common that the test cell should produce a tested device every 1–3 s (referred to as the throughput rate). If this rate is maintained over the manufacturing life of an IC – usually consisting of millions of devices – then test costs can be con-tained at levels which make it profitable to manufacture.

There are two methods used to increase throughput. The first is to absolutely minimize the time required to test a single device. This relates to test methods and testability methods that we have covered, and they are the main concerns of this book. On the other hand, the ATE also plays an important role in minimizing the time required to test a device. The ATE generally implements very specialized instrumentation, and data transport and processing mechanisms to minimize the time overhead of the test process. The second method to increase throughput is to test multiple devices in parallel on one piece of ATE. Typical SOC devices are tested with as few as two "sites", up to as many as 32 or 64 "sites". DRAMs and Flash memories, which have very long test execution times, often employ many hundreds of test sites in parallel and run on highly specialized test equipment.

1.3 HDLs in Digital System Test

Although hardware description languages have become an inevitable part of digital system design process, they have not benefited the digital system testing. In spite of this, there are many areas that hardware description language can influence and improve testing, some of which are described below and extensively used in this book.

1.3.1 Hardware Modeling

Analyzing a circuit for test purposes requires a hardware model. Depending on the type of analysis, different types of models may be required. Since HDLs can represent hardware at various levels of abstraction, and HDL simulators can handle mixed level descriptions, using an HDL for representing a circuit for analyzing it for test will benefit the corresponding test methods.

An HDL application at the transistor level is to analyze an HDL model of a transistor level circuit for propagation of transistor-level faults to the upper logic level. Going higher than the transistor level, for generating good circuit response, an HDL model at the behavioral level can be simulated to efficiently produce good circuit response vectors for test purposes. In simulations for studying fault effects, gate level descriptions can be used.

For testability purpose, describing the CUT and its testability hardware in an HDL produces a simulation model that can be analyzed for the effectiveness of the testability method. Where a BIST technique is being analyzed, a model of CUT and its associated BIST are useful for configuring the BIST hardware and measuring its effectiveness. There are other situations that HDL hardware models are used for issues related to test that will be seen in the chapters that follow.

1.3.2 Developing Test Methods

With the model of a CUT being available in an HDL, we can use procedural constructs and capabilities of an HDL, and use it as a software programming language to process the CUT model and perform tasks that are related to test. One such example is the use of an HDL for generating random test vectors, applying them to the CUT, and sorting the test vectors according to their effectiveness in terms of detecting faults.

Injecting faults in a circuit and applying tests to check if the test vectors detect the injected fault is another example of such applications. In applications like these, we use HDL facilities for hardware description and modeling to model the CUT, and we use software-like features of the HDL for writing programs for testing and implementation of test methods. The software-like parts go in HDL testbenches and primarily use procedural language constructs. For creating an interface between these two uses, we have developed HDL functions that can interact with an HDL model of a CUT from an HDL's procedural environment. In the Verilog HDL, these functions are implemented in procedural language interface (PLI).

Text input and output facilities, and handling external files are capabilities in HDLs that make them usable as a programming language.

1.3.3 Virtual Testers

A tester is an equipment that uses a device's test interface to apply test vectors to it, and reads back its response. A virtual tester does the exact same thing, except that it is a program and not an equipment [8]. It is very important for a designer to be able to plan testing of his or her circuit before the design is built. A virtual tester developed in an HDL acting on an HDL model of a circuit being designed is an important tool that enables a design engineer to think about test issues early in the design process. With this, test strategies can be developed as a design is being developed.

A virtual tester is an HDL testbench that mimics a tester, with the additional capability of being able to manipulate the CUT for studying its testability or changing parameters for making it a more testable circuit. To properly play its role as a tester, a virtual tester can be allowed only to interact with the CUT through its test ports. We take advantage of HDL capabilities as virtual testers for developing test strategies for scan-based designs in Chaps. 7 and 9.

1.3.4 Testability Hardware Evaluation

Another application of HDLs is in evaluating hardware that is to be added to a CUT for making it testable, by another device, or by itself. The latter case is BIST, whose design and configuration can greatly benefit by the use of HDLs. In the design of BISTs, a template BIST architecture is inserted in the CUT, and its parameters are adjusted for better coverage of faults and lower test time. An HDL testbech can take responsibility for instantiating a CUT and its associated preliminary BIST, and simulating and adjusting BIST register lengths and configurations for a more efficient testing of the CUT. In this book, we are using Verilog in Chap. 9 for BIST evaluation and configuration.

1.3.5 Protocol Aware ATE

A recent addition to ATE industry is the use of HDLs for programming a tester for testing a DUT. Instead of using predefined bit patterns to test a CUT, specialized hardware (typically FPGA-based) is embedded in the tester hardware to directly interpret HDL commands, construct data input patterns on-the-fly, and adopt to variations in CUT output timing and data content. The concept and implementation of Protocol Aware-based tester instrumentation is explained in more detail in Appendix A.

The topics discussed above on use of HDLs in test are extensively used in chapters of this book. For the start, Chap. 2 presents the Verilog HDL and discusses some of the ways it can contribute to the digital system test methodology.

1.4 ATE Architecture and Instrumentation[1]

As noted earlier, the goal of manufacturing test is to perform electrical fault testing as quickly as possible on as many devices in parallel as is practical. ATE hardware typically contains instrumentation which, at a very high level, performs the same function as benchtop instrumentation that might be found in an engineering lab. The types of instruments in the test can roughly be categorized into four groups.

1.4.1 Digital Stimulus and Measure Instruments

Digital stimulus and measure instruments are instruments which source data patterns to the DUT and then verify that the output data pattern is correct. Typically, these are architected around a large pattern memory, up to 64Meg locations (or "vectors") deep. The stimulus data are then "formatted" by shaping the data bit with a rising or falling edge which occur as specific times relative to the start of each bit cycle. Lastly, the data is driven to the device from a buffer that has programmable voltage levels. Data produced by the device are conversely received by a voltage comparator with programmable thresholds and then latched (or "strobed") at a specific point in time relative to the start of the bit cycle and then compared in real time to expected data from the pattern memory. Digital instrumentation is generally differentiated based on the maximum data rate that can be achieved, the timing accuracy of the drive, and compare strobes and the cost of the hardware. General-purpose digital instruments typically have 64–256 digital channels per instrument card and can have data rates in excess of 1 GVector per second and timing accuracy of less than 100 ps. Low-cost instrumentation used for scan testing or for cost-sensitive devices typically operates at rates of 100–200 MVectors per second with timing accuracy of approximately 1 ns. Specialized digital instruments built for high-speed SERializer/DESserializer (SERDES) applications can operate at rates in excess of 10 GVectors per second [9, 10].

1.4.2 DC Instrumentation

DC instrumentations are instruments which are used to either power the DUT or to perform DC parametric testing of power management components such as embedded voltage regulators. All DC options can source voltage and measure current, and some can also force current and measure voltage. These options are generally segmented based on power capability, accuracy, and cost. Very high density DC cards can have hundreds of source/measure channels operating at power levels of less than 1 W, while very high power channels used for testing of large processors can deliver much higher power levels to the DUT. It is not uncommon for high-end server processors to draw many hundreds of watts of power while performing scan test.

1.4.3 AC Instrumentation

AC instrumentation primarily consists of arbitrary waveform generators (AWGs) and waveform digitizers used to test AC functionality such as audio and video performance, intermediate frequency (IF) frequency testing of RF systems, or linearity testing of analog to digital converters

---

[1]This section has been provided by Ken Lanier of Teradyne.

(ADCs) and digital to analog converters (DACs). This instrumentation is typically differentiated based on waveform fidelity (noise and distortion levels) and frequency range. High-end audio converters can required THD and SNR levels in the range of –120 dBc, while linearity testing of high-resolution ADCs and DACs can require accuracy of several parts per million (ppm).

1.4.4   RF Instrumentation

RF instruments are fundamentally meant to perform continuous waveform (CW) testing of RF components such as mixers, low-noise amplifiers (LNAs), and modulation/demodulation components for devices such as mobile phone or local area network (LAN) transceivers [11, 12]. The instruments typically have the ability to measure bidirectional signal power for scattering parameter (S-parameter) measurement, and to perform very high-fidelity waveform modulation and demodulation in order to measure the accuracy of data constellations and to perform end-to-end testing of an RF transceiver using embedded digital data. RF instrumentation is typically segmented based on frequency range (typically several gigahertz) and waveform fidelity.

1.4.5   ATE

What makes ATE unique from bench instrumentation is the density and pin count of the instrumentation, and the speed at which it can setup and perform measurements.

Typical ATE systems will accommodate several thousand pins of digital and analog resources inside a card cage referred to as the "Test Head" in order to perform multi-site testing. In addition, the need to keep all of the instruments in close proximity to the DUT (within several inches – in order to maximize signal fidelity) requires very careful power and thermal management. Most large-scale ATEs incorporate integrated liquid cooling systems to allow a large number of high-density channel cards to be placed within a space typically less than one-half to one square meter. Instruments in a typically configured tester for SOC applications can dissipate between 10 and 40 KW in this space.

Whereas typical laboratory instruments are controlled through standard data busses such as Ethernet, GPIB, or PCI, ATE systems use specialized (usually proprietary) data busses to minimize system latency. ATE is measured on a parameter called "Device Limited Test Time." That is, the time the ATE itself adds to the test process above and beyond what was minimally required to test the device. For example, if a scan pattern is to be executed at a rate of 10 MVectors per second for 100,000 Vectors, then the test time for that pattern should be 10 ms. If the tester takes 2 ms to setup the test, start it, and log the results, then it has added 20% overhead. Furthermore, if that test is done on two devices in parallel and 3 ms is added, then it has achieved a certain level of parallel test efficiency (PTE) which must also be maximized.

ATE from different vendors is differentiated on measurement capability and device throughput. The architecture of the tester must achieve certain goals:

- *Minimize hardware setup time*: To achieve this, hardware setups such as voltage levels and timing are stored in local memory on each instrument. At the start of the test, the test controller broadcasts one command to all instruments to recall and apply a specific setup on all pins in parallel.
- *Optimize instrument handshaking*: Often, instruments in a tester must communicate with each other in real time. For example, a digital pattern may run for 1000 vectors to write to device

Fig. 1.11 UltraFLEX ATE
(Courtesy of Teradyne)

registers and then pause while a power supply current measurement is taken by a DC instrument. Most ATEs provide a separate synchronization bus that allows the digital hardware to start the measurement with zero overhead by allowing a synchronization command for the DC instrument to be embedded in the appropriate step in the pattern.

- *Minimize data post processing time*: In many cases, the tester will collect data from the DUT – from an ADC, for example – that must be postprocessed to determine the quality of the part. A typical example is to perform spectral analysis of digitized data with a fast Fourier transform (FFT) to calculate noise and distortion. This processing cannot be done on the main tester controller because it would slow down the operation of the entire machine. Some ATE architectures include entirely separate computers with a dedicated data bus which will download data from an instrument and perform the postprocessing math as a background process while the ATE is performing subsequent tests.

Figure 1.11 shows a typical ATE test system. The Test Head can be seen on the front of the machine, suspended by a "manipulator" which allows the Head to be positioned at any angle in order to be docked to the equipment that is handling the device packages or moving a wafer to different die positions for testing. This is the same equipment that is shown in the background of Fig. 1.6.

The UltraFLEX test system features a hybrid air/liquid cooled test head with universal slots. The system is available with a 24-slot standard capacity (-SC) test head and a 36-slot high-capacity (-HC) test head. The UltraFLEX system offers digital speed and pin count needed for multi-site test applications of SoC and SiP devices operating at greater than 200 MHz. The UltraFLEX system also has a mainframe cabinet containing the power distribution unit, heat exchanger for liquid cooling, air-cooling resources, clock reference, and an integrated manipulator. There is also cabinet space for mounting additional third-party rack instrumentation.

1.5 Summary

This chapter discussed some of the very basic concepts of test. Perhaps the most important question this chapter should have answered is why test is needed. We tried to do this by discussing a design flow and showing that manufacturing is just another step in making a product, and just like all the steps from design to final production its output needs to be tested. We showed manufacturing test as a step in HDL-based design and showed how much of the same tools used in the earlier phases of design could be used in postmanufacturing test.

Another question this chapter should have answered is why we need methods for test and methods for making circuits testable. We tried to answer these questions by showing complexity of test and the time it takes to test every component that is produced, and its impact on the cost of the final product. In effect, test and testability methods are software and hardware solutions for reducing test time and doing test of a final product more thoroughly.

This chapter also discussed the role of ATEs in test. Regardless of how testable our circuit is, and how good our test sets are, the final testing by an ATE requires an efficient and fast ATE. Furthermore, our test and testability methods cannot be devised and used independent of the equipment that our product is tested on. Understanding capabilities of ATEs and what they can and cannot do is important.

In addition to all the above, this chapter showed how hardware description languages can be used in all aspects of digital system testing. Devising a test strategy, generating test, evaluating how much testing we need, and deciding on the hardware for making our circuit testable are all test-related tasks that HDLs are well suited to be useful for. This chapter tried to highlight the importance of HDLs in test, and showed why it is important for a design engineer to understand test and a test engineer to understand design flow. HDLs can be regarded as a link between design and test.

References

1. Kurup P, Abbasi T (1995) Logic synthesis using synopsis. Kluwer, Boston
2. Jha NK, Gupta SK (2003) Testing of digital systems. Cambridge University Press, Cambridge, UK
3. Nadeau-Dostie B (ed) (2000) Design for at-speed test. In: Diagnosis and measurement. Kluwer, Boston
4. Abramovici M, Breuer MA, Friedman AD (1994) Digital systems testing and testable design. IEEE Press, Piscataway, NJ (revised printing)
5. Willaims MJY, Angell JB (1973) Enhancing testability of large-scale integrated circuits via test points and additional logic. IEEE Trans Comput C-22(1):46–60
6. Eichelberger EB, Lindbloom E, Waicukauski JA, Williams TW (1991) Structured logic testing. Prentice-Hall, Englewood Cliffs, NJ
7. Wang L-T, Wu C-W, Wen X (2006) VLSI test principles and architectures: design for testability. Morgan Kaufmann, San Francisco
8. Sehgal A, Iyengar V, Chakrabarty K (2004) SOC test planning using virtual test access architectures. IEEE TransVLSI Syst 12 (12):1263–1276
9. Mak TM, Tripp MJ (2005) Device testing, U.S. Patent No. 6,885,209, April 26
10. Laquai B, Cai Y (2001) Testing gigabit multilane serdes interfaces with passive jitter injection filters. In: Proceedings of IEEE International Test Conference, pp. 297–304, October (2001)
11. Agilent Technologies (2005) Spectrum analyzer basics, AN 150 5952–0292, App. Note, 2005
12. Agilent Technologies (2006) Fundamentals of RF and microwave noise figure measurement, AN 57–1 5952–8255E, App. Note, 2006
13. Navabi Z (2006) Verilog digital system design, Second Edition. McGraw Hill Company, N.Y, New York
14. Navabi Z (2007) VHDL: Modular design and synthesis of cores and systems, Third Edition. McGraw Hill-Professional, N.Y, New York
15. Navabi Z (2006) Embedded core design with FPGAs. McGraw Hill-Professional, N.Y, New York

Chapter 2
Verilog HDL for Design and Test

In Chapter 1, we discussed the basics of test and presented ways in which hardware description languages (HDLs) could be used to improve various aspects of digital system testing. The emphasis of this chapter is on Verilog that is a popular HDL for design. The purpose is to give an introduction of the language while elaborating on ways it can be used for improving methodologies related to digital system testing. After, the basic concepts of HDL modeling, the main aspects of describing combinational and sequential circuits using different levels of abstraction, and the semantics of simulation in Verilog language are expressed. Then, we get into the testbench techniques and virtual tester development, which are heavily utilized in the presentation of test techniques in the rest of this book. Finally, a brief introduction to the procedural language interface (PLI) of Verilog and the basics of implementing test programs in PLI is given. The examples we present in this chapter for illustrating Verilog language and modeling features are used in the rest of this book as circuits that are to be tested. The HDL codes for such examples are presented here. Verilog coding techniques for gate-level components that we use for describing our netlists in the chapters that follow are also shown here.

2.1   Motivations of Using HDLs for Developing Test Methods

Generally speaking, tools and methodologies design and test engineers use are different, and there has always been a gap between design and test tools and methods. This gap results in inconsistencies in the process of design and test, such as designs that are hard to test or the time needed to convert design to the format compatible for testing. On the other hand, we have seen in new design methodologies that incorporating test in design must start from the beginning of the design process [1, 2]. It is desirable to bring testing in the hands of designers, which certainly requires that testing is applied at the level and with the language of the designers. This way, designers will be able to combine design and test phases.

Using RT-level HDLs in test and DFT, helps advancing test methods to RTL, and at the same time alleviates the need for the use of software languages and reformatting designs for the evaluation and application of test techniques. Furthermore, actual test data can be applied to post-manufacturing model of a component, while keeping other component models at the design level, and still simulating in the same environment and keeping the same testbench. This also allows reuse of design test data, and migration of testbenches from the design stage to post-manufacturing test. In a mixed-level design, these advantages make it possible to test a single component described at the gate level while leaving others in RTL or even at the system level.

On the other hand, when we try to develop test methods in an HDL environment, we are confronted with the limitations of HDL simulation tools. Such limitations include the overhead that test methods put on the simulation speed and the inability to describe complex data structures. PLI provides a library of C language functions that can directly access data within an instantiated

Verilog HDL data structure and overcomes the HDL limitations. With PLI, the advantages of doing testable hardware design in an HDL and having a software environment for the manipulation and evaluation of designs can be achieved at the same time. Therefore, not only the design core and its testbench can be developed in a uniform programing environment, but also all the facilities of software programing (such as complex data structures and utilization of functions) are available. PLI provides the necessary accesses to the internal data structure of the compiled design, so test methods can be performed in such a mixed environment more easily and without having to mingle with the original design.

In this book, by means of the PLI interface, a mixed HDL/PLI test environment is proposed and the implementations of several test applications are exercised. In the sections that follow, a brief description of HDL coding and using testbench techniques combined with PLI utilities for developing test methods are given.

2.2 Using Verilog in Design

For decades, HDLs have been used to model the hardware designs as an IEEE standard [3]. Using HDLs and their simulators, digital designers are capable of partitioning their designs into components that work concurrently and are able to communicate with each other. HDL simulators can simulate the design in the presence of the real hardware delays and can imitate concurrency by switching between design parts in small time slots called "delta" delays [4]. In the following subsections, the basic features of Verilog HDL for simulation and synthesis are described.

2.2.1 Using Verilog for Simulation

The basic structure of Verilog in which all hardware components and testbenches are described is called a *module*. Language constructs, in accordance to Verilog syntax and semantics form the inside of a module. These constructs are designed to facilitate the description of hardware components for simulation, synthesis, and specification of testbenches to specify test data and monitor circuit responses. A module that encloses a design's description can be described to test the module under design, in which case it is regarded as the testbench of the design. Figure 2.1 shows a simulation

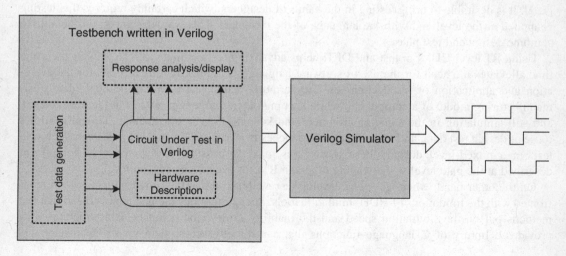

Fig. 2.1 Simulation in Verilog

model that consists of a design with a Verilog testbench. Verilog constructs (shown by dotted lines) of the Verilog model being tested are responsible for the description of its hardware, while language constructs used in a testbench are in charge of providing appropriate input data or applying data stored in a text file to the module being tested, and analysis or display of its outputs. Simulation output is generated in the form of a waveform for visual inspection or data files for record or for machine readability.

2.2.2 Using Verilog for Synthesis

After a design passes basic the functional validations, it must be synthesized into a netlist of components of a target library. The target library is the specification of the hardware that the design is being synthesized to. Verilog constructs used in the Verilog description of a design for its verification or those for timing checks and timing specifications are not synthesizable. A Verilog design that is to be synthesized must use language constructs that have a clear hardware correspondence.

Figure 2.2 shows a block diagram specifying the synthesis process. Circuit being synthesized and specification of the target library are the inputs of a synthesis tool. The outputs of synthesis are a netlist of components of the target library, and timing specification and other physical details of the synthesized design. Often synthesis tools have an option to generate this netlist in Verilog.

2.2.2.1 Postsynthesis Simulation

When the netlist is provided by the synthesis tool that uses Verilog for the description of the netlist components (Fig. 2.3), the same testbench prepared for the pre-synthesis simulation can be used with this gate-level description. This simulation, which is often regarded as post-synthesis simulation, uses timing information generated by the synthesis tool and yields simulation results with detailed timing.

Since the same testbench of the high-level design is applied to the gate-level description, the resulted waveform or printed data must be the same. This can be seen when comparing Fig. 2.1 with Fig. 2.3, while the only difference is that the post-synthesis simulation includes timing details.

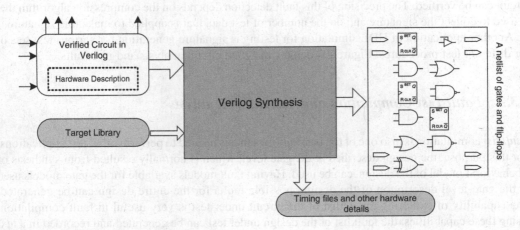

Fig. 2.2 Synthesis of a Verilog design

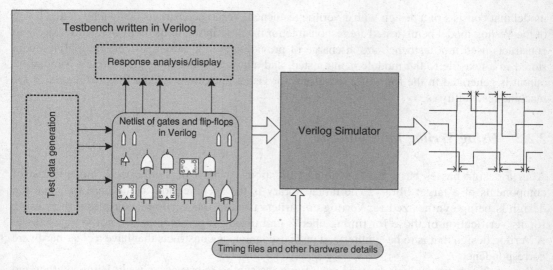

Fig. 2.3 Postsynthesis simulation in Verilog

2.3 Using Verilog in Test

As mentioned, HDL capabilities can be utilized to enhance exercising existing test methods and to develop new ones with little effort. The subsections that follow illustrate some possible usages of Verilog in the test of digital systems.

2.3.1 Good Circuit Analysis

An important tool in testing is one that generates good circuit responses from a circuit's golden model. This response is to be to compared with responses from faulty circuits. By applying testbench data to the golden model, it is possible to record the good behavior of the circuit for future use. The golden signatures can also be generated this way. A signature is the result of an accumulative compression on all the outputs of the golden model. Later, when checking if a circuit is faulty or not, the same input data and the same signature collection algorithm must be applied to the design under test. By comparing the obtained signature with the recorded golden signature, the presence or absence of faults in the circuit can be verified. The precision of this fault detection depends on the compression algorithm that is used to collect the signature and on the number of test data that is applied to make this signature.

Another application of HDL simulation for testing is signature generation for various test sets or for different test procedures. Figure 2.4 depicts the good circuit analysis and its results.

2.3.2 Fault List Compilation and Testability Analysis

Fault list compilation is also one of the basic utilities that is needed to perform other test applications. For this purpose, the design described at the gate level, which is normally resulted from synthesis of a behavioral model of the design, can be used. Having fault models available for the gate models used in the gate-level description of the design, possible faults for the entire design can be generated. The capability of exploring the netlist of the circuit under test is very useful in fault compilation. Using these capabilities, the fault list of the design under test can be generated and recorded in a text

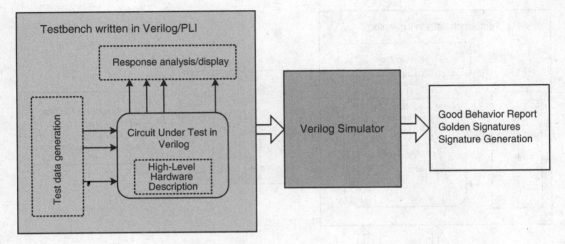

Fig. 2.4 Good circuit analysis using Verilog

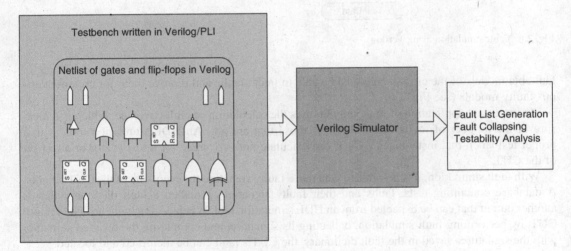

Fig. 2.5 Fault list compilation and testability measurement using Verilog

file as the fault list (Fig. 2.5). In order to reduce test time, fault collapsing, which is also implementable in the HDL environment, is performed.

Certain test applications, such as test generation or testability hardware insertion methods, need measurements to estimate how testable their internal nodes are. Methods used for fault compilations can also be used for applications such as this.

2.3.3 Fault Simulation

As mentioned, an HDL environment is able to generate a list of faults. This list can be used in an HDL simulation environment for fault simulation of the circuit under test. To complement the facilities that the HDL and its environment provide, we have developed Verilog PLI functions for producing fault models of a CUT for the purpose of fault simulation. The PLI functions inject faults in the good circuit model to create a faulty model of the CUT.

Assuming test data and the fault list and a mechanism for fault injection (FI) are available, fault simulation can be implemented in an HDL testbench. This testbench needs to instantiate golden and

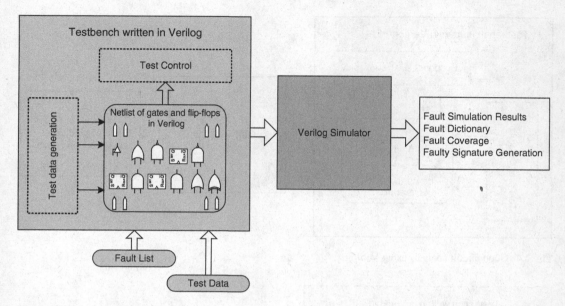

Fig. 2.6 Fault simulation using Verilog

faultable models of the circuit, and must be able to inject faults and remove them for creating various faulty models (see Fig. 2.6).

An important application of fault simulation is the calculation of fault coverage, which is a measure of the number of faults detected versus those that are not. An HDL simulation tool, with a proper testbench that instantiates a CUT, can calculate fault coverage for a test vector or a test set of the CUT.

With fault simulation, it is possible to generate a faulty signature for every one of the CUT's faults. A database containing tests, faults and their faulty signatures is called a fault dictionary that is another output that can be expected from an HDL simulation tool. When dealing with an actual faulty CUT, by performing fault simulation, collecting its signature, and comparing the resulted signature with the signatures saved in the fault dictionary, the CUT's fault can be identified and located.

2.3.4 Test Generation

Another application of Verilog PLI for test applications is test generation. The same netlist that was used for fault simulation is instantiated in the testbench of Fig. 2.7. This environment is able to inject a fault, generate some kind of random or pseudo random test data, and check if the test vector detects the injected fault. We can also find the number of undetected faults that a test vector detects. The result can be a collection of test vectors that detect a good number of circuit faults. This collection is a test set produced by HDL simulation of CUT.

2.3.5 Testability Hardware Design

Efficient design of hardware that makes a design testable is possible in an HDL environment. By means of the testability measurements and other information provided by simulating a design, we can decide on the type and the place of the testability hardware that we intend to insert into the original design. After that, by applying test generation and fault simulation applications provided in this environment, a proper

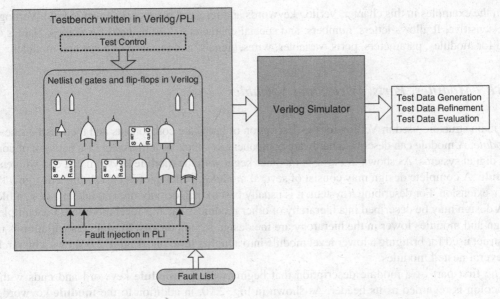

Fig. 2.7 Test generation using Verilog

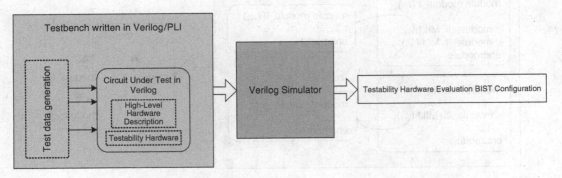

Fig. 2.8 Testability hardware design using Verilog

test set can be found for the new circuit, and the testbench can act as a virtual tester for the DFT-inserted circuit. In this case, various testability factors of the new circuit, such as new testability measurements, fault coverage, test time, and even power consumption estimation during test, can be obtained.

Along with this DFT evaluation, changing the configuration of the testability hardware is also possible. For this purpose, the important parameters of the DFT, such as the place for inserting test points, the length and the number of scan chains, and the number of clocks in the BIST circuit, can be changed until the best possible configuration is obtained (Fig. 2.8).

2.4 Basic Structures of Verilog

As mentioned, all the design and test processes described in this book are implemented in Verilog. The following subsections cover the basics of this language and the rules of describing designs in various levels of abstraction. For more details on HDL modeling and testbenches, the reader is encouraged to refer to [5, 6].

In the examples in this chapter, Verilog keywords and reserved words are shown in bold. Verilog is case sensitive. It allows letters, numbers, and special character "_" to be used for names. Names are used for modules, parameters, ports, variables, wires, signals, and instance of gates and modules.

2.4.1 Modules, Ports, Wires, and Variables

The main structure used in Verilog for the description of hardware components and their testbenches is a *module*. A module can describe a hardware component as simple as a transistor or a network of complex digital systems. As shown in Fig. 2.9, modules begin with the **module** keyword and end with **endmodule**. A complete design may consist of several modules. A design file describing a design takes the *.v* extension. For describing a system, it is usually best to include only one module in a design file.

A design may be described in a hierarchy of other modules. The top-level module is the complete design and modules lower in the hierarchy are the design's components. Module instantiation is the construct used for bringing a lower level module into a higher level one. Figure 2.9 shows a hierarchy of several nested modules.

The first part of a module description that begins with the **module** keyword and ends with a semicolon is regarded as its header. As shown in Fig. 2.10, in addition to the **module** keyword, a

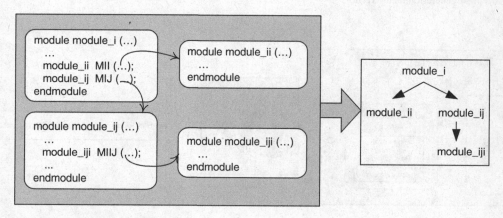

Fig. 2.9 Module outline and hierarchy

```
module acircuit (input a, b, input[7:0] av, bv, output w, output[7:0] wv);
    wire d, c;
    wire [7:0] dv;
    reg e;
    reg [7:0] ev;

    assign d = a & b;
    assign dv = av & bv;
    assign w [6:0] = av [7:1] & dv [7:1];
    assign cv[7] = d ^ bv[3];

    always @(av,bv,a,b) begin
        ev = {av[3:0],bv[7:4]}
        e = a | b;
    end

    assign wv = ev;

endmodule
```

Fig. 2.10 Port, wire, and variable declaration

module header includes the module name and list of its ports. Port declarations specifying the mode of a port (i.e. input, output, etc.), and its length can be included in the header or as separate declarations in the body of the module. Module declarations appear after the module header. A port may be **input**, **output**, or **inout**. The latter type is used for bidirectional input/output lines. The size of a multibit port comes in a pair of numbers separated by a colon and bracketed by square brackets. The number on the left of the colon is the index of the left most bit of the vector, and that on the right is the index of the right most bit of the vector.

In addition to ports not declared in the module header, this part can include declaration of signals used inside the module or temporary variables. Wires (that are called **net** in Verilog) are declared by their types, **wire**, **wand**, or **wor**; and variables are declared as **reg**. Wires are used for interconnections and have properties of actual signals in a hardware component. Variables are used for behavioral descriptions and are similar to variables in software languages. Figure 2.10 shows several wire and variable declarations.

Wires represent simple interconnection wires, busses, and simple gate or complex logical expression outputs. When wires are used on the left-hand side of **assign** statements, they represent outputs of logical structures. Wires can be used in scalar or vector form. Multiple concurrent assignments to a **net** are allowed and the value that the wire receives is the resolution of all concurrent assignments to the **net**. Figure 2.10 includes several examples of wires used on the right and left hand sides of **assign** statements.

In contrast to a **net**, a **reg** variable type does not represent an actual wire and is primarily used as variables are used in a software language. In Verilog, we use a **reg** type variable for temporary variables, intermediate values, and storage of data. A **reg** type variable can only be used in a procedural body of Verilog. Multiple concurrent assignments to a **reg** should be avoided.

In the vector form, inputs, outputs, wires, and variables may be used as a complete vector, part of a vector, or a bit of the vector. The latter two are referred to as part-select and bit-select. Examples of part-select and bit-select on right and left hand sides of an **assign** statement are shown in Fig. 2.10. The statement that assigns the *ev* **reg**, besides part-select indexing, illustrates concatenation of *av[3:0]* and *bv[7:4]* and assigning the result to *ev*. This structure especially is useful to model swapping and shifting operations.

2.4.2 Levels of Abstraction

Operation of a module can be described at the gate level, using Boolean expressions, at the behavioral level, or a mixture of various levels of abstraction. Figure 2.11 shows three ways the same operation can be described. Module *simple_1a* uses Verilog's gate primitives, *simple_1b* uses concurrent statements, and *simple_1c* uses a procedural statement. Module *simple_1a* describes instantiation of three gate primitives of Verilog. In contrast, *simple_1b* uses Boolean expressions to describe the same functions for the outputs of the circuit. The third description, *simple_1c*, uses a conditional **if** statement inside a procedural statement to generate proper function on one output, and uses a procedural Boolean function for forming the other circuit output.

2.4.3 Logic Value System

Verilog uses a 4-value logic value system. Values in this system are 0, 1, Z, and X. Value 0 is for logical 0 which in most cases represents a path to ground (Gnd). Value 1 is logical 1 and it represents a path to supply (Vdd). Value Z is for float, and X is used for uninitialized, undefined, undriven, unknown, and value conflicts. Values Z and X are used for wired-logic, busses, initialization values, tristate structures, and switch-level logic.

```
module simple_1a (input i1, i2, i3, output w1,w2);
   wire c1;
   nor g1(c1, i1, i2);
   and g2 (w1, c1, i3);
   xor g3(w2, i1, i2, i3);
endmodule
```

```
module simple_1b (input i1, i2, i3, output w1, w2);
   assign w1 = i3 & ~ (i1 | i2);
   assign w2 = i1 ^ i2 ^ i3;
endmodule
```

```
module simple_1c (input i1, i2, i3, output w1, w2);
   reg w1, w2;
   always @ (i1, i2, i3) begin
      if (i1 | i2) w1 = 0; else w1 = i3;
      w2 = i1 ^ i2 ^ i3;
   end
endmodule
```

Fig. 2.11 Module definition alternatives

A gate input, or a variable or signal in an expression on the right-hand side of an assignment can take any of the four logic values. Output of a two-valued primitive gate can only take 0, 1, and X while output of a tristate gate or a transistor primitive can also take a Z value. A right-hand-side expression can evaluate to any of the four logic values and can thus assign 0, 1, Z, or X to its left-hand-side **net** or **reg**.

2.5 Combinational Circuits

A combinational circuit can be represented by its gate-level structure, its Boolean functionality, or description of its behavior. At the gate level, interconnection of its gates are shown; at the functional level, Boolean expressions representing its outputs are written; and at the behavioral level a software-like procedural description represents its functionality. At the beginning of this section, implementation of a NAND gate using primitive transistors of Verilog as a glance to transistor-level design is illustrated. Afterward, the implementation of a 2-to-1 multiplexer is described in various levels of abstraction to cover important concepts in combinational circuits. Examples for combining various forms of descriptions and instantiation of existing components are also shown.

2.5.1 Transistor-level Description

Verilog has primitives for unidirectional and bidirectional MOS and CMOS structures [7]. As an example of instantiation of primitive transistors of Verilog, consider the two-input CMOS NAND gate shown in Fig. 2.12.

The Verilog code of Fig. 2.13 describes this CMOS NAND gate. Logically, nMOS transistors in a CMOS structure push 0 into the output of the gate. Therefore, in the Verilog code of the CMOS NAND, input to output direction of nMOS transistors are from *Gnd* toward *y*. Likewise, nMOS

Fig. 2.12 CMOS NAND gate

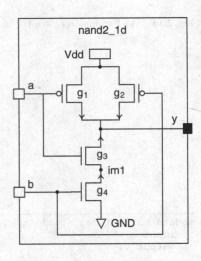

```
module cmos_nand ( input a, b, output y );
    wire im1;
    supply1 vdd;
    supply0 gnd;

    pmos #(4, 5)
        g1 (y, vdd, a),
        g2 (y, vdd, b);
    nmos #(3, 4)
        g3 (im1, gnd, b),
        g4 (w, im1, a);
endmodule
```

Fig. 2.13 CMOS NAND Verilog description

transistors push a 1 value into *y*, and therefore, their inputs are considered the *Vdd* node and their outputs are connected to the *y* node. The *im1* signal is an intermediate **net** and is explicitly declared.

2.5.2 *Gate-level Description*

We use the multiplexer circuit of Fig. 2.14 to illustrate how primitive gates are used in a design. The description shown in Fig. 2.15 corresponds to this circuit. The module description has inputs and outputs according to the schematic of Fig. 2.14.

The statement that begins in Line 6 and ends in Line 8 instantiates two **and** primitives. The construct that follows the primitive name specifies 0-to-1 and 1-to-0 propagation delays for the instantiated primitive ($t_{\rho lh} = 2$, $t_{\rho hl} = 4$). This part is optional and if eliminated, 0 values are assumed $t_{\rho lh}$ and $t_{\rho hl}$ delays.

Line 7 shows inputs and outputs of one of the two instances of the **and** primitive. The output is *im1* and inputs are module input ports *a* and *b*. The port list on Line 7 must be followed by a comma if other instances of the same primitive are to follow, otherwise a semicolon should be used, like the end of Line 9. Line 8 specifies input and output ports of the other instance of the **and** primitive. Line 10 is for instantiation of the **or** primitive at the output of the majority gate. The output of this gate is

Fig. 2.14 2-to-1 multiplexer circuit

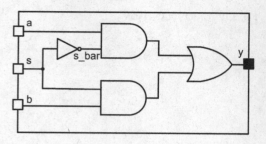

```
module mux2to1 ( a, b, s, y );
   input a, b, s;
   output y;

   not #(1,1) (s_bar, s);        //Line 05
   and #(2,4)                    //Line 06
      ( im1, a, s_bar ),         //Line 07
      ( im2, b, s );             //Line 08
   or  #(3,5) ( y, im1, im2);    //Line 09

endmodule
```

Fig. 2.15 Verilog code for the multiplexer circuit

y that comes first in the port list, and is followed by inputs of the gate. In this example, intermediate signals for interconnection of gates are *im1*, *im2*, and *s_bar*. Scalar interconnecting wires need not be explicitly declared in Verilog. The two **and** instances could be written as two separate statements, like instantiation of the **or** primitive. If we were to specify different delay values for the two instances of the **and** primitive, we had to have two separate primitive instantiation statements.

2.5.3 Equation-level Description

At a higher level than gates and transistors, a combinational circuit may be described by the use of Boolean, logical, and arithmetic expressions. For this purpose, the Verilog concurrent **assign** statement is used. Table 2.1 shows Verilog operators that can be used with **assign** statements.

Figure 2.16 shows a 2-to-1 multiplexer using a conditional operator. The expression shown reads as follows: **if** *s* is 1, then *y* is *i1* **else** it becomes *i0*.

If there is more than one **assign** statement, because of the concurrency property of Verilog, the order in which they appear in module is not important. These statements are sensitive to events on their right-hand sides. When a change of value occurs on any of the right-hand-side **net** or variables, the statement is evaluated and the resulting value is scheduled for the left-hand side **net**.

2.5.4 Procedural Level Description

At a higher level of abstraction than describing hardware with gates and expressions, Verilog provides constructs for procedural description of hardware. Unlike gate instantiations and **assign**

Table 2.1 Verilog operators

Bitwise operators	&	|	^	~	~^	^~	
Reduction operators	&	~&	|	~|	·^	~^	^~
Arithmetic operators	+	–	*	/	%		
Logical operators	&&	||	!				
Compare operators	<	>	<=	>=	++		
Shift operators	>>	<<					
Concatenation operators	{}	$\{n\{\}\}$					
Conditional operators	?:						

```
module mux2_1 (input [3:0] i0, i1, input s, output [3:0]y );
    assign y = s ? i1 : i0;
endmodule
```

Fig. 2.16 A 2-to-1 Multiplexer using condition operator

statements that correspond to concurrent substructures of a hardware component, procedural statements describe the hardware by its behavior. Also, unlike concurrent statements that appear directly in a module body, procedural statements must be enclosed in procedural blocks before they can be put inside a module.

The main procedural block in Verilog is the **always** block. This is considered a concurrent statement that runs concurrent with all other statements in a module. Within this statement, procedural statements like **if-else** and **case** statements are used and are executed sequentially. If there are more than one procedural statement inside a procedural block, they must be bracketed by **begin** and **end** keywords.

Unlike assignments in concurrent bodies that model driving logic for left-hand-side wires, assignments in procedural blocks are assignments of values to variables that hold their assigned values until a different value is assigned to them. A variable used on the left hand side of a procedural assignment must be declared as **reg**.

An event control statement is considered a procedural statement, and is used inside an **always** block. This statement begins with an at-sign, and in its simplest form, includes a list of variables in the set of parenthesis that follow the at-sign, e.g., @ $(v1, v2,...)$.

When the flow of the program execution within an **always** block reaches an event-control statement, the execution halts (suspends) until an event occurs on one of the variables in the enclosed list of variables. If an event-control statement appears at the beginning of an **always** block, the variable list it contains is referred to as the *sensitivity list* of the **always** block. For combinational circuit modeling, all variables that are read inside a procedural block must appear on its sensitivity list.

2.5.4.1 Multiplexer Example

As an example of a procedural block, consider the 2-to-1 multiplexer of Fig. 2.17. This example uses an **if-else** construct to set y to *i0* or *i1* depending on the value of *s*. As in the previous examples, all circuit variables that participate in the determination of value of y appear on the sensitivity list of the **always** block. Also since y appears on the left-hand side of a procedural assignment, it is declared as **reg**.

The **if-else** statement shown in Fig. 2.17 has a condition part that uses an equality operator. If the condition is true (i.e., *s* is 0), the block of statements that follow it will be taken, otherwise the block of statements after the **else** is taken. In both cases, the block of statements must be bracketed by **begin** and **end** keywords if there are more than one statement in a block.

```
module mux2_1 (input i0, i1, output reg s, y );
   always @( i0, i1, s ) begin
      if ( s==1'b0 )
         y = i0;
      else
         y = i1;
   end
endmodule
```

Fig. 2.17 Procedural multiplexer

```
module alu_4bit (input [3:0] a, b, input [1:0] f, output reg [3:0] y );
   always @ ( a or b or f ) begin
      case ( f )
         2'b00 : y = a + b;
         2'b01 : y = a - b;
         2'b10 : y = a & b;
         2'b11 : y = a ^ b;
         default: y = 4'b0000;
      endcase
   end
endmodule
```

Fig. 2.18 Procedural ALU

2.5.4.2 Procedural ALU Example

The **if-else** statement, used in the previous example, is easy to use, descriptive, and expandable. However, when many choices exist, a **case** statement which is more structured may be a better choice. The ALU description of Fig. 2.18 uses a **case** statement to describe an ALU with add, subtract, AND, and XOR functions. The **case** statement shown in the **always** block uses f to select one of ALU functions in the **case** alternatives. The last alternative is the **default** alternative that is taken when f does not match any of the alternatives that appear before it. This is necessary to make sure that unspecified input values (here, those that contain X and/or Z) cause the assignment of the default value to the output and do not leave it unspecified.

2.5.5 Instantiating Other Modules

We have shown how primitive gates can be instantiated in a module and wired with other parts of the module. The same applies to instantiating a module within another. For regular structures, Verilog provides repetition constructs for instantiating multiple copies of the same module, primitive, or set of constructs. Examples in this section illustrate some of these capabilities.

2.5.5.1 ALU Example Using Adder

The ALU of Fig. 2.18 starts from scratch and implements every function it needs inside the module. If we have a situation that we need to use a specific design from a given library, or we have a function that is too complex to be repeated everywhere it is used, we can make it into a module and instantiate it when we need to use it.

```
module ALU_Adder (input [7:0] a,b, input addsub, // Line 01
                   output gt, zero, co, output [7:0] r );
   wire [7:0] b_bbar;
   add_8bit ADD (a, b_bbar, addsub, r, co);            // Line 04
   assign b_bbar = addsub ? ~b : b;                    // Line 05
   assign gt = (a>b);
   assign zero = (r == 0);
endmodule
```

Fig. 2.19 ALU Verilog code using instantiating an adder

Figure 2.19 shows another version of the above ALU circuit. In this new version, addition is handled by instantiation of a predesigned adder (*add_8bit*). Instantiation of a component, such as *add_8bit* in the above example, starts with the component name, an instance name (*ADD*), and the port connection list. The latter part decides how local variables of a module are mapped to the ports of the component being instantiated. The above example uses an ordered list, in which a local variable, e.g., *b_bbar*, takes the same position as the port of the component it is connecting to, e.g., *b*. Alternatively, a named port connection such as that shown below can be used.

$$add\_8bit\ ADD\ (.a(a),.b(b\_bbar),.ci(addsub),.s(r),.co(co));$$

Using this format allows port connections to be made in any order. Each connection begins with a dot, followed by the name of the port of the instantiated component, e.g. *b*, and followed by a set of parenthesis enclosing the local variable that is connected to the instantiated component, e.g. *b_bbar*. This format is less error-prone than the ordered connection.

2.5.5.2 Iterative Instantiation

Verilog uses the **generate** statement for describing regular structures that are composed of smaller sub-components. An example is a large memory array or a systolic array multiplier. In such cases, a cell unit of the array is described, and by means of several **generate** statements, it is repeated in several directions to cover the entire array of the hardware.

Here, we show the description of a parametric n-bit AND gate using this construct. Obviously, *n*-input gates can be easily obtained by using vector inputs and outputs for Verilog primitives. However, the example shown in Fig. 2.20 besides illustrating the iterative **generate** statement of Verilog, introduces the structure of components that are used in this book to describe gate-level circuits for test applications. This description is chosen due to the PLI requirements for implementing test applications that are discussed later.

The code of Fig. 2.20 uses the **parameter** construct to prepare parametric size and delays for this AND gate. In the body of this module on Line 8, a variable for generating *n* instances of **and** primitive is declared using the **genvar** declaration. The **generate** statement that begins on Line 10 loops *n* times to generate an instance of the **and** gate in every iteration. Together, the *and_0* instance and the **generate** statement make enough **and** gates to AND together bits 0 to n-1 of input vector *in*. This is done by use of the intermediate wire, *mwire*. Since the resulted *and_n* must represent a bitwise function, *mwire* **net** is declared to accumulate bit-by-bit AND results. Line 9 shows the first two bits of the *in* input vector ANDed using the **and** primitive, and the result is collected in bit 0 of *mwire*. After that, each instanced **and** in the **generate** statement takes the next bit from *in* and ANDs it with the calculated bit of *mwire* to generate the next bit of *mwire*. The resulted hardware for this parametric *and_n* gate, is concatenation of 2-input **and** primitives that AND all bits of the *in* input vector.

The complete component library for test applications of this book can be found in Appendix B.

```
module and_n
  #(parameter n = 2, tphl = 1, tplh = 1)(out,in);

  input [n-1:0] in;
  output out;
  wire [n-2:0] mwire;

  genvar i;                                        //Line 08
  and and_0 (mwire [0], in [0], in [1]);           //Line 09
  generate                                         //Line 10
    for (i=1; i <= n-2; i=i+1) begin : AND_N       //Line 11
      and inst (mwire [i], mwire [i-1], in [i+1]); //Line 12
    end
  endgenerate

  bufif1 #(tplh, tphl) inst(out, mwire [n-2], 1'b1); //Line 16

endmodule
```

Fig. 2.20 Using iterative instantiation for test primitive AND gate

2.6 Sequential Circuits

As with any digital circuit, a sequential circuit can be described in Verilog by the use of gates, Boolean expressions, or behavioral constructs (e.g., the **always** statement). While gate-level descriptions enable a more detailed description of timing and delays because of complexity of clocking and register and flip-flop controls, these circuits are usually described by the use of procedural **always** blocks. This section shows various ways sequential circuits are described in Verilog.

2.6.1 Registers and Shift Registers

Figure 2.21 shows an 8-bit register with *set* and *reset* inputs that are synchronized with the clock. The *set* input puts all 1s in the register, and the *reset* input resets it to all 0s. The sensitivity list of the procedural statement shown includes **posedge** of *clk*. This **always** statement only wakes up when *clk* makes a 0 to 1 transition. When this statement does wake up, the value of *d* is put into *q*. Obviously, this behavior implements a rising-edge register. Instead of **posedge**, the use of **negedge** would implement a falling-edge register.

In order to provide procedural description for shift registers the concatenation construct can be used as shown in Fig. 2.22. This partial code, that can be used in the body of an **always** statement like that of Fig. 2.21, does a left-shift if *shift_left* is 1, and right shifts, otherwise.

2.6.2 State Machine Coding

Along with simple sequential circuits, such as registers, shift registers, and counters, Verilog constructs enable the designer to model finite state machines of any type. State machines can be modeled as Moore or Mealy machines. In both cases, based on the current state of the sequential circuit and its input, the next state is decided. The difference is in the determination of outputs. Unlike a Moore machine that has outputs that are only determined by the current state of the machine, in a

```
module register (input [7:0] d, input clk, set, reset, output reg [7:0] q);
    always @ ( posedge clk ) begin
        if ( set )
            q <= 8'b1;
        else if ( reset )
            q <= 8'b0;
        else
            q <= d;
    end
endmodule
```

Fig. 2.21 An 8-bit register

```
if ( shift_left )
    q <= {q[6:0], s_in};
else
    q <= {s_in, q[7:1]};
```

Fig. 2.22 Concatenation for a 8-bit shift register

Mealy machine, the outputs are declared regarding the state the machine is in as well as the inputs of the circuit. This makes Mealy outputs not fully synchronized with the circuit clock.

This section shows coding for state machines and introduces the Huffman coding style. The example we use is a Residue-5 divider. The coding styles used here apply to such controllers and are used in later sections of this chapter to describe the controller of a simple adding machine. It must be mentioned that the Residue-5 example presented here is one of the test cases for the application of test methods in this book. Simpler and more detailed examples can be found in [6].

2.6.2.1 Residue-5 Divider

The Residue-5 divider is a circuit that performs the integral division modulo-5 on the sequences coming on its input. For this purpose, the circuit divides the first received input by five and stores the remainder. For the next data on the input port, the circuit adds the new value to the stored remainder, divides the result by 5, and stores the new remainder. This circuit can be modeled using a finite state machine. The remainder stored in this circuit shows its internal state and its output. State diagram for the Residue-5 divider using 2-bit input x is depicted in Figs. 2.23 and 2.24. For the sake of readability, Fig. 2.23 just includes arcs related to two states.

The machine has five states that are labeled, *Zero*, *One*, *Two*, *Three*, and *Four*; each of which shows the resulted Residue-5 remainder. In the Moore state machine modeling, the output depends just on the current state, so in Fig. 2.23 the output is defined for each state. In addition to the x input, the machine has a *reset* input that forces the machine into its *Zero* state. The resetting of the machine is synchronized with the circuit clock.

2.6.2.2 The Moore Implementation of Residue-5 in Verilog

The Verilog code of the Moore machine of Fig. 2.24 is shown in Fig. 2.25. After the declaration of inputs and outputs of this module, **parameter** declaration declares five states of the machine as 3-bit parameters. The square-brackets following the **parameter** keyword specify the size of parameters being declared. Following parameter declarations in the code of Fig. 2.25, the 3-bit *current* **reg** type

Fig. 2.23 A part of
Residue-5 Moore state
machine

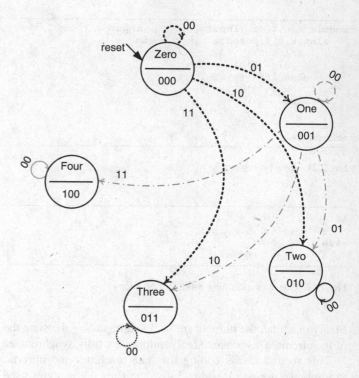

Fig. 2.24 Complete Residue-5
Moore state machine

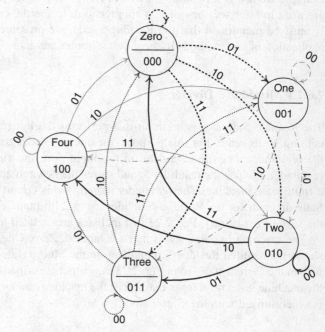

variable is declared. This variable holds the current state of the state machine. The body of the code
of this circuit has an **always** block and an **assign** statement.

The **assign** statement shown in Fig. 2.25 puts the proper value on the output regarding the current
state. This statement is concurrent with the **always** block that is responsible for making the state
transitions. The **always** block used in the module of Fig. 2.25 describes state transitions of the state

```verilog
module residue5(input clk, reset, input[1:0] x, output[2:0] out);
reg[2:0] current;
parameter Zero = 3'b000, One = 3'b001, Two = 3'b010,
          Three = 3'b011, Four = 3'b100;
    always @(posedge clk) begin
        if(reset == 1)
            current <= Zero;
        else
            case(current)
            Zero: case(x)
                    2'b00:  current <= Zero;
                    2'b01:  current <= One;
                    2'b10:  current <= Two;
                    2'b11:  current <= Three;
                  endcase
            One: case(x)
                    2'b00:  current <= One;
                    2'b01:  current <= Two;
                    2'b10:  current <= Three;
                    2'b11:  current <= Four;
                  endcase
            Two: case(x)
                    2'b00:  current <= Two;
                    2'b01:  current <= Three;
                    2'b10:  current <= Four;
                    2'b11:  current <= Zero;
                  endcase
            Three: case(x)
                    2'b00:  current <= Three;
                    2'b01:  current <= Four;
                    2'b10:  current <= Zero;
                    2'b11:  current <= One;
                  endcase
            Four: case(x)
                    2'b00:  current <= Four;
                    2'b01:  current <= Zero;
                    2'b10:  current <= One;
                    2'b11:  current <= Two;
                  endcase
            default: current <= Zero;
          endcase
    end
    assign out = current;
endmodule
```

Fig. 2.25 Moore machine Verilog code

diagram of Fig. 2.24. The main task of this procedural block is to inspect input conditions (values on *reset* and *x*) during the present state of the machine defined by *current* and set values into *current* for the next state of the machine.

The flow into the **always** block begins with the positive edge of *clk*. Since all activities of this machine are synchronized with the clock, only *clk* appears on the sensitivity list of the **always** block. Upon entry into this block, the *reset* input is checked and if it is active, *current* is set to *Zero* (*Zero* is a declared **parameter** and its value is 0). The value put into *current* in this pass through the **always** block gets checked in the next pass with the next edge of the clock. Therefore, assignments to *current* are regarded as the next-state assignment. When such an assignment is made, the **case** statement skips the rest of the code of the **always** block, and this **always** block will next be entered with the next positive edge of *clk*. Upon entry into the **always** block, if *reset* is not 1, program flow reaches the **case** statement that checks the value of *current* against the five states of the machine.

Figure 2.26 shows the Verilog code of the *Two* state and its diagram from the state diagram of Fig. 2.24. As shown, the **case** alternative that corresponds to the *Two* state specifies the next values for that state. Determination of the next state is based on the value of *x*. If *x* is 1, the next state becomes *Three*, and if *x* is 2, the next state becomes *Four*, and so on. As shown in the **assign** statement in Fig. 2.25, the output bits of this circuit are taken directly from the *current* register.

This same machine can be described in Verilog in several different forms. A finite state machine can also be described as a Mealy machine. As mentioned, in this case the output depends not only on the current state, but also on the input of the circuit. In Mealy machines, the output becomes available one cycle sooner than that of a Moore machine, causing fewer states than Moore.

2.6.2.3 Huffman Coding Style

The Huffman model for a digital system characterizes it as a combinational block with feedbacks through an array of registers. Verilog coding of digital systems, according to the Huffman model, uses an **always** statement for describing the register part and another concurrent statement for describing the combinational part. This model of representing a digital component is very useful for test purposes, as we see in the chapters that follow.

We describe the state machine of Fig. 2.24 to illustrate this style of coding. Figure 2.27 shows the combinational and register part partitioning that we use for describing this machine.

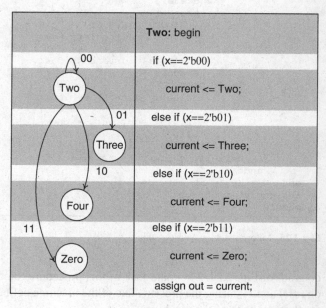

Fig. 2.26 Next values from state *two*

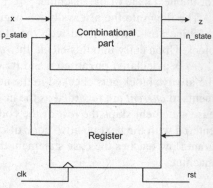

Fig. 2.27 Huffman partitioning of Residue-5 divider

```
module residue5_huffman(input clk, rst, input[1:0] x, output[2:0] out);
reg[2:0] n_state, p_state;
parameter Zero = 3'b000, One = 3'b001, Two = 3'b010,
          Three = 3'b011, Four = 3'b100;
    always@(p_state, x) begin
        n_state = Zero;
        case(p_state)
            Zero:… n_state = …
            One:… n_state = …
            Two:… n_state = …
            Three:… n_state = …
            Four:… n_state = …
            default:…
        endcase
    end// Combinational part
    always@(posedge clk, posedge rst) begin
        if(rst)
            p_state = Zero;
        else
            p_state = n_state;
    end// Register part

    assign out = p_state;
endmodule
```

Fig. 2.28 Verilog Huffman coding style

The *combinational* block uses *x* and *p_state* as input and generates *out* and *n_state*. The *register* block clocks *n_state* into *p_state*, and resets *p_state* when *rst* is active.

Figure 2.28 shows the Verilog code of Fig. 2.24 according to the partitioning of Fig. 2.27. As shown, parameter declaration declares the states of the machine. Following this declaration, *n_state* and *p_state* variables are declared as 3-bit **reg**s that hold values corresponding to the five states of the Moore Residue-5 divider. The *combinational* **always** block follows this **reg** declaration. Since this is purely a combinational block, it is sensitive to all its inputs, namely, *x* and *p_state*. Immediately following the block heading, *n_state* is set to its inactive or reset value. This is done so that this variable is always reset with the clock to make sure it does not retain its old value. Note that retaining old values implies latches, which is not what we want in our combinational block.

The body of the combinational **always** block of Fig. 2.28 contains a **case** statement that uses the *p_state* input of the **always** block for its **case** expression. This expression is checked against the states of the Moore machine. As in the other styles discussed before, this **case** statement has **case** alternatives for all of the states. For brevity, the statements in the **case** alternatives are not shown. These statements set the *n_state* variable using the same procedure as setting the *current* variable in Fig. 2.25. In a block corresponding to a **case** alternative, based on input values, *n_state* is assigned values. Unlike the other style where *current* is used both for the present and next states, here we use two different variables, *p_state* and *n_state*.

The next procedural block shown in Fig. 2.28 handles the register part of the Huffman model of Fig. 2.27. In this part, *n_state* is treated as the register input and *p_state* as its output. On the positive edge of the clock, *p_state* is either set to the *Zero* state (000) or is loaded with contents of *n_state*. Together, *combinational* and *register* blocks describe our state machine in a very modular fashion.

As with the other style we presented, a separate **assign** statement (or any other concurrent statement) is used for the assignment of values to the output. The advantage of this style of coding is in its modularity and defined tasks of each block. State transitions are handled by the *combinational* block and clocking is done by the *register* block. Changes in clocking, resetting, enabling, or presetting the machine only affect the coding of the *register* block. In this code, the a synchronous resetting is applied.

2.7   A Complete Example (Adding Machine)

In this section, the complete RTL design of a simple CPU is described. Although this design has the structure of a simple CPU, since its ALU actually just performs adding operation, we refer to it as *Adding Machine*. In this part, almost all Verilog constructs explained in this chapter are exercised. Furthermore, the basics of RTL design and datapath and controller partitioning are introduced. Later, this Adding Machine is used as one of the test cases in this book for demonstrating test methods.

2.7.1   Control/Data Partitioning

The first step in an RT-level design is the partitioning of the design into a data part and a control part. The data part consists of data components and the bussing structure of the design, and the control part is usually a state machine generating control signals that control the flow of data in the data part [8].

Figure 2.29 shows a general sketch of an RT-level design that is partitioned into its data and control parts. As shown in this figure, a processor is divided into *datapath* and *controller* parts. The datapath has storage elements (registers) to store intermediate data, handles transfer of data between its storage components, and performs arithmetic or logical operations on data that it stores. The datapath also has communication lines for transfer of data; these lines are referred to as *busses*. Activities in the datapath include reading from and writing into data registers, bus communications, and distributing control signals generated by the controller to the individual data components.

The controller commands the datapath to perform proper operation(s) according to the instruction it is executing. Control signals carry these commands from the controller to the datapath. Control signals are generated by the controller state machine that, at all times, knows the status of the task that is being executed and the sort of the information that is stored in datapath registers. Controller is the thinking part of a design.

2.7.2   Adding Machine Specification

The design of Adding Machine begins with the specification of the design, including the number of general purpose registers and the instruction format. The machine has two 8-bit external data buses (input bus and output bus) and a 6-bit address bus. The address bus connects to the memory in order to address locations that are being read from or written into. Data read from the memory

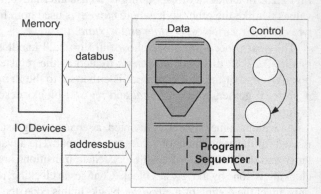

Fig. 2.29 Control/data partitioning for Adding Machine

Table 2.2 Adding machine instruction set

Opcode	Instruction	Instruction class	Description
00	add immd	Arithmetic	AC ← AC + immd
01	lda adr	Data-transfer	AC ← Mem [adr]
10	sta adr	Data-transfer	Mem [adr] ← AC
11	jmp adr	Control-flow	PC ← adr

are instructions and instruction operands, and data written into the memory are instruction results and temporary information. Adding Machine also communicates with its IO devices through its external busses. The address bus addresses a specific device or a device register while the data bus contains data that is to be written or read from the device.

Each instruction of Adding Machine is 8 bits wide, and occupies a memory word. The instruction format of the machine has an explicit operand (immediate data or memory location the address of which is specified in the instruction) and an implicit operand. Adding Machine has four instructions, divided into three classes of arithmetic (**add**), data transfer (**lda, sta**), and control-flow instructions (**jmp**).

Adding Machine instructions are described below. A tabular list and summary of this instruction set is shown in Table 2.2.

- **add** *immd*: adds the *immd* data with an 8-bit register named accumulator (*AC*) and stores the result back in *AC*.
- **lda** *adr*: reads the content of the memory location addressed by *adr* and writes it into *AC*.
- **sta** *adr*: writes the content of *AC* into the memory location addressed by *adr*.
- **jmp** *adr*: jump to the memory location addressed by *adr*.

2.7.3 CPU Implementation

In the following subsections, the Verilog implementation of the Adding Machine in register transfer level of abstraction is described.

2.7.3.1 Datapath Design

As mentioned, Adding Machine has an 8-bit register called accumulator (*AC*). All data transfers and arithmetic instructions use *AC* as an operand. In a real CPU, there may be multiple accumulators or an array of registers that is referred to as a register file.

To store the instruction that is read from the memory, a register is used at the output of the memory unit called instruction register (*IR*). The program counter (*PC*) is implemented as a counter that is incremented for program sequencing. Using these registers, the implementation of datapath is shown in Fig. 2.30. The input data bus connects to the input of *IR* in order to bring the instruction read from the memory into this register. Similarly, this bus connects to *AC* to bring data read from the memory into the *AC* register. The control signal for loading *IR* and *AC* are *ld_ir* and *ld_ac*, respectively. *PC* has three control signals *ld_pc*, *inc_pc*, and *clr_pc* to load, increment, and clear it, respectively. The right most 6 bits of *IR* connect to the input of *PC* for the execution of the *jmp* instruction. When a bus has more than one source driving it, e.g., *IR* and *PC* driving *adr*_bus, a multiplexer and control signals from the controller select the source.

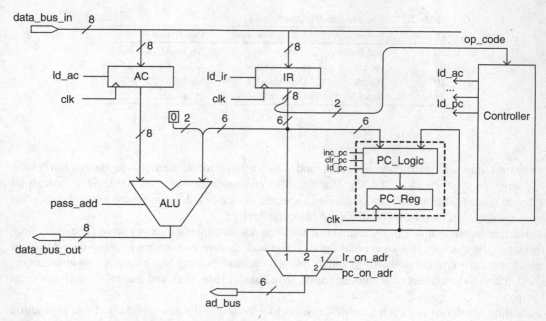

Fig. 2.30 Adding machine multicycle datapath

2.7.3.2   Controller Design

After the design of the datapath and figuring control signals and their role in activities in the datapath, the design of the controller becomes a simple matter. The block diagram of this part is shown in Fig. 2.31.

The controller of our Adding Machine has four states, *Reset*, *Fetch*, *Decode*, and *Execute*. As the machine cycles through these states, various control signals are issued. In state *Reset*, for example, the *clr_pc* control signal is issued. State *Fetch* issues *pc_on_adr*, *rd_mem*, *ld_ir*, and *inc_pc* to read memory from the present *PC* location, route it to *IR*, load it into *IR*, and increment *PC* for the next memory fetch. Depending on *op_code* bits, that are the controller inputs, the *Execute* state of the controller issues control signals for the execution of **lda**, **sta**, **add**, and **jmp** instructions. The *Decode* state is a simple wait state.

The next section discusses details of the controller signals and their role in execution of these instructions. As before, our processor description has a datapath and a control component. The controller is described using a state machine coding style. At the end, the description of our small example is completed by wiring datapath and controller in a top-level Verilog module.

2.7.3.3   Datapath HDL Description

Datapath components of Adding Machine are described by **always** and **assign** statements according to their functionalities described above. Afterward, these modules are instantiated into the datapath module. Figure 2.32 shows the Verilog code of the datapath. Structure and signal names in this description are according to those shown in Fig. 2.30.

2.7.3.4   Controller HDL Description

The controller code for our Adding Machine example is shown in Fig. 2.33. This code corresponds to the right-hand side control block in Fig. 2.29 which is shown in more details in Fig. 2.31.

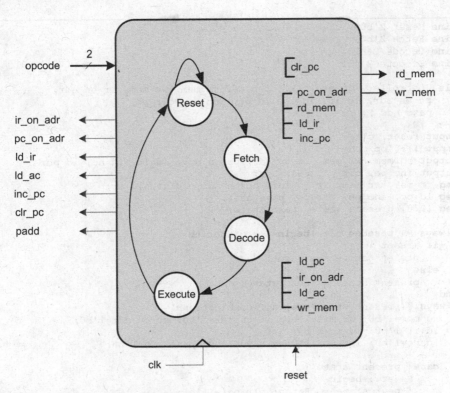

Fig. 2.31 Simple CPU Adding Machine multicycle controller

```
module DataPath ( clk, ir_on_adr, pc_on_adr, ld_ir, ld_ac, ld_pc, inc_pc,
          clr_pc, pass_add, adr_bus, op_code, data_bus_in, data_bus_out);

    input clk, ir_on_adr, pc_on_adr, ld_ir, ld_ac, ld_pc, inc_pc, clr_pc,
        pass_add;
    output [5:0] adr_bus;
    output [1:0] op_code;
    input [7:0] data_bus_in;
    output [7:0] data_bus_out;

    wire [7:0] ir_out;
    wire [5:0] pc_out;
    wire [7:0] a_side;

    IR ir( data_bus_in, ld_ir, clk, ir_out );
    PC pc( ir_out[5:0], ld_pc, inc_pc, clr_pc, clk, pc_out );
    AC ac( data_bus_in, ld_ac, clk, a_side );
    ALU alu( a_side, {2'b00,ir_out[5:0]}, pass_add, data_bus_out );

    assign adr_bus = ir_on_adr ? ir_out[5:0] : pc_on_adr ? pc_out : 6'b0;
    assign op_code = ir_out[7:6];
endmodule
```

Fig. 2.32 Datapath HDL description

In addition to *clk* and *reset*, the controller has the *op_code* input that is driven by *IR* and comes to the controller from the *DataPath* module (see Fig. 2.30).

The sequencing of control states is implemented by a Huffman style Verilog code. In this style, an always block (*registering*) handles the assignment of values to *present_state*, and another **always**

```verilog
`define Reset 2'b00
`define Fetch 2'b01
`define Decode 2'b10
`define Execute 2'b11

module Controller (reset, clk, op_code, rd_mem, wr_mem, ir_on_adr,
        pc_on_adr, ld_ir, ld_ac, ld_pc, inc_pc, clr_pc,
        pass_add );

   input reset, clk;
   input [1:0]op_code;
   output rd_mem, wr_mem, ir_on_adr, pc_on_adr, ld_ir, ld_ac, ld_pc;
   output inc_pc, clr_pc, pass_add;
   reg rd_mem, wr_mem, ir_on_adr, pc_on_adr, ld_ir, ld_ac;
   reg ld_pc, inc_pc, clr_pc, pass_add;
   reg [1:0] present_state, next_state;

   always @( posedge clk )begin : registering
      if (reset )
         present_state <= `Reset;
      else
         present_state <= next_state;
   end
   always @(present_state) begin : combinational
      rd_mem=1'b0; wr_mem=1'b0; ir_on_adr=1'b0; pc_on_adr=1'b0;
      ld_ir=1'b0; ld_ac=1'b0;
      ld_pc=1'b0; inc_pc=1'b0; clr_pc=1'b0; pass_add=1'b0;

      case( present_state )
         `Reset : begin
            next_state = `Fetch; clr_pc = 1'b1;
         end
         `Fetch : begin
            next_state = Decode ; pc_on_adr=1'b1; rd_mem=1'b1;
            ld_ir=1'b1; inc_pc=1;
         end
         Decode : begin
            next_state = `Execute;
         end
         `Execute: begin
            next_state = `Fetch;
            case( op_code )
               2'b00: begin // lda
                  ir_on_adr=1'b1; rd_mem=1'b1; ld_ac=1'b1;
               end
               2'b01: begin // sta
                  ir_on_adr=1'b1; pass_add = 1'b0;
                  wr_mem=1'b1;
               end
               2'b10: ld_pc=1'b1; // jmp
               2'b11: begin // add
                  pass_add=1'b1; ld_ac=1'b1;
               end
            endcase
         end
      endcase
   end
endmodule
```

Fig. 2.33 Controller HDL description

statement (*combinational*) uses this register output as the input of a combinational logic determining *next_state*. This combinational block also sets values to control signals that are outputs of the controller.

In the body of the *combinational* **always** block, a **case** statement checks *present_state* against the states of the machine (*Reset, Fetch, Decode*, and *Execute*), and activates the proper control signals.

The *Reset* state activates *clr_pc* to clear *PC* and sets *Fetch* as the next state of the machine. In the *Fetch* state, *pc_on_adr*, *rd_mem*, *ld_ir*, and *inc_pc* become active, and *Decode* is set to become the next state of the machine. By activating *pc_on_adr* and *rd_mem*, the *PC* output goes on the memory address and a read operation is issued. Assuming the memory responds in the same clock, contents of memory at the *PC* address will be put on *data_bus_in*. This bus is connected to the input of *IR* and issuance of *ld_ir* loads its contents into this register. The next state of the controller is *Decode* that makes the new contents of *IR* available for the controller. In the *Execute* state, a newly fetched instruction in *IR* decides on control signals to issue to execute the instruction.

In the *Execute* state, *op_code* is used in a **case** expression to decide on control signals to issue depending on the opcode of the fetched instruction. The **case** alternatives in this statement are four *op_code* values of 00, 01, 10, and 11 that correspond to **lda, sta, jmp**, and **add** instructions.

For **lda**, *ir_on_adr*, *rd_mem*, and *ld_ac* are issued. These control signals cause the address from *IR* to be placed on the *adr_bus* address bus, memory read to take place and data from memory to be loaded into *AC*.

The controller executes the **sta** instruction by issuing *pass_add*, *ir_on_adr*, and *wr_mem*. As shown in Fig. 2.33, these signals take contents of *AC* to the input bus of the memory (i.e., *data_bus_out*), and *wr_mem* causes the writing into the memory to take place. Note that *pass_add* causes *AC* to pass through *ALU* unchanged. The **jmp** instruction is executed by enabling *PC* load input, which takes the jump address from *IR* (see Fig. 2.33).

The last instruction of this machine is **add**, for execution of which, *pass_add* and *ld_ac* are issued. This instruction adds data in the upper 6 bits of *IR* with *AC* and loads the result into *AC*.

2.7.3.5 The Complete HDL Design

The top-level module for our Adding Machine example is shown in Fig. 2.34. In the *CPU* module shown, *DataPath* and *Controller* modules are instantiated. Port connections of the *Controller*

```
module CPU( reset,clk,adr_bus,rd_mem,wr_mem,data_bus_in,data_bus_out );
input reset;
input clk;
input [7:0]data_bus_in;
output [5:0]adr_bus;
output rd_mem;
output wr_mem;
output[7:0]data_bus_out;
wire ir_on_adr, pc_on_adr, ld_ir, ld_ac, ld_pc, inc_pc, clr_pc, pass_add;
wire [1:0] op_code;

Controller cu ( reset, clk, op_code, rd_mem, wr_mem, ir_on_adr, pc_on_adr,
               ld_ir, ld_ac, ld_pc, inc_pc, clr_pc,  pass_add );

DataPath dp ( clk, ir_on_adr, pc_on_adr, ld_ir, ld_ac, ld_pc, inc_pc,
          clr_pc, pass_add, adr_bus, op_code, data_bus_in, data_bus_out );

endmodule
```

Fig. 2.34 Adding Machine top-level module

include its output control signals, the *op_code* input from *DataPath* and the *reset* external input. Port connections of *DataPath* consist of *adr_bus* and *data_bus_*in and *data_bus_out* external busses, *op_code* output, and control signal inputs.

2.8   Testbench Techniques

The previous sections described Verilog for designing combinational and sequential circuits, as well as complete systems. This section discusses about testbenches and their role in simulation. However, the primary intention of this part is to show how testbench techniques could help us to develop test environments and virtual testers for digital circuit testing. This section shows how Verilog language constructs can be used for the application of data to a module under test, and how module responses can be displayed and checked.

A Verilog testbench is a Verilog module that instantiates a module under test (MUT), applies data to it and monitors its output. Because a testbench is in Verilog, it can go from one simulation environment to another. A module and its corresponding testbench form a simulation model in which MUT is tested regardless of what simulation environment is used.

Based on these considerations, testbenches could play a very important role in the development of test applications in HDL environments. Therefore, a test designer must understand testbenches and language constructs that are used for testing a design module. The basics of testbench techniques in Verilog HDL are discussed in this section, and more complete testbenches to develop test applications are illustrated in the next chapters.

2.8.1   Testbench Techniques

All that a testbench covers can be categorized in instantiating a module, applying generated or existing data to the inputs of the MUT, delay management, and then collecting the responses of the circuit and, if required, comparing them with the expected responses. Therefore, testbench techniques can be categorized in order to answer the following questions: 1) How is the data generated or provided, 2) How are the circuit responses getting reported, 3) What are data generation and response collection sensitive, and 4) What language constructs are to be used to manage the termination of a testbench?

Answers to the above questions are discussed in the rest of this section, and for preparing for the materials that follow Short answer for the above questions are given in the following.

1. The methods to provide data include *deterministic* – assigning a specific data to inputs, *arithmetic* – for example, using a counter to provide new data, *periodic* – toggling the value of a signal in certain periods, *random* – for example, using **$random** task function of Verilog, and *Text IO* – reading data from a stored text file, e.g. using **$fscanf** or **$fread**.
2. To report the circuit responses, Verilog display utilities such, as **$display** or **$monitor** can be used. These tasks, show the results in the simulator's console. Another way is to use *Text IO* to record the responses in text files for future references, e.g., using **$fdisplay** or **$fwrite**.
3. It is important to decide on the conditions that test data are applied to a design under test, and conditions for collection of its responses. Various choices for such conditions are: a) End of a delay, which can abe based on different time slots, equal time slots, or random amount of delay, and b) Change of a signal which is appropriate to make handshaking and synchronization between the testbench and the design under test.

4. While applying data and collecting responses, the duration of running a testbench must also be specified. The methods to manage the end time of a testbench include **$stop**, **$finish** or managing iterations using **repeat** or **for** construct.

Examples of the above items will be seen in the testbenches that are discussed in the following sections for testing combinational and sequential circuits.

2.8.2   A Simple Combinational Testbench

Developing a testbench for a combinational circuit is straightforward; however, selection of data and how much testing should be done depends on the MUT and its functionality. Previously, a simple ALU was described (Fig. 2.18) that we use here to test, and its header is repeated in Fig. 2.35 for reference. The *alu_4bit* module is a four function ALU. Data inputs are *a* and *b*, and its function input is *f*.

A testbench for *alu_4bit* is shown in Fig. 2.36. Variables corresponding to inputs and outputs of the MUT are declared in the testbench. Variables connecting to the inputs are declared as **reg** and outputs as **wire**. Instantiation of *alu_4bit*, shown in the testbench, associates local **regs** and **wires** with the ports of this module.

Variables that are associated with the inputs of *alu_4bit* have been given initial values when declared. Application of data to the *b* input is done in an **initial** statement. For the first 60 ns and every 20 ns, a new value is assigned to *b*, and after 20 ns the testbench finishes the simulation. This last 20 ns wait, allows effects of the last input change to be shown in the simulation run results.

Application of data to the *f* input of *alu_4bit* is done in an **always** statement. Starting with the initial value of 0, *f* is incremented by 1 every 23 ns. The **$finish** statement in the **initial** block is reached at 80 ns. At this time, all active procedural blocks stop and simulation terminates. Simulation

```verilog
module alu_4bit (input [3:0] a, b, input [1:0] f, output reg [3:0] y );

//...

endmodule
```

Fig. 2.35 alu_4bit module declaration

```verilog
module   test_alu_4bit;
   reg   [3:0] a=4'b1011, b=4'b0110;
   reg   [1:0] f=2'b00;
   wire  [3:0] y;

   alu_4bit MUT( a, b, f, y);

   initial begin
     #20 b=4'b1011;
     #20 b=4'b1110;
     #20 b=4'b1110;
     #20 $finish;
   end
   always #23 f = f + 1;

endmodule
```

Fig. 2.36 Testbench for *alu_4bit*

control tasks are **$stop** and **$finish**. The first time the flow of a procedural block reaches such a task, simulation stops or finishes. A stopped simulation can be resumed, but a finished one cannot. In this example, the data generation for *b* is deterministic, and its data application condition is based on different time slots (we used 20 ns intervals). For the *f* input, data generation is arithmetic, and data application is based on equal time slots (periodic 23 ns).

2.8.3   A Simple Sequential Testbench

Test of sequential circuits involves synchronization of clock with other data inputs. We use the *residue5* module as an example here. As shown in the header of this circuit, repeated in Fig. 2.37 for reference, it has a clock input, a reset, data input, and output.

Figure 2.38 shows a testbench for the Residue-5 circuit. As before, variables corresponding to the ports of MUT are declared in the testbench. When the *residue5* module is instantiated, these variables are connected to its actual ports.

The **initial** block of this testbench generates a positive pulse on *rst* that begins at 13 ns and ends at 63 ns. The timing is so chosen to cover at least one positive clock edge so that the synchronous *rst* input can initialize the states of the Residue-5 circuit. The *d_in* data input begins with value X and is initialized to 2'b01 while *rst* is 1.

In addition to the **initial** block, *test_residue5* module includes two **always** blocks that generate data on *d_in* and *clk*. Clock is given a periodic signal that toggles every 11 ns. The Residue-5 *d_in* input is assigned a new value every 37 ns. In order to reduce the chance of changing several inputs at the same time, we usually use prime numbers for the timing of sequential circuit inputs.

```
module residue5(input clk, reset, input[1:0] x, output[2:0] out);
    reg[2:0] current;
//...
endmodule
```

Fig. 2.37 *Residue-5* sequential circuit

```
module test_residue5;
    reg clk, rst;
    reg [1:0] d_in;
    wire [2:0] d_out;

    residue5 MUT ( clk, rst, d_in, d_out );

    initial begin
        clk=1'b0
    end
    initial begin
        #13 rst=1'b1;
        #19 d_in = 2'b01;
        #31 rst=0'b0;
        #330 $finish;
    end
    always #37 d_in = d_in+1;
    always #11 clk = ~clk;

endmodule
```

Fig. 2.38   A testbench for the *residue5* module

Instead of initializing **reg** variables when they are declared, we have used an **initial** block for this purpose. It is important to initialize variables, like the *clk* clock, for which their old values are used for determining their new values. If not done so, *clk* would start with value X and complementing it would never change its value. The **always** block shown generates a periodic signal with a period of 22 ns to provide a free running clock.

The waveform generated on *d_in* may or may not be able to test the whole functionality of this state machine. However, periods of *clk* and *d_in*, and the testbench duration can be changed to make this happen.

2.8.4 Limiting Data Sets

Instead of setting a simulation time limit, a testbench can put a limit on the number of data put on inputs of a MUT. This will also be able to stop simulation from running forever.

Figure 2.39 shows a testbench for our MUT that uses **$random** to generate random data on the *x* input of the circuit. The **repeat** statements in the **initial** blocks shown cause *clock* to toggle 13 times every 5 ns, and *x* to receive a random data 10 times every 7 ns. Instead of a deterministic set of data to guarantee a deterministic test state, random data are used here. This strategy makes it easier to generate data, but due to unpredictable inputs, makes the analysis of circuit responses more difficult. In large circuits, using random data is more useful, and is usually more appropriate to set data inputs and not control signals. The testbench of Fig. 2.39 stops at 70 ns.

2.8.5 Synchronized Data and Response Handling

The previous examples of testbenches for MUT used independent timings for the clock and data. Where several sets of data are to be applied, synchronization of data with the system clock becomes difficult. Furthermore, changing the clock frequency would require changing the timing of all data inputs of the module being tested.

The testbench of Fig. 2.40 uses an event control statement to synchronize data applied to *x* with the clock that is generated in the testbench. The *clock* signal is generated in an **initial** statement using the **repeat** construct. An **always** statement is used for generation of random data on *x*. This loop waits for the positive edge of *clock* and 3 ns after the clock edge, and a new random data is generated for *x*. The stable data after the positive edge of the clock will be used by *residue5* on the

```
module test_residue5;
   reg   reset=1, clock=0;
   reg [1:0] x;
   wire [2:0] z;

   residue5 MUT (clock, reset, x, z);

   initial #24 reset=1'b0;
   initial repeat(13) #5 clock=~clock;
   initial repeat(10) #7 x=$random;

endmodule
```

Fig. 2.39 Testbench using **repeat** to limit data sets

```
module test_residue5;
    reg   reset=1, clock=0;
    reg  [1:0] x;
    wire [2:0] z;

    residue5 MUT ( clock, reset, x, z );

    initial #24 reset=0;
    initial repeat(13) #5 clock=~clock;
    always @(posedge clock) #3 x=$random;
    initial forever @(posedge clock) #1 $displayb(z);
    always @(z) $display("Output changes at %t to %b", $time, z);
    initial $monitor("New state is %d and occurs at %t", MUT.current, $time);
endmodule
```

Fig. 2.40 Synchronizing data with clock

next leading edge of the clock. This technique of data application guarantees that changing of data and clock do not coincide.

In this testbench, 1 ns after the positive edge of the clock, that is when the circuit output is supposed to have its new stable value, the z output is displayed using the **$display** task. This method is appropriate for behavioral simulation, but when dealing with synthesized circuit which includes internal delays, calculating the exact time in which response is ready would not be easy and reliable. A more convenient way to display new output values is to wait for an event on the output z, which means that it has received a new value. This can be complemented by displaying the time of change using the **$time**, task.

Using hierarchical naming, this testbench can be used for displaying internal variables and signals of MUT. The **initial** statement containing **$monitor** is responsible for displaying *MUT, current*, which is the current state of *residue5* addressed by its hierarchical name. The **initial** statement starts **$monitor** in the background. Display occurs when the task is started and when an event occurs on one of the variables in the task arguments. The **%b**, **%d**, and **%t** format specifications in this testbench cause the related signals to be reported as binary, decimal, and in time unit, respectively.

2.8.6 Random Time Intervals

We have shown how **$random** can be used for generation of random data. The testbench we are discussing in this section uses random delays for assigning values to x.

Figure 2.41 shows a testbench for the Residue-5 circuit that uses **$random** for its delay control. As shown, the *running* **initial** statement applies appropriate initial values to inputs of the MUT. In this procedural block, nonblocking assignments cause intra-assignment delay values to be regarded as absolute timing values. Then, the testbench waits for 13 complete clock pulses before it finishes the simulation. As shown, an **always** block concurrent with the *running* block continuously generates clock pulses of 5 ns duration.

Also concurrent with these blocks is another **always** block that generates random data on t, and uses t to delay the assignment of random values to x. This block generates data on the x input for as long as the **$finish** statement in the *running* block is not reached. Assume that it is desirable to check if the state machine of *residue5* ever meets state *Three* or not. The last **always** block in this testbench waits on observing 2'b11 on the internal state of the Reside-5 circuit (the *current* **reg**), and if it is found, it will be reported.

```
module test_residue5;
   reg reset, clock;
   reg [1:0] x;
   wire [2:0] z;

   residue5 MUT ( clock, reset, x, z );

   initial begin :running
      clock = 1'b0; x = 1'b0;
      reset = 1'b1; reset = #7 1'b0;
      repeat (13) begin
         @( posedge clock );
         @( negedge clock );
      end
      #5;
      $finish;
   end

   always #5 clock=~clock;
   always begin
      t = $random;
      #(t) x=$random;
   end

   always begin
      wait (MUT.current == 2'b11);
      $display("state is 2'b11");
   end
endmodule
```

Fig. 2.41 Testbench using random time intervals

2.8.7 Text IO

Input and output from external files are discussed here. In VHDL, this is referred to as Text IO, and we use the same terminology here. The input side of Text IO means that instead of generating test data, a testbench can apply data to the MUT from a pre recorded text file. This is equivalent to a *stored vector* testing that is done by an ATE. In this book, using this type of providing data is very common.

Figure 2.42 shows a testbench that uses Text IO to read data and expected output of the MUT. Three file pointers *dataFile, responseFile*, and *reportFile* of type **integer** are declared and are assigned in the first **initial** block to three physical text files "*Res5.dat*," "*Res5.rsp*," and "*Res5.rpt*," respectively. This assignment is performed by using the **$fopen** task function. The second argument in these statements shows the mode of opening file, which could be "*r*" as read, "*w*" as write, and "*a*" as append.

The next **initial** block is responsible for reading data and the expected output from the related text files, managing the required delay and then collecting the responses of the MUT and comparing them with the expected values. All of the mentioned processes continue until the end of one or more of the input files is reached; this condition is checked with the **$feof** task function in the condition part of the **while** statement.

The data reading from the input files can be done using **$fscanf** or **$fread**. In this case, **$fscanf** is used. This task function has an integer return value which shows if the reading was successful or not. Therefore, variable *status* of type **integer** should be declared and used here. After reading and applying data from *dataFile* to *d_in*, and reading the expected response from *responseFile*, the testbench waits for the **posedge** of the clock to make sure that the input is affected, and the internal state of the circuit has been changed. Then after a very short delay, the expected and the actual responses of MUT

```
module test_residue5;
   reg  rst, clk;
   reg [1:0] d_in;
   wire [2:0] d_out;
   reg [2:0] expected_out;
   integer dataFile, responseFile, reportFile, status;

   residue5 MUT (clk, rst, d_in, d_out);

   initial begin
     clk = 0;
     rst = 1'b1; #7; rst = 1'b0;
     dataFile = $fopen("Res5.dat", "r");
     responseFile = $fopen("Res5.rsp", "r");
     reportFile = $fopen("Res5.rpt", "w");
   end

   always #5 clock=~clock;

   initial begin
     while((!$feof(dataFile) && (!$feof(responseFile))
     begin
         status = $fscanf(dataFile, "%b\n",d_in);
         status = $fscanf(responseFile, "%b\n",expected_out);
         @posedge(clk);
         #1; if(expected_out == d_out)
                 $display("correct output = %d", d_out);
             else
                 $fdisplay(reportFile, "wrong output.. d_out= %b\t
                           expected_out = %b\n", d_out, expected_out);
     end
     #1;
     $finish;
   end

endmodule
```

Fig. 2.42 Testbench using text IO

can be compared. Since the internal state of the circuit changes right at the **posedge** of the clock, this 1 ns delay guarantees that we are not looking at the previous state of the circuit. The correctness of results is reported on the console of the simulator using **$display** or **$monitor**, and in case that they are not equal, it can be reported in a text file using **$fprintf**, **$fwrite**, or **$fdisplay**. Concurrent with the mentioned **initial** blocks, an **always** block generates a periodic clock with a 10 ns period.

2.8.8 Simulation Code Coverage

A good testbench that can verify the correctness of a design should guarantee that is able to exercise most of the design under test and especially its critical parts. The percentage of the statements, blocks, paths, etc., in a design that are covered using a testbench is the *code coverage* of that testbench. Most of the simulation environments provide tools to estimate the code coverage for testbenches. During the compilation part in an HDL simulator, the kind of required code coverage can be specified; then, the simulator calculates the specified type of code coverage for the instantiated design. If the resulted code coverage is less than expected, it can be decided that the testbench is not a good quality testbench. The parameters that most of the HDL simulators support for code coverage include *statement coverage*, *condition coverage*, *block coverage*, and *branch coverage*.

Code coverage matrices measure how much of the design a testbench covers. On the other hand, if we want to estimate how much of the possible design faults this testbench covers, we must apply this testbench to the post-synthesis model of the design, and simulate it. Reports generated by this simulation are called fault coverage. We may think, or hope, that high-level code coverage and low-level fault coverage somehow correspond. Although this correspondence is very weak, but using code coverage we can have a sense of how good a testbench would be for gate-level fault simulation. The advantage of using high-level simulation is that it is much faster than the gate-level simulation. Therefore, the fast behavioral testbench can be performed as an estimation of a good testbench; it can get matured in this level of simulation at a lower cost and then get adjusted for covering more faults.

As an example, Figs. 2.43 and 2.44 show the Verilog code of a *Comparator* and its block diagram on which various code coverages are depicted.

```verilog
module Comparator (input a, b, output a_gtoreq_b, a_lt_b);

    always @ ( a, b ) begin
        if ( a < b ) begin
            a_gtoreq_b = 0;
            a_lt_b = 1;
        end
        else begin
            a_gtoreq_b = 1;
            a_lt_b = 0;
        end
    end
endmodule
```

Fig. 2.43 Behavioral code of a comparator

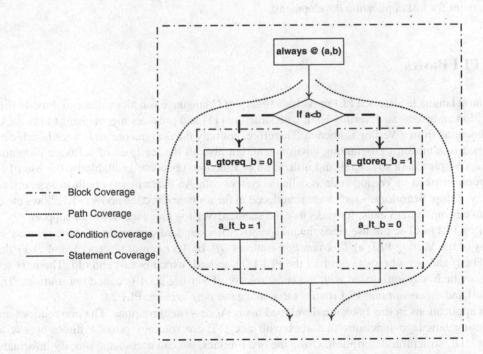

Fig. 2.44 Types of code coverage for a comparator

In these figures, output *a_gtoreq_b* becomes 1 when the first input is greater than or equal to the second input, and *a_lt_b* checks if *a* is less than *b* or not. Figure 2.44 shows the block diagram related to this code, and various types of code coverages are specified with different line styles.

Condition coverage means how many of the edges of all the conditions in the code will be visited with the related testbench. For example in Fig. 2.44, if the testbench is such that *a* is always less than *b*, then the left-hand side branch of the condition is never covered.

In this block diagram, *statements* are identified using solid lines. *Statement coverage* specifies how many of these statements can be examined by the testbench.

Two curved dotted lines on the sides of the block diagram show two *paths* in this code. *Path coverage* shows how many paths in a design are covered using a certain testbench. For example, in a **case** statement, the code branches out to many paths and they converge at the same place, and there might be some paths from the divergence to the convergence point that have not been examined.

Finally, the dash-and-dot lines in Fig. 2.44 represent the *blocks* of the code in which their coverage can be calculated using the *block coverage* option of the simulator. To this point, we have discussed the HDL techniques to develop testbenches useful for HDL design. However, as mentioned at the beginning of this chapter, the main objective of using HDLs and testbenches in this book is utilizing their facilities to implement existing test applications and developing new ones. As mentioned in Sect. 2.2, an HDL environment can provide utilities to develop fault simulation, test generation, DFT evaluation and configuration, and various other test applications. However, some facilities are required for test purposes that Verilog HDL basic constructs do not provide. For example, Verilog is not able to model a defective wire without making changes in the components of the original design [9]. In addition, a number of test utilities such as fault compilation and testability measurements need to explore the gate-level netlist of a design at a reasonable cost. In the standard Verilog language, this cannot easily be done since it does not have mechanisms for creating software-like structures. Fortunately, these drawbacks of HDL environment are compensated for by using the PLI of Verilog. PLI also has other capabilities that facilitate integration of design and test [10]. The following section briefly introduces PLI and its features and illustrates how it can be useful for providing a convenient environment for test application development.

2.9 PLI Basics

Procedural language interface PLI provides a library of C language functions that can directly access data within an instantiated Verilog HDL data structure [11] and provides mechanisms to invoke C or C++ functions from a Verilog testbench. Therefore, not only the design core and its testbench can be developed in a uniform programing environment, but also all the facilities of software programing (such as complex data structures and utilization of functions) become available by the use of PLI. A function invoked in Verilog code is called a system call. An example of a built-in system call is **$display**, **$stop**, **$random**, which were introduced in the testbench section above. PLI allows the user to create custom system calls, for tasks that the standard Verilog language does not support.

Verilog PLI has been in use since the mid-1980s. This standard comprises of three primary generations of the Verilog PLI: a) *Task/function* routines (*tf*), b) *Access* routines (*acc*), and c) *VPI* routines. The *tf* and *acc* libraries construct the PLI 1.0 standard, which is vast and old. The next set of routines, which was introduced with the latest release of Verilog 2001 is called *vpi* routines. These are small and down-to-point PLI routines that make the new version, PLI 2.0.

Test applications in this book are developed using Access (*acc*) routines. The *acc* routines are C programing language functions that start with **acc_**. These routines provide direct access to a Verilog HDL structural description. Using the *acc* routines, we can access and modify information, such as delay and logic values on various objects in a Verilog HDL description. More information about these routines can be found in the next subsection.

2.9.1 Access Routines

Access routines are C programing language routines that provide procedural access to information within Verilog-HDL. Access routines perform one of two operations, read or write. Using read operations, certain data and information can be obtained about particular objects in the circuit directly from its internal data structure. The objects that access routines can perform read operations for, included module instances, module ports, module paths, intermodule paths, top-level modules, primitive instances, primitive terminals, **net**s, **reg**s, **parameter**s, **specparam**s, timing checks, named events, integer, and real and time variables. Write operations replace new data or information for objects in the circuit by directly changing the related variables into the internal data structures. Access routines can write to intermodule paths, module paths, primitive instances, timing checks, register logic values, and sequential UDP logic values.

According to the operation performed by access routines, they are classified into six categories: 1) *Fetch* routines return a variety of information about different objects in the design hierarchy, 2) *Handle* routines return handles – the pointer to an object PLI in the data structure, to a variety of objects in the design hierarchy, 3) *Modify* routines alter the values of a variety of objects in the design hierarchy, 4) *Next* routines when used inside a loop construct can find each object of a given type that is related to a particular reference object in the design hierarchy; for example, ports of a module, the instantiated modules within it – which are called its children, or the module which instantiated this module – which is called its parent, 5) *Utility* routines perform a variety of operations, such as initializing and configuring the access routine environment, and 6) *Vcl* or Value Change Link (*VCL*) allows a PLI application to monitor the value changes of selected objects. VCL can monitor value changes for events, scalar and vector registers, scalar nets, bit-selects of expanded vector nets, and unexpanded vector nets. On the other hand, VCL cannot extract information about the following objects: bit-selects of unexpanded vector nets or registers, part-selects, and memories.

2.9.2 Steps for HDL/PLI Implementation

Figure 2.45 shows the general view of implementing and running test programs in a mixed HDL/PLI environment. All test applications in this book are implemented based on this block diagram.

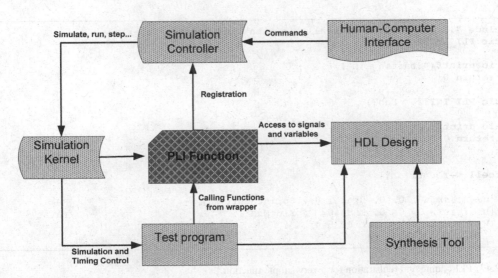

Fig. 2.45 The general view of running test programs in mixed HDL/PLI environment

The main part of this block diagram is the PLI function; the PLI functions should be written and compiled using a C compiler. A number of examples for writing PLI functions are given in the following. After completing the C code for the PLI function using the *acc* routines, the provided function must register its system tasks and functions with the HDL simulator. Registering each system task and function must be performed by filling the entries of an array of *s_tfcell* structures shown in Fig. 2.46. The resulted *struct* must take place at the end of the C code which implements the PLI function.

To make these steps more clear, a simple PLI function (perhaps the simplest possible) is shown in Fig. 2.47. This is just a simple PLI call for printing a message. The first line of this code includes the **veriuser.h** header to be able to use the **io_printf** function. **veriuser.h** and **acc_user.h** are two header files in the directory of the HDL simulator installation path, and should be included in the C code of the PLI function to have access to the routines of these libraries.

The last part of this C code performs the registration of these PLI functions with the HDL simulator. This code varies from one simulator to another, and we have shown this for Mentor Graphics's *ModelSim* simulator. As depicted in this code, there must be one entry for each declared function in this code, and the last entry of the **veriusertfs** must always be zero. The first field in each entry shows the type of the declared function, which for common uses we usually set it as **usertask**. This value means that the registered task does not return any value. The last field declares the name that the PLI function will be invoked with in the HDL testbench (notice the *$* character at the beginning of these names). The fifth field is the name of the C function that describes the PLI function.

```
typedef struct t_tfcell
{
    PLI_INT16 type;      /* USERTASK, USERFUNCTION, or USERREALFUNCTION  */
    PLI_INT16 data;      /* passed as data argument of callback function */
    p_tffn    checktf;   /* argument checking callback function          */
    p_tffn    sizetf;    /* function return size callback function       */
    p_tffn    calltf;    /* task or function call callback function      */
    p_tffn    misctf;    /* miscellaneous reason callback function       */
    char *    tfname;    /* name of system task or function              */
}
```

Fig. 2.46 *t_tfcell* struct for registering the PLI function with the HDL simulator

```
#include "..\\HDLsimulatorInstallationPath\\include\\veriuser.h"
static PLI_INT32   Start()
{
    io_printf("Starting...\n");
    return 0;
}
static PLI_INT32   End()
{
    io_printf("Ending...\n");
    return 0;
}

s_tfcell veriusertfs[] =
{
    {usertask, 0, 0, 0, Start, 0, "$printStart"},
    {usertask, 0, 0, 0, End, 0, "$printEnd"},
    {0} /*last entry must be 0 */
};
```

Fig. 2.47 PLI coding and registration for a very simple function

```
module verilog_test();

initial
    $printStart();
    //   . . .
    //   . . .
    $printEnd();
    $stop();
endmodule
```

Fig. 2.48 PLI function call in HDL testbench

After providing this C code and compiling it with the C compiler, it must be built to generate a Dynamic Linked Library (*.dll1*) file. When the C part is done, we get to the HDL simulator part. The resulted *.dll1* file must be placed in the working directory of the HDL simulator. For invoking the prepared PLI function, the pseudo code in Fig. 2.48 can be used.

In order to simulate this testbench in the presence of the PLI.*dll* in ModelSim simulation environment, the following command must be performed in the simulator console.

vsim −c −pli dllFileName TestbenchName

In this line, **vsim** is the simulation command, **-c** is for the command mode, **-pli** means in the presence of the *.dll* file the name of which appears next, and finally the name of the top-level module or the testbench must be declared. In order to link more than one PLI *.dll* file to the HDL project, the following command should be used.

vsim −c −pli dllFileName_1 −pli dllFileName_2 … −pli dllFileName_n TestbenchName

By running this command, the simulation of the testbench and designs added to the project is done, and it can be run like any other normal HDL project. As a result of running this testbench, the following lines will be printed on the simulator console.

Starting…
Ending…

In the next subsection, the implementation of fault injection and removal as more complex PLI functions and also a very important part of test applications are discussed.

2.9.3 Fault Injection in the HDL/PLI Environment

The most important utilities for implementing most of test algorithms are fault injection (FI) and fault removal (FR) functions. As mentioned, PLI provides mechanisms for reading and writing **net** and **reg** values. Therefore, we can force and release values in the data structures corresponding to **net**s, which give us the capabilities for FI and FR on and from circuit lines. In PLI, a **handle** is a pointer to a specific object in the design hierarchy. **handles** give information about a unique instance of a special object to *acc* routines. They contain useful information such as how and where we can find data about the object. For reading and writing information about an object, most *acc* routines require a handle argument. For each input argument of a PLI function, a variable of type **handle** will be used.

The FI and FR processes are done simply by using the **acc_set_value** PLI routine that sets the desired value on the target wire or removes the value from it. In order to implement PLI *InjectFault* and *RemoveFault*, there are two **structs** named **s_setval_value** and **s_setval_delay**, for which several fields must be set. However, the most important fields that need to be mentioned are the **model** field in **acc_setval_delay** and the **value** field of **acc_setval_value**. Figure 2.49 depicts these two **structs** for one of the input ports of an AND gate.

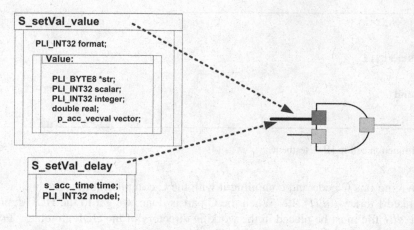

Fig. 2.49 PLI structures for stuck-at fault injection and fault removal

```
static PLI_INT32 injectFault ()
{
    arg1 = acc_handle_by_name ((char*) acc_fetch_tfarg_str(1), null);
    arg2 = acc_handle_tfarg(2);

    //get value of arg2 into value_of_arg2
    value_of_arg2.format = accScalarVal;
    acc_fetch_value (arg2, "%%", &value_of_arg2);

    //prepare a data
    value_for_arg1.format = accScalarVal;
    value_for_arg1.value.scalar = Value_for_arg3.value.scalar;

    //prepare delay mode
    delay_of_arg1.model = accForceFlag;
    delay_of_arg1.time.type = accSimTime;
    delay_of_arg1.time.low = 0;
    delay_of_arg1.time.high = 0;

    //Put it in arg1
    acc_set_value (arg1, &value_for_arg1, &delay_for_arg1);

    acc_close();
    return 0;
}
```

Fig. 2.50 Fault injection PLI code

In FI, the model field must be defined as **accForceFlag**. This means that the desired value will be forced on the wire until it is removed by calling a PLI function for FR. During this time, the wire will not take values assigned to it by the Verilog simulator. The C code for the PLI FI is shown in Fig. 2.50.

The *removeFault* function sets the **s_setval_delay** model field to **accReleaseFlag**. Once this is done, values coming from HDL simulator will again appear on the wire. In other words, by putting the desired fault value on a variable of type **s_setvalue_value** and setting the model field of a variable of type **s_setvalue_delay** to **accForceFlag** or **accReleaseFlag**, FI and FR can be achieved. Figure 2.51 shows that the faulty value of the selected wire is applied by the PLI **$InjectFault** function. Only after calling **$RemoveFault** for that wire, it will accept the normal values, propagated to it by the HDL testbench.

Figures 2.52 and 2.53, respectively illustrate the usage of inject and remove fault on the wires of a full adder and the resulted waveform.

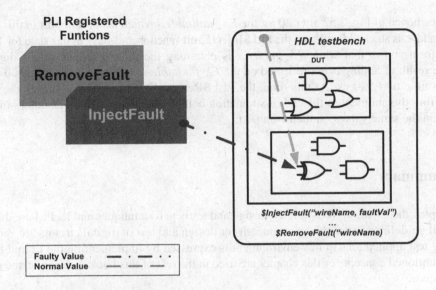

Fig. 2.51 Stuck-at fault injection and fault removal mechanism

```
module testbench();
reg a, b, cin;
wire sum_f, sum_g;
wire co_f, co_g;

    FA FA_ golden (a, b, cin, sum_g, co_g);
    FA FA_faultable (a, b, cin, sum_f, co_f);

    initial begin
        #20;
        $InjectFault("testbench.FA_faultable.s", 1'b0);
        repeat(10) begin
            #150;
            {a,b,cin} = $random();
        end
        $RemoveFault("testbench.FA_faultable.s");
        repeat(10) begin
            #150;
            {a,b,cin} = $random();
        end
        $stop;
    end
endmodule
```

Fig. 2.52 Fault injection and removal for a full adder – testbench

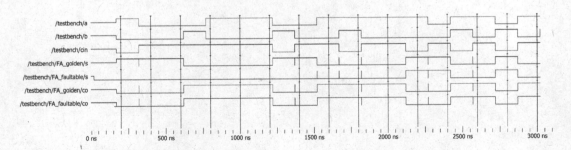

Fig. 2.53 Fault injection and removal for a full adder – waveform

In the testbench of Fig. 2.52 after 20 ns, the *FA_ faultable.s* which is the *sum* port in this instance of the full adder, is stuck to 0 utilizing the PLI **$InjectFault** function and stays in this state for 1,500 ns. The waveform shows that while *FA_golden.s* is obtaining the related output values during this period, the result of adding is not reflected on the *FA_faultable.s*, and its value is always 0 until the simulation time of 1,520 ns. At this time, the PLI **$RemoveFault** function, removes the injected fault and from this moment to the end of simulation both faultable and golden instances of the full adder obtain the same values on their *sum* port.

2.10   Summary

In this chapter, the basics of Verilog HDL design and testbench techniques and its PLI are discussed. The overall guidelines to use this environment for design and test of digital circuits are shown and developing test applications in this environment is expressed by implementing the FI and FR utilities. All mentioned concepts of this chapter are used in the rest of this book to describe and enhance test techniques.

References

1. The International Technology Roadmap for Semiconductors (ITRS) website. (2007) [Online]. Available: http://www.itrs.net/
2. Ungar LY, Ambler T (2007) "Economics of Built-In Self-Test," *ITC*
3. IEEE Std 1364-2001, IEEE Standard Verilog Language Reference Manual, SH94921-TBR (print) SS94921-TBR (electronic), ISBN 0-7381-2827-9 (print and electronic), 2001
4. Stephen B, Zvonko V (2002) *Fundamentals of digital Logic with Verilog design*, McGraw-Hill; ISBN: 0-07-283878-7
5. Zainalabedin N (2006) *Verilog digital system design: RT level synthesis, testbench, and verification*, McGraw Hill, ISBN: 0-07-144564-1
6. Zainalabedin N (2007) *Embedded core design with FPGAs*, McGraw Hill, ISBN: 978-0071474818
7. Neil H.E. Weste and David Harris, CMOS VLSI Design: *A Circuits and Systems Perspective (3rd Edition)*, Addison Wesley; 3rd edition (May 11, 2004), ISBN: 0321149017
8. Patterson DA, Hennessy JL, Ashenden PJ, Larus JR, Sorin DJ *Computer Organization and Design: The Hardware/Software Interface, Third Edition*, Morgan Kaufmann; 3 edition (August 2, 2004), ISBN: 1558606041
9. Hesscot CJ, Ness DC, Lilja DJ (2005) "A methodology for stochastic fault simulation in vlsi processor architectures," *MoBs*
10. Riahi PA, Navabi Z, Lombardi F (2005) "Simulating Faults of Combinational IP Core-based SOCs in a PLI Environment," *DFT*
11. IEEE Std 1364–2001, IEEE Standard Procedural Language Interface Reference Manual, clause 20 through clause 25

Chapter 3
Fault and Defect Modeling

A model of a physical object or model of a phenomenon is a representation of the object or phenomenon that is used for the specific purpose of analyzing the behavior of the object, studying the phenomenon, or studying the effect of the object or phenomenon on its environment or surroundings. A computer simulation program uses a model. The information we obtain from running a simulation program depends on the model that is used for simulation. For example, a model of a circuit can be developed for predicting its temperature radiation; its logical behavior, or its behavior of when the circuit becomes faulty.

Just as the binary logic value system, containing 0 and 1, is used as a simplified model for complex line values in digital systems, a simplified value system is needed to model faults on circuit lines. Such a fault model should be simple and should be able to facilitate analysis of faulty behavior of a digital system.

This chapter discusses fault models and issues that are related to fault analysis and simulation of a circuit that is faulty. We try to set a solid background for the reader in understanding faults and circuit fault models, as this background is crucial in understanding the materials in the rest of this book. We discuss various fault models, ways of reducing faults for faster analysis, and simulation issues for generating fault lists.

3.1 Fault Modeling

This chapter is on fault and defect modeling. Before we discuss various ways in which such modeling can be done, we need to give a clear definition of various terms that are used in this regard.

As discussed in Chap. 1, a *defect* in an electronic system refers to a flaw in the actual hardware.

A *fault*, on the other hand, is a representation of a defect and is used in computer programs for analyzing defects in electronic components.

An *error* is caused by a defect, and it happens when a defect in hardware causes a line or a gate output to have a wrong value. In a computer simulation program, an error is defined as an observed fault.

A *failure* occurs when a defect causes a misbehavior in the functionality of a system that cannot be reversed or recovered. In a computer simulation program, change in the intended functionality of a system due to an existing fault is referred to as a failure.

In order to be able to predict the behavior of a defective system, or generate tests to find the defective parts, or in general, for any type of computer analysis of defects, a good fault model is needed. A good fault model is one that has a close correspondence with the actual defects, it is easy to represent in a computer program, and it is as brief as possible for an optimized computer processing time.

Z. Navabi, *Digital System Test and Testable Design: Using HDL Models and Architectures*,
DOI 10.1007/978-1-4419-7548-5_3, © Springer Science+Business Media, LLC 2011

At the end of this section, we present our fault model that satisfies these requirements, and Sect. 3.2 that follows this section presents the details of such a fault model. However, before we narrow down our choices of faults to the model that we use, we need to explain various faults, their applications, short comings, and the kind of hardware description that they apply to. Such categorization of fault is explained here.

3.1.1 Fault Abstraction

As defects occur in a circuit, their effects may be seen at various levels of abstraction [1–4]. For example, a physical defect of a short between an nMOS transistor gate and its source may be seen at the switch, gate, RT, or system-level. Figure 3.1 is a graphical representation of physical, switch, gate, RT, and system abstraction levels.

At the physical level, mask misalignment in layout (Fig. 3.2) may cause a short between gate of a transistor and its source [5]. As shown in Fig. 3.3, such a defect has different switch level behaviors depending on how the transistor is used [6, 7]. In Fig. 3.3a, this defect translates to the T_2 switch always transistors staying open. Thus, this defect becomes a switch open fault in T_2.

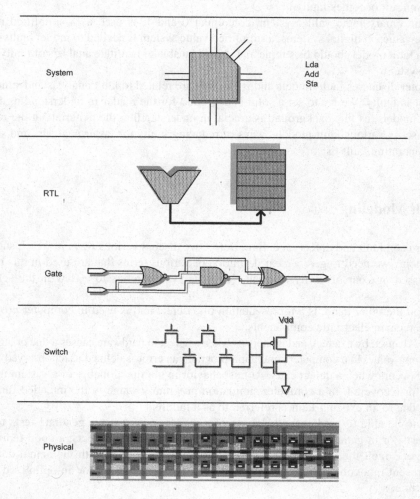

Fig. 3.1 Abstraction levels

Fig. 3.2 Mask misalignment

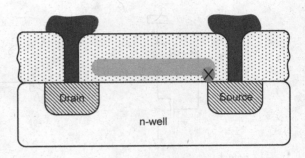

Fig. 3.3 Switch level faults

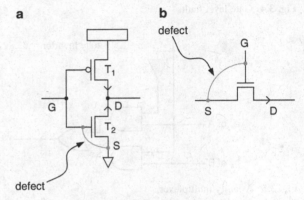

On the other hand in Fig. 3.3b, the same defect translates to switch level fault of switch output being open when it is supposed to be 0. In this scenario, the switch transmits a 1 on its source correctly to its drain.

The defect of Fig. 3.2 causes different gate-level faults depending on the type of the gate, its structure, and the conditions that the gate is used in. Gate input driving strengths, output load capacitance, and other inputs and outputs use conditions affect how a switch fault translates to a gate fault. Figure 3.4 shows a possible inverter gate fault caused by the defect of Fig. 3.2.

The same defect discussed above can cause various RT-level faults depending on the RT-level component function and its logic structure. Some examples of RT-level faults are an adder producing the wrong result, a multiplexer selecting a wrong source, or a bus ORing several of its sources instead of selecting one. As an example of an RT-level fault and how a gate-level fault translates to such a fault consider the circuit of Fig. 3.5.

Figure 3.5 shows a bit of a multiplexer selecting vector A or B depending on s being 0 or 1. This multiplexer uses AND-OR CMOS gate structure, and let us assume that its CMOS inverter is the faulty inverter of Fig. 3.4. If s is 0, according to Fig. 3.5, the inverter output (line i) is 1 and the A input of the multiplexer is selected to go on the W output. If s changes from 0 to 1, the faulty inverter output (line i) becomes Z, and since the AND gate that this inverter is driving is a CMOS gate, the open input causes it to retain its old value of 1. Since s is 1, both i and j lines are 1. This input combination causes w_k to become the OR result of a_k and b_k. This means that when s changes from 0 to 1, instead of W becoming B, it becomes A OR B. This is an RT-level fault that is caused by the defect of Fig. 3.2.

The same defect that has resulted in switch level faults of Fig. 3.3, gate-level fault of Fig. 3.4, and RT-level fault of Fig. 3.5 can result in various forms of system-level faults. As in other abstraction levels, the exact form of the system fault depends on the type of the system and the condition in which the faulty RT-level component is used. A system-level fault in a CPU may be execution of a wrong instruction, or in a NoC, a system fault may be a NoC switch directing data to a wrong destination.

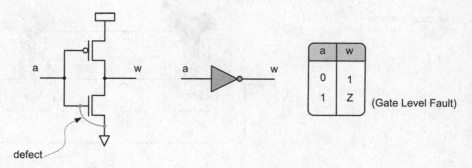

Fig. 3.4 Gate level faults

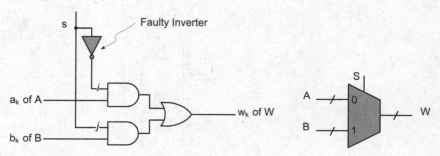

Fig. 3.5 A faulty multiplexer

Fig. 3.6 A system-level fault, executing an instruction incorrectly

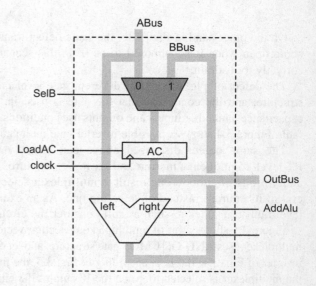

Consider Fig. 3.6 as a partial datapath of a CPU. This datapath section includes an ALU, an accumulator, three busses, and a multiplexer that we assume is the defective multiplexer of Fig. 3.5. Let us assume that instruction *lda* followed by *add*, as described below, are to be executed in this datapath.

$$lda : AC \leftarrow ABus$$

$$add : AC \leftarrow ABus + AC.$$

For *lda*, *SelB* becomes 0, the faulty multiplexer functions correctly, and data on *ABus* is correctly loaded into *AC*, which becomes available on *OutBus*. Following this instruction, *add* is to execute, a requirement of which is the ALU input *AddAlu* to become 1. In order to route the output of the

ALU into *AC*, *SelB* input of the multiplexer must also become 1. Because of the faulty multiplexer, *SelB* of 1 causes *ABus* ORed with *BBus* to appear on the multiplexer output. This will cause the accumulator to be loaded with

$$AC \leftarrow (ABus + AC) \,|\, ABus$$

instead of

$$AC \leftarrow ABus + AC.$$

This example shows that the gate-source short-defect of Fig. 3.2 can lead to an instruction fault at the system-level in a CPU.

As another system-level fault consider the NoC switch of Fig. 3.7. The switch router shown here selects source *A* or *B* for the switch destination *D*. When the router selects *A*, correct data from *A* will appear on *D*. However, when port *B* is selected, the faulty multiplexer places data on port *A* ORed with that of port *B* on the input of destination *D*. This is because of the faulty behavior of the multiplexer of Fig. 3.5 and is contrary to the correct operation of the switch that should route data on *B* to port *D*.

Examples presented above show how a low-level defect appears as faults at various abstraction levels. We can always deduce how a low-level defect appears at a higher level, but the opposite of this is not necessarily true. For example, referring to Fig. 3.6, there may be many reasons why an *add* instruction ORes one of the add operands with the result. Such a system fault cannot be traced to a specific lower level fault, even to a fault at its most immediate lower abstraction level.

3.1.2 Functional Faults

A fault that affects functionality of a system is said to be a *functional fault* [5, 7]. A gate-level functional fault causes a faulty gate to have a different truth table than the non-faulty gate. For example, a fault in a NAND gate may change its functionality to become like a NOR or an inverter. At the system-level, a system-level functional fault causes the system to perform a function different than what the system was originally designed for. For example, in a processor, a functional fault may cause an add instruction to execute as subtraction, or fetching of data or instruction to be done from a wrong location.

Faults discussed in Sect. 3.1.1 were all functional faults. For example, the inverter of Fig. 3.4 has a different truth table than a good inverter, as shown in Fig. 3.8.

At a higher level of abstraction, consider the multiplexer of Fig. 3.5. Because a faulty inverter is used in this circuit, the overall functionality of the multiplexer is faulty. Without looking at the internal structure or circuitry of this circuit, we can just say that the multiplexer has a functional fault. The good and faulty functions of the multiplexer are shown in Fig. 3.9.

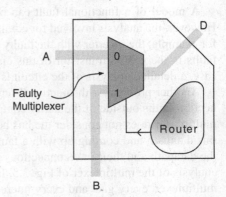

Fig. 3.7 NoC switch with faulty multiplexer

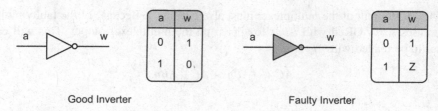

Good Inverter   Faulty Inverter

Fig. 3.8 Good and faulty inverters

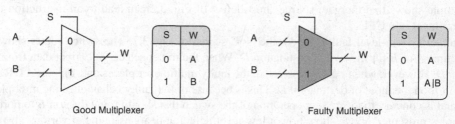

Good Multiplexer   Faulty Multiplexer

Fig. 3.9 Good and faulty multiplexers

As another functional fault, at a higher level of abstraction than the above two faults, consider the section of the datapath depicted in Fig. 3.6. As in the above cases, we only look at the datapath section from the outside, and see that under certain conditions, the datapath functions incorrectly. This again is regarded as a functional fault at the system-level.

Good and faulty partial Verilog codes of this datapath section are shown in Fig. 3.10. Notice that the details of the datapath such as the multiplexer, ALU, and their interconnections are not described here, and only its overall functionality as it relates to the inputs and outputs of this system is mentioned.

It is worth mentioning again that a functional fault of a component at a certain level of abstraction (e.g., gate-level) ignores lower level details of the component (e.g., transistor-level), and only considers the input–output behavior of the component. At an abstraction level, the exact cause of the faulty behavior is irrelevant.

3.1.3 Structural Faults

In Sect. 3.1.1, we had a faulty transistor that was used in an inverter that made the inverter function incorrectly. The inverter was used in a multiplexer that made it faulty, and the multiplexer was used in a datapath that made the datapath function incorrectly. By a careful analysis of the source of the fault, we were able to model the incorrect functionality of the faulty component.

A model of a functional fault can be a very accurate representation of the fault at that level. However, the analysis involved for coming up with the model may be a very involved process. Take, for example, the inverter with the faulty transistor (Fig. 3.4). Other faults in the nMOS transistor, or faults in the pMOS transistor of this circuit may lead to many different truth tables, and for each case, a detailed analysis of the circuit is needed.

Another problem with functional representation of a fault is that it does not consider interconnection faults outside of the faulty component. For example, in Fig. 3.6, the functional fault of the multiplexer does not consider the bus between the multiplexer and the accumulator. For analyzing this datapath and coming up with a fault model for the datapath, faults for every sub-structure of the datapath and their interconnections must be considered. A similar statement can be said about analysis of the multiplexer of Fig. 3.5. In this case, in order to come up with the faulty behavior of multiplexer, every gate and every interconnection must be analyzed for their faulty behavior.

```
always @(posedge clk) begin
   if (loadAC == 1 ) begin
      . . .
      if (selB == 0 )
         AC <= Abus;
      else if ((selB == 1 ) && ( AddALU == 1))
         AC <= AC + Abus;
      . . .
      end
   end
assign outBus = AC;
                                                                                Good Behavior
```

```
always @(posedge clk) begin
   if (loadAC == 1 )  begin
      . . .
      If (selB == 0 )
         AC <= Abus;
      else if ((selB == 1 ) && ( AddALU == 1))
         AC <= ( AC + Abus ) | Abus;
      . . .
      end
   end
assign outBus = AC;
                                                                                Faulty Behavior
```

Fig. 3.10 Good and faulty behavior of the datapath section of Fig. 3.6

Complexity of analysis for extracting a faulty model for a component (e.g., the multiplexer) by analyzing its sub-components (e.g., inverter, AND, and OR gates) and their interconnections (e.g., line i in Fig. 3.5) can be simplified by only considering interconnection faults, and lumping faults on the two ends of an interconnection and those of the interconnection itself into faults belonging only to the interconnection.

This fault model is called structural fault model and assumes that components forming a hardware module are fault free, and only the interconnection of the components may be faulty [5, 7, 8].

Referring to Fig. 3.4, gate-level *structural fault* for this component means that the inverter is good and lines a and w have the potential of having faults. If this inverter is used in a gate-level circuit, such as the multiplexer of Fig. 3.5 (a version of which is shown in Fig. 3.11), the inverter fault can be structurally modeled at the gate-level as line i being permanently stuck-at-1. This is justified because, when s is 0, gate capacitance of G_2 charges to 1. When s becomes 1, the faulty inverter produces a Z on its output, which causes charge on line i to remain at 1.

In addition to modeling the faulty behavior of G_1, line i stuck-at-1 also models certain faults of the G_2 NAND gate. Extending this concept to all gates of circuit of Fig. 3.11, all gate and interconnection faults can be structurally modeled by faults on various lines of this circuit.

Furthermore, extending a similar concept to higher abstraction levels, structural RT-level faults are defined as faults on busses interconnecting various RT-level components. Figure 3.12 shows candidates for RT-level structure faults in the partial datapath of Fig. 3.6.

At a higher level, when several systems are wired together, system-level structural faults are only considered on interconnections of such systems. For example, in a NoC consisting of several switches and processing elements, faulty switches, faulty processing elements, and faulty interconnections are all modeled by structural faults on interconnections. Figure 3.13 uses three switches providing communication between six processing elements. In this circuit, candidates for structural system-level faults are marked by boxed F's. As in other abstraction levels, we assume all nine PEs and switches are fault free.

Fig. 3.11 Inverter fault becomes
fault on line i

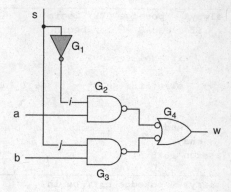

Fig. 3.12 Candidates for
RT-level structural faults

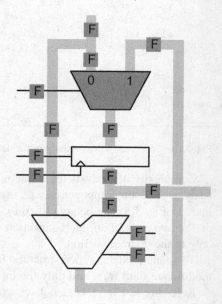

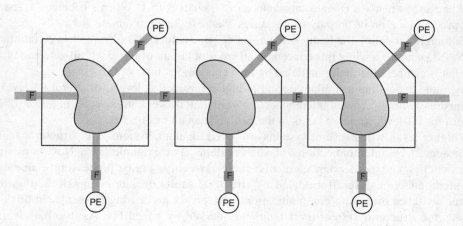

Fig. 3.13 NoC structural faults

3.2 Structural Gate Level Faults

Section 3.1.3 defined structural fault model as a model that lumps faults of the components between which an interconnection is used and faults of the interconnection itself, into faults belonging to the interconnect. The focus of this section is on gate-level structural faults. At the gate-level, an interconnection is a simple wire between two gates, called a line. In order to comply with this definition with regard to inputs, outputs, and multi-branch fanout systems, we introduce three new components: inputs, outputs, and fanouts, in addition to the standard gates of a gate-level circuit.

Figure 3.14 shows the multiplexer of Fig. 3.11 with the addition of these new components, I_1, I_2, I_3, O_1, and $FO1$. As shown, in this circuit, only a line connects two components together. l_1 is between I_1 and $FO1$ and l_2 is between $FO1$ and G_1.

The rest of this section discusses structural faults as related to nine lines of this circuit.

3.2.1 Recognizing Faults

Although materials presented in this sub section apply to structural and functional faults as well, we are presenting them here to be used in the proceeding sub-sections on structural gate-level faults. A structural fault within a component is recognized if it changes the functionality of the component. For example, in the multiplexer of Fig. 3.14, if l_4 is permanently stuck-at-1 (see discussion related to Fig. 3.11), the Boolean function at output w becomes

$$w_{\text{Faulty}} = a + b.s$$

Since this function is different than the good behavior of the circuit,

$$w_{\text{Good}} = a.\overline{s} + b.s$$

then, fault on l_4 can be recognized.

Figure 3.15 shows Karnaugh maps of the faulty and good w outputs. Input combinations where the two Karnaugh maps entries are different designate input combinations that recognize the faulty circuit from the good one. In this example, $abs = 101$ produces a 1 for the faulty circuit and 0 for the good circuit, and therefore recognizes l_4 stuck-at-1. In the materials that follow, we will only show the faulty Karnaugh maps, with shaded boxes for conflicts with the good circuit.

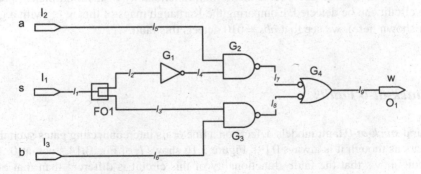

Fig. 3.14 Structural gate-level circuit

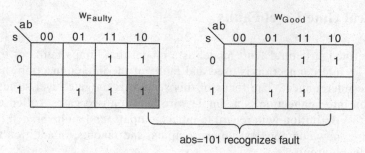

Fig. 3.15 Faulty and good Karnaugh maps

The topic discussed here is called *Boolean Difference*, which will be dealt with in detail in the chapter on test generation.

In Chap. 1, test set, test vector, and other related terms were formally defined. In this chapter, we use the term "test" very frequently. To avoid ambiguities, we define test as follows: A set of values at inputs of a faulty circuit that produces an output different than the good circuit output is called a test for the fault of the faulty circuit.

3.2.2 Stuck-open Faults

A familiar stuck-open fault discussed in the previous sections is l_4 stuck-open (Fig. 3.14). This structural fault accounts for defects on gates G_1, G_2 or line l_4. The faulty behavior of multiplexer in presence of this fault is different from the normal multiplexer output, and therefore, it is detectable.

As another example, consider line l_3 in Fig. 3.14 having a stuck-open fault. Note that l_3 and l_2 are branches of a fanout whose stem is l_1. Since structural faults on l_3 account for defects on l_3, $FO1$, and G_3, structural fault models of l_3 do not necessarily mean a bad fanout stem (l_1) or other fanout branches (l_2).

l_3 broken open causes the G_3 gate to retain its charge forever, causing it to act as an inverter for inverting the b input. If this is the only fault in the circuit, the faulty multiplexer output becomes

$$l_3 \text{ stuck-open} : w_{\text{Faulty}} = a.\overline{s} + b$$

Since this functionality is different from the normal functionality of the multiplexer, presence of a fault in the circuit can be detected. Comparing the Karnaugh maps of this w_{Faulty} with a good multiplexer (not shown here), we see that $abs = 010$ detects this fault.

3.2.3 Stuck-at-0 Faults

The structural *stuck-at-0* fault models defects on a line or its interconnecting gates such that the line value appears as though it is always 0 [9]. Figure 3.16 shows l_3 of Fig. 3.14 stuck-at-0. The output in this circuit shows that the faulty functionality of this circuit is different than that of the good circuit behavior.

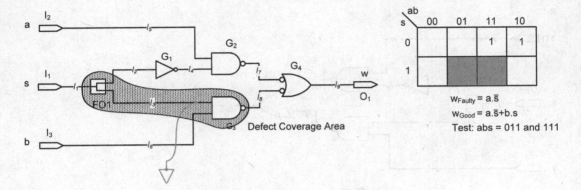

Fig. 3.16 Stuck-at-0 fault

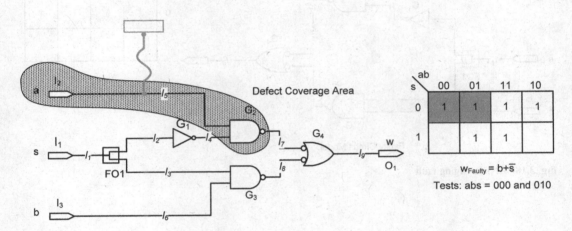

Fig. 3.17 Stuck-at-1 fault

3.2.4 Stuck-at-1 Faults

The structural stuck-at-1 fault models structural and functional defects that cause a line to appear as though it is pulled by a supply voltage and never changes [9]. In Fig. 3.17, the stuck-at-1 fault models defects on the I_2 input pin, G_2 NAND gate, and line l_5, which cause this line to look like it is pulled to a supply voltage. Faulty output expression, faulty Karnaugh map, and tests that detect this fault are also shown in this figure.

3.2.5 Bridging Faults

Defects on neighboring gates and lines may cause a bridging effect on two lines. Two types of bridging faults are AND bridging and OR bridging [10]. In an *AND-bridging*, two bridged lines appear as if they are forming an AND function which feeds all destinations of the bridged lines.

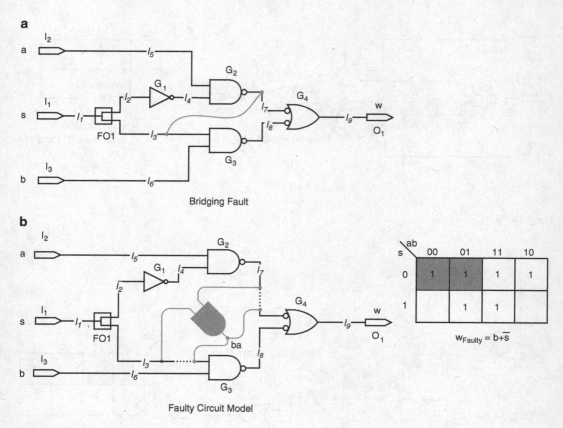

Fig. 3.18 AND-bridging fault

3.2.5.1 AND-bridging Faults

Figure 3.18a shows a bridging fault between l_7 and l_3, and Fig. 3.18b shows the AND-bridging fault circuit model and the faulty output of the multiplexer. The bridging output, *ba*, feeds both gates that the bridged lines used to feed. Based on this circuit, the faulty *w* becomes

$$w_{\text{Faulty}} = b + \overline{s}$$

and tests that detect this fault are

$$abs = 000 \text{ and } 010.$$

3.2.5.2 OR-bridging Faults

Depending on the logic behind and after a bridging fault, the fault may be modeled by an OR function instead of an AND function. Figure 3.19 shows an *OR-bridging* fault corresponding to the bridging fault of Fig. 3.18a. The faulty multiplexer output is also shown in this figure. The Karnaugh map shown here indicates that this fault can be detected by *abs* = 010.

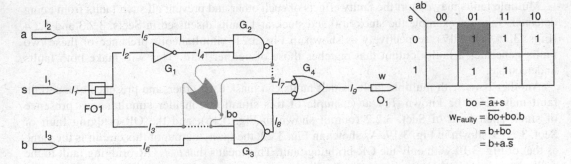

Fig. 3.19 OR-bridging fault

$$bo = \overline{a} + s$$
$$w_{Faulty} = \overline{bo} + bo.b$$
$$= b + \overline{bo}$$
$$= b + a.\overline{s}$$

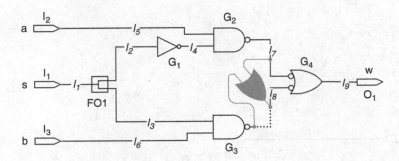

Fig. 3.20 Modeling a state-dependent fault

3.2.6 State-dependent Faults

Bridging faults between two lines affect the destination of both participating lines. In a different situation, a fault on a line may depend on a certain value on another line. While the latter line value affects that of the former, no faulty value appears on the latter line. This type of fault is referred to as a *state-dependent* fault.

Figure 3.20 models a state-dependent fault on line l_8 that depends on the value of line l_7. If l_7 is 0, l_8 has its normal value; however, when l_7 becomes 1, l_8 input of G_4 stays at 1 regardless of the output value of G_3. Note that the only difference between the circuitry modeling this fault and that of an OR-bridging fault is that in the latter case, the output of the modeling OR gate replaces destinations of l_7 and l_8, while in the former case, only l_8 is affected. The faulty output of this circuit becomes $w_{Faulty} = a.\overline{s}$, which is detectable by $abs = 011$ or 111.

Another type of a state-dependent fault is when the dependent line is affected by value 0 of the affecting line. This case is modeled by an AND gate, the output of which is used to model the faulty line.

3.2.7 Multiple Faults

Any of the faults discussed above and any number of such faults can happen in a circuit simultaneously. A major problem with multiple faults is that the combination of all types and many faults of each type become too many cases to deal with for finding the fault, generating test for the fault, or simply analyzing the faulty circuit [11, 12].

Multiple faults may distort the faulty effects of each other and prevent all such faults from being detected. Take for example the stuck-at-0 and stuck-at-1 faults discussed in Sects. 3.2.3 and 3.2.4 (Figs. 3.16 and 3.17), respectively as shown in Fig. 3.21, simultaneous presence of these two faults generates a faulty output that matches those of neither fault. This will make both faults undetectable.

Another issue with multiple faults is that faults can mask each other, and presence of a certain fault may never be known. For an example of this situation consider simultaneous presence of stuck-open fault of Sect. 3.2.2 (output shown in Fig. 3.15), and the OR-bridging fault of Sect. 3.2.5.2 shown in Fig. 3.19. As shown in Fig. 3.22, the faulty output of this circuit is the same as that of Fig. 3.19 with only the OR-bridging fault. This means that l_3–l_7 OR-bridging fault in the multiplexer masks the l_4 stuck-open fault. Thus, in presence of the former fault, the latter cannot be detected.

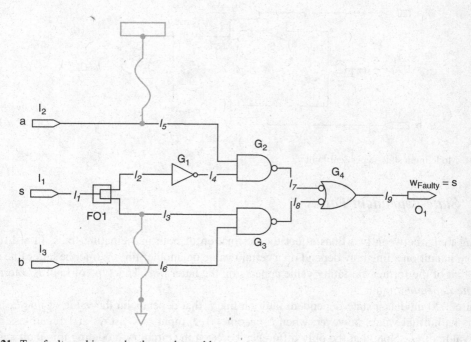

Fig. 3.21 Two faults making each other undetectable

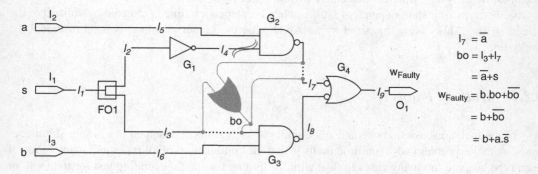

Fig. 3.22 Multiple faults (stuck-open and OR-bridging)

3.2.8 Single Stuck-at Structural Faults

Our search for a manageable fault model has narrowed down our choices to structural gate-level fault models. Choices for the types of structural faults are still too many, and multiple faults make the situation even harder to handle. Further simplifications are needed for a fault model that allows handling of complex algorithms processing millions of gates.

The first step in the simplification ahead of us is to select stuck-at fault model among other fault types, and the second step is to assume single fault and disregard multiple faults. Justifications for these simplifications are presented below.

3.2.8.1 Stuck-at Faults

Stuck-at faults are stuck-at-0 and stuck-at-1 which we will refer to as SA0 and SA1. These faults account for many transistor-level faults and can model other fault types.

We use the CMOS NAND gate of Fig. 3.23 to show how various transistor-level faults are translated to stuck-at faults at the gate-level [6]. To provide proper interfacing, and study the effect of faulty logic values on the surrounding circuits of this NAND gate, three CMOS inverters are used in inputs and outputs of this circuit. We use Verilog modeling and a Verilog testbench for analyzing transistor faults in the NAND gate.

Figure 3.24 shows the Verilog code of the CMOS NAND gate. This description uses *tranif0* and *tranif1* Verilog bidirectional switches for the nMOS and pMOS transistors. Gate capacitances of a and b inputs are modeled by the Verilog *trireg* construct with 50 time units representing the capacitance. For a more realistic representation of events in this circuit, 3 and 4 time unit delays are used for nMOS and pMOS transistors, respectively.

Inverters in Fig. 3.23 are modeled similar to the NAND gate using bidirectional switches with delay parameters. Figure 3.25 shows the corresponding Verilog code.

The Verilog code of circuit shown in Fig. 3.23 for study of faults in the NAND gate of Fig. 3.24 is shown in Fig. 3.26. This circuit uses an instance of the NAND gate of Fig. 3.24 and three instances of the inverter of Fig. 3.25.

The *GoldenCircuit* module along with the *FaultyCircuit* module will be used in a testbench to apply data to inputs and compare outputs. Verilog code of this testbench is shown in Fig. 3.27. This testbench applies 11, 00, 01, 10, ... to a and b inputs of circuit of Fig. 3.23 every 35 ns.

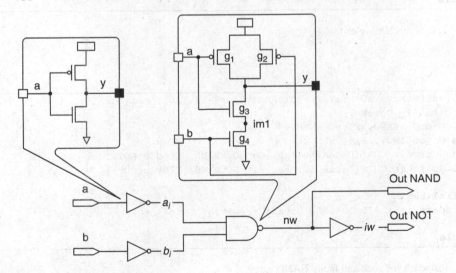

Fig. 3.23 Studying effects of transistor faults

```
module CMOSnand (input a, b, output y)
   supply0 Gnd;
   supply1 Vdd;
   trireg #( 0, 0, 50 ) aa, bb;
   assign aa = a, bb = b;
   wire im1;
   tranif0 #(4) g1 (y, Vdd, aa );
   tranif0 #(4) g2 (y, Vdd, bb );
   tranif1 #(3) g3 (y, im1, aa );
   tranif1 #(3) g4 ( im1, Gnd, bb );
endmodule
```

Fig. 3.24 CMOS NAND gate using bidirectional switches

```
module CMOSnot (input a, output y);
   supply0 Gnd;
   supply1 Vdd;
   trireg #( 0, 0, 50 ) aa;
   assign aa = a;
   tranif0 #(4) g1 ( y, Vdd, aa );
   tranif1 #(3) g4 ( y, Gnd, aa );
endmodule
```

Fig. 3.25 Verilog code for interface inverters

```
module GoldenCircuit (input a, b, output nw, iw);
   wire ai, bi;
   CMOSnot  U1 (a, ai);
   CMOSnot  U2 (b, bi);
   CMOSnot  U3 (nw, iw);
   CMOSnand U4 (ai, bi, nw);
endmodule
```

Fig. 3.26 Test circuit Verilog code

```
module FaultAnalysis0 (); // FaultAnalysis1  // FaultAnalysis2
   reg DataA=1, DataB=1;
   wire GoodOutNAND, FltyOutNAND;
   wire GoodOutNOT, FltyOutNOT;
   GoldenCircuit CUA (DataA, DataB, GoodOutNAND, GoodOutNOT);
   FaultyCircuit CUT (DataA, DataB, FltyOutNAND, FltyOutNOT);

   initial begin
      repeat (15) #35 {DataA, DataB} = {DataA, DataB} + 1;
   end
endmodule
```

Fig. 3.27 Testbench for good and faulty NAND gates

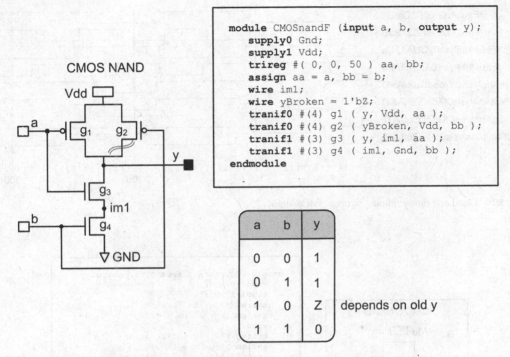

Fig. 3.28 Modeling open pMOS drain

The *FaultyCircuit* module instantiated in the testbench is similar to the *GoldenCircuit* of Fig. 3.26 except that it instantiates *CMOSnandF*, which is the faulty version of *CMOSnand*. Several faulty NAND modules (*CMOSnandF*) and their corresponding waveforms produced by this testbench will be shown next.

Open pMOS drain. The first fault to consider is drain of one of the NAND pMOS gates is stuck-open. The transistor circuit, corresponding truth table, and the Verilog code that models this circuit are shown in Fig. 3.28.

Simulation of this circuit along with the good NAND gate in the environment of Fig. 3.23, and using the testbench of Fig. 3.27, generates the waveform shown in Fig. 3.29. The *DataA* and *DataB* inputs are those of the input inverters in Fig. 3.23. The third and fourth waveforms are those of the inputs of the good and faulty NAND gates. As for the outputs, X and Z values are interpreted as 1 or 0 by the logic that follows a gate output. Therefore, instead of directly looking at the NAND outputs, we look at how they are interpreted by the inverters that follow them. Therefore, the signals to compare the waveform shown are *GoodOutNOT* and *FltyOutNOT*.

The *GoodOutNOT* output is the AND result of *a* and *b*, which is different than *FltyOutNOT*. On the other hand, except for a short delay, *FltyOutNOT* is logically the same as the *a* input. Considering the AND function on this output, this value can only be justified if input *b* of the faulty NAND is stuck-at-1. Therefore, the fault shown in Fig. 3.28 is modeled by *b:SA1*.

Grounded nMOS Gate. Another transistor-level fault to consider is the case of upper nMOS gate of the NAND being grounded. The circuit modeling this fault, its truth table, and the corresponding Verilog code for simulation are shown in Fig. 3.30. The Verilog code shows *aBroken* that is set to 0 is used at the gate of g_3.

Figure 3.31 shows the simulation run of this faulty NAND gate after going through the output inverter. Since this output is always 0, the transistor fault of Fig. 3.30 can be interpreted either as one of the NAND inputs being SA0, or its output being SA1, i.e., *a:SA0*, *b:SA0*, or *w:SA1*.

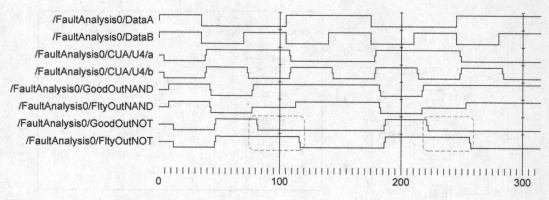

Fig. 3.29 Good and faulty outputs of open pMOS drain

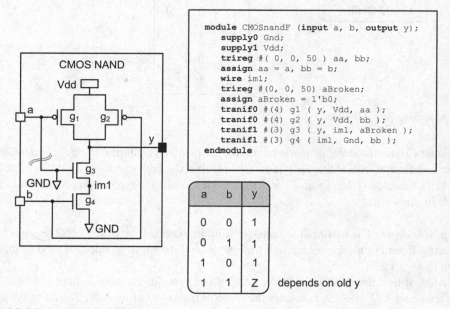

```
module CMOSnandF (input a, b, output y);
    supply0 Gnd;
    supply1 Vdd;
    trireg #( 0, 0, 50 ) aa, bb;
    assign aa = a, bb = b;
    wire im1;
    trireg #(0, 0, 50) aBroken;
    assign aBroken = 1'b0;
    tranif0 #(4) g1 ( y, Vdd, aa );
    tranif0 #(4) g2 ( y, Vdd, bb );
    tranif1 #(3) g3 ( y, im1, aBroken );
    tranif1 #(3) g4 ( im1, Gnd, bb );
endmodule
```

a	b	y
0	0	1
0	1	1
1	0	1
1	1	Z

depends on old y

Fig. 3.30 Modeling grounded nMOS gate

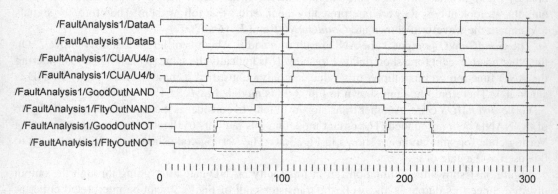

Fig. 3.31 Grounded nMOS gate fault simulation

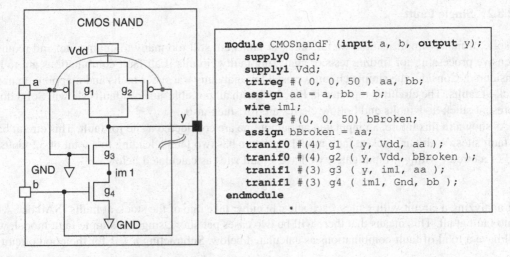

Fig. 3.32 Modeling broken pMOS gate, bridged to another

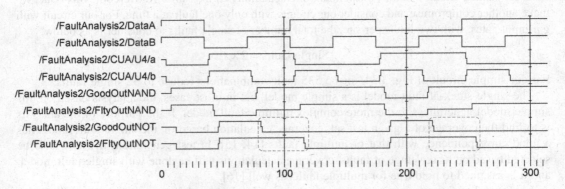

Fig. 3.33 Wrong input driving pMOS Gate

Broken pMOS gate bridged. Another fault to study is the case that a pMOS gate is broken from its own input and bridged to the other NAND input. This case is shown in Fig. 3.32. The Verilog code shown here uses *bBroken*, that is connected to *a*, for the Gate input of g_2 pMOS transistor.

Simulating the faulty circuit of Fig. 3.32, with the testbench of Fig. 3.27, reveals that the fault shown can be modeled by *b* input of NAND gate stuck-at-1 (*b*:SA1). As shown in Fig. 3.33, *FltyOutNOT* is the same as the *a* input with a slight delay.

Broken pMOS gate pulled. Another case that we analyzed was when the Gate input of g_2 is pulled to supply instead of being driven by the *b* input of the NAND circuitry. Simulation of this case reveals that this situation can also be modeled by *b*:SA1.

The above examples show that most transistor-level faults can be approximated by stuck-at faults. In addition, as we will discuss in Sect. 3.3, other fault types, e.g., bridging and state dependent, can also be detected by tests for stuck-at faults. Furthermore, transient faults can be modeled by temporary stuck-at faults [13].

Because of this, the stuck-at fault model is the most general and will be used for logic circuit fault modeling.

3.2.8.2   Single Fault

In spite of many simplifications so far, stuck-at faults are still too many to enumerate, and require intensive processing for finding tests or analyzing faulty circuits if all fault combinations are to be considered. Consider the simple circuit we used for analyzing transistor faults and mapping them to stuck-at faults. The circuit is repeated in Fig. 3.34 with all possible stuck-at faults shown. Solid dots represent stuck-at-1 faults and hollow circles are for stuck-at-0.

As shown in this figure, an input and an output line are considered a site for fault. This circuit has 16 fault sites. With our stuck-at fault model, each site has two faults, leading to a total of 32 faults.

For a circuit with g n-input gates, we have s fault sites as calculated below.

$$s = g \times (n+1)$$

For analyzing a circuit with s sites, each site can either take one of the stuck-at faults (SA0 or SA1) or no fault at all. This means that there will be two cases per site. Using this simple fault model, we will have a total of fault combinations as calculated below. Subtracting a 1 is for the good circuit.

$$\text{Fault combinations} = 2^s - 1$$

With this, for our simple circuit of Fig. 3.34, there will be a total of $12^{16} - 1 = 65{,}535$ fault combinations. Even for a small circuit like this, this amount of calculation is impossible. To overcome this issue, we make another compromise and consider our circuits with only one fault at a time. For our circuit with g n-input gates, with two faults per site, the total number of single faults to consider is as below:

$$\text{Single faults} = 2 \times s$$

For our simple circuit of Fig. 3.34, the 65,535 fault combinations reduce to 32 single faults.

The single-stuck-at fault model is a simple model, but in most cases, the analysis based on this simple model generalizes to the more complex multiple fault model. If the purpose of the analysis is to find the coverage of a given test set, coverage calculation based on the single fault model has a good correspondence with that of multiple faults [14, 15]. In test generation, if a test is to be selected based on its coverage of faults, coverage calculations will be done with single fault model, and it is assumed to hold true for multiple faults as well [16].

In physical testing of circuits, we assume that circuits are tested frequent enough that faults are detected and identified as they occur, before we run into multiple fault situations. Although this assumption may be unrealistic for complete systems, it is a realistic situation for sub-modules of a system tested independently.

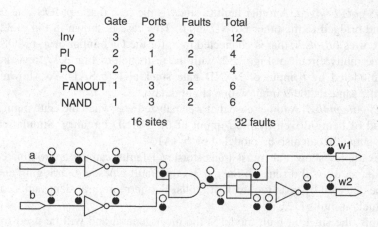

	Gate	Ports	Faults	Total
Inv	3	2	2	12
PI	2	1	2	4
PO	2	1	2	4
FANOUT	1	3	2	6
NAND	1	3	2	6
	16 sites			32 faults

Fig. 3.34 Stuck-at faults in s gate-level circuit

Another reason single faults are preferred over multiple faults is that multiple faults can mask each other and prevent them from being detected or distinguished. This situation was discussed in Sect. 3.2.7.

Considering the discussions in this section (3.2.8), although not perfect, and, in some cases, over simplified, the single stuck-at fault model remains the most used fault model for testing electronic equipment today. Most algorithms and test methods are based on this fault model and are good approximations for other fault types.

3.2.9 Detecting Single Stuck-at Faults

For detecting a fault, finding Boolean function of the faulty circuit and comparing it with the good circuit function, as discussed in Sect. 3.2.1, applies to functional faults and structural faults of any type. However, a simpler method exists for detecting single stuck-at faults that we explain here.

A stuck-at fault to be detected must be activated and then propagated. To activate a fault, primary input values must be adjusted such that the faulty line receives a value different than its faulty value. Once a fault is activated, input values must be adjusted such that the fault effect can propagate to a primary output. A fault effect on a line propagates to a primary output when toggling the value of line also toggles the output [17].

As an example, consider the SA1 fault on l_4 of Fig. 3.35. In order to activate this fault, a 0 must reach this line ($l_4 = 0/1$, 0: good value, 1: faulty value). This can be achieved by $s = 1$. Now that l_4: SA1 is activated, it must be propagated to the output. For this to happen, l_5 must become 1, so that faulty value of l_4 propagates to the output of G_2. Thus $a = 1$.

This makes $l_7 = 1/0$. For l_7 output of G_2 to propagate to l_9, and thus w, l_8 must be 1; otherwise, $l_8 = 0$ forces $l_9 = 1$ regardless of value of l_7. Making $l_8 = 1$ can be achieved by a 0 on b. This makes $l_9 = 0/1$, which shows that w becomes 0 if there is no fault in the circuit, and it becomes 1 in presence of l_4:SA1. The input combination $abs = 101$ detects l_4:SA1.

The process discussed above is the basis of many test generation algorithms that will be presented in chapter on test generation. We are only presenting this to help us present materials in the rest of this chapter.

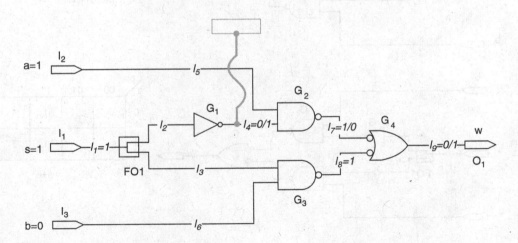

Fig. 3.35 Detecting stuck-at faults

3.3 Issues Related to Gate Level Faults

Certain definitions, methods, and terms that relate to functional and structural gate-level faults are discussed here. This material provides a background for Sect. 3.4 on reducing number of faults.

3.3.1 Detecting Bridging Faults

For justifying the use of the stuck-at fault model, in an earlier section, we indicated that bridging faults could be detected by stuck-at faults. This section elaborates on this issue and presents several illustrating examples. An AND-bridging fault between lines i and j can be detected by a test that puts a 0 on i, a 1 on j, and detects a SA0 on j. Similarly, an OR-bridging fault between lines i and j can be detected by a test that puts a 1 on i, a 0 on j, and detects a SA1 on j. In both cases (AND and OR bridging), the test should detect the stuck-at fault on line that is receiving a faulty value due to the bridging fault.

Figure 3.36 shows two faulty versions of our multiplexer example. One circuit has an AND-bridging between lines l_5 and l_6, and the other has an OR-bridging between these two lines. Faulty functions are shown, and gray boxes in the Karnaugh maps show conflicts with the good multiplexer, which designate input combinations detecting faults.

As directed above, to detect the AND-bridging fault of Fig. 3.36a, we set l_5 to 1, l_6 to 0, and detect stuck-at-0 on l_5. A SA0 on l_5 is activated by $a = 1$, propagated to l_7 by $s = 0$, and propagated to w by $b = 0$. Therefore, $abs = 100$ detects fault of Fig. 3.36a, which represents one of the gray boxes in this figure. If we start with l_5 equal to 0 and l_6 equal to 1, a similar analysis finds the other test for the above-mentioned AND-bridging fault that is $abs = 011$.

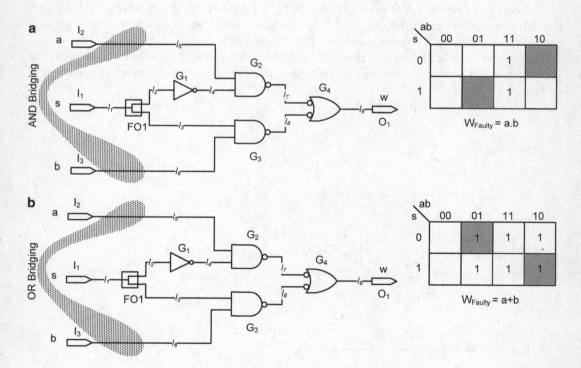

Fig. 3.36 Detecting bridging faults

In order to detect the OR-bridging fault of Fig. 3.36b, we start with $l_5 = 1$, $l_6 = 0$, and detect SA1 on l_6. This translates to $abs = 100$. Starting with $l_5 = 0$ and $l_6 = 1$, the test for the OR-bridging faults becomes $abs = 101$.

3.3.2 Undetectable Faults

A fault is *undetectable* if the circuit in presence of the fault produces the same output as the fault-free circuit. Figure 3.37 shows an AND-bridging fault between l_7 and l_8 that produces the same output as the good multiplexer output and is therefore undetectable.

3.3.3 Redundant Faults

Redundancies in logic circuits happen for multi-output circuit minimization, reliability purposes, static hazard removal, or are unintentional. In any case, circuit redundancies provide conditions that some faults in the circuit go undetected. Such faults are referred to as redundant faults. See for example the circuit in Fig. 3.38. Gate G_5 removes the potential hazard shown in the Karnaugh map. A stuck-at-1

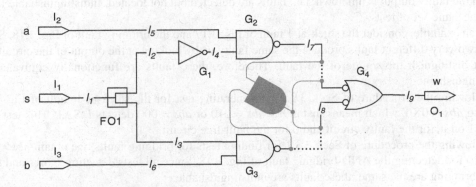

Fig. 3.37 Undetectable AND-bridging fault

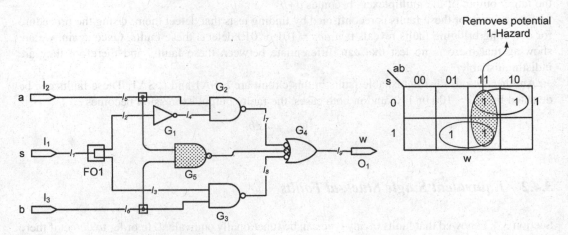

Fig. 3.38 Redundant fault, due to hazard removal

fault at the output of this gate causes no change in the circuit output, and therefore is undetectable. Not all faults in the redundant parts of a circuit are necessarily undetectable. For example, a stuck-at-0 at the output of G_5 causes w to become permanently 1, which makes this fault a detectable one.

3.4 Fault Collapsing

Thus far, we have been able to reduce fault types, simplify our fault model, and accept certain approximations in fault coverage and detection. These have been for the purpose of simplifying algorithms dealing with faults, and reducing processing time of processes that deal with faults. This section presents another step in this direction. Normally, we are to reduce our processing time by reducing the number of faults that we deal with. We discuss indistinguishable faults, fault equivalence, and methods of removing unnecessary faults [18].

3.4.1 Indistinguishable Faults

When several faults produce the same faulty outputs, for every input, they are said to be *indistinguishable*. In this case, detecting that a fault has occurred is possible, but exactly which fault has caused the faulty output is unknown, i.e., faults are detected but not located. Indistinguishable faults have the same set of tests.

As an example, consider the stuck-at-1 fault of Fig. 3.17 and the bridging fault of Fig. 3.18. Since these two very different faults produce the same faulty w, by looking at the output of the circuit, we cannot distinguish the source of the fault. Therefore, these faults are functionally equivalent or indistinguishable.

Following the procedure of Sect. 3.2.9 for generating test for the stuck-at-1 fault of Fig. 3.17 leads to $abs = 0X0$, which means that either $abs = 010$ or $abs = 000$ detects l_5:SA1. This test produces 1 on w for the faulty circuit and 0 for the fault-free circuit.

Following the procedure of Sect. 3.3.1 for finding tests for bridging faults, we obtain $abs = 000$ and 010 for detecting the AND-bridging fault of Fig. 3.18. Since all tests detecting l_5:*SA1* and l_7–l_3: AND-bridging are the same, these faults are indistinguishable.

Another set of indistinguishable faults in the multiplexer circuit are the OR-bridging fault between l_5 and l_6 and the OR-bridging fault between l_5 and l_4, as shown in Fig. 3.39. In both cases, the faulty output of the multiplexers becomes $a+b$.

Equivalency of these faults is reconfirmed by finding tests that detect them. Using the procedure for detecting bridging faults reveals that $abs = 101$ or 010 detects these faults. Once again, we are showing that there is no test that can differentiate between these faults, and therefore they are indistinguishable.

Another set of indistinguishable faults in this circuit are l_2:SA1 and l_7:SA1. These faults can be detected by $abs = 100$ or 110, and in both cases, the faulty output expression becomes

$$w_{\text{Faulty}} = s.b.$$

3.4.2 Equivalent Single Stuck-at Faults

Section 3.4.1 showed that faults of any type can be functionally equivalent. In order to detect if there are equivalent faults in a circuit, faulty functions of all types of faults must be calculated and

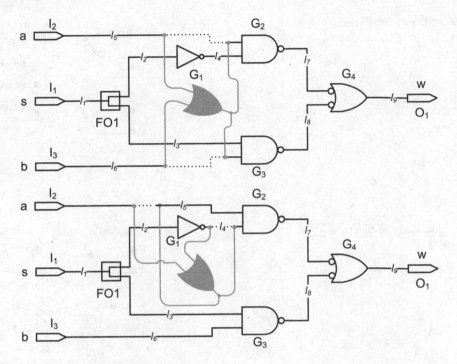

Fig. 3.39 Two indistinguishable OR-bridging faults

compared. Alternatively, tests can be generated for all faults, and faults with the same set of tests are equivalent. Obviously, with the large number of faults in a circuit, either alternative involves a prohibitively intensive computation. Even if we limit ourselves to stuck-at faults, an exhaustive search for all equivalent faults is a hard process.

Instead of looking for functional equivalency, if we limit ourselves to structural local fault equivalence, the process will be significantly simplified. In structural local fault equivalence, only faults on inputs and outputs of simple logic gates are checked for equivalency. Although this is not exact, and there will still be equivalent stuck-at faults that are not detected, the methods used for this purpose are simple and can be applied to arbitrary large circuits [19].

3.4.3 Gate-oriented Fault Collapsing

The simplest gate-oriented fault collapsing method involves a local processing of circuit gates from circuit inputs toward the outputs. This is based on the fact that stuck-at faults on ports of a logic gate can be reduced independently, by a local gate-based process.

3.4.3.1 Gate Faults

For a single gate fault collapsing case, consider six stuck-at faults on ports of a 2-input AND gate. These faults can be reduced to four without considering where the gate inputs and outputs are connected to.

This example is illustrated in Fig. 3.40. Originally, a SA0 (hollow circle) and a SA1 (solid dot) are placed on AND gate inputs and output.

Fig. 3.40 AND gate local fault collapsing

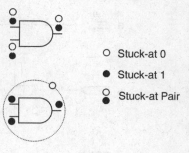

Fig. 3.41 Float and fixed fault in primitive gates

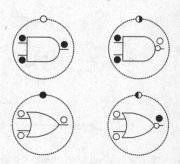

The SA0 faults on the AND gate inputs have the same effect on the output of this gate, and thus are equivalent. Furthermore, the SA0 fault on the output of this gate affects the logic that this output is connected to no different than the input SA0 faults. Therefore, the output SA0 fault of this AND gate is also indistinguishable from its input SA0 faults. This means that the three SA0 faults on the ports of an AND gate form an equivalent class, the members of which cannot be distinguished from one another. As such, we only need to select only one member of this equivalent class, which reduces the AND gate faults to only four. Graphically (as shown in Fig. 3.40), we represent this case by an n-input AND gate having $n+1$ SA1 fixed faults on its inputs and output, and a float SA0 fault that can be placed on any of its ports.

Figure 3.41 shows extension of the AND gate fault collapsing rules to other basic gates. An n-input NOR gate has n fixed SA0 faults on its inputs, a fixed SA1 fault on its output, and a float fault that appears as SA1 if placed on an input, and becomes SA0 if placed on the output of the gate.

3.4.3.2 Gate-oriented Fault Collapsing Procedure

Using gate fault properties discussed above, a simple fault collapsing procedure is presented here. In this procedure, first a pair of faults is placed on every port of every component. Then, line faults are reduced to only one of a kind. Line faults move forward next to gate that they drive. For gate faults, gates closer to the input are treated first. Fixed gate faults are kept, and their float faults are placed on their outputs. Output faults move forward close to the gates they are driving. In general, in this procedure, like bubbles, faults move forward on a line as close as they can get to the output of the circuit. Figure 3.42 shows these rules for lines, inverters, basic gates, and fanouts.

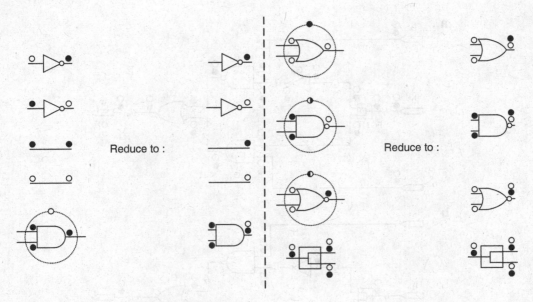

Fig. 3.42 Rules for local gate-oriented fault collapsing

Figure 3.43 shows application of this procedure to the multiplexer circuit example of the previous sections. Originally, all faults are placed on all ports, including inputs, output, and fanout. This will put a pair of SA0, SA1 faults on each end of a circuit line. Line faults will be reduced first.

For example, four faults on the two ends of l_3 become two and will be placed next to G_3. Repeating this for all circuit lines results in Fig. 3.43b. We then treat gate faults moving from the input side of the circuit toward its output.

According to Fig. 3.42, the G_3 NAND gate requires a SA1 on each of its inputs, and SA0 and SA1 on its output. This means that the SA0 faults on l_3 and l_6 inputs of G_3 are not needed and can be removed. As for the output of G_3, the required SA0 and SA1 faults are already placed on l_8 at the input of G_4. Before going forward, the inverter that is at the same distance to circuit inputs as G_3 must be treated. Gate G_2 SA0 and SA1 faults move to the output of this gate and become SA1 and SA0. These faults move on l_4 close to G_2 and collapse into faults already at this location. We then follow this procedure for G_2, which leads to keeping a SA1 on its l_5 input, a SA1 on its l_4 input, and SA0 and SA1 on its l_7 output. As before, output faults move to the inputs of the next gate. The last level in this circuit is that of G_4. Considering that this is also a NAND, we remove SA0 faults at its inputs and keep the SA0 and SA1 faults at its output. Figure 3.43c shows faults in the multiplexer after collapsing.

3.4.4 Line-oriented Fault Collapsing

A simpler fault collapsing can be achieved when we view faults on lines instead of those on the gate ports. This approach is based on the classic concepts and is only different from the previous material in its view of placing faults. Instead of initially arranging all possible faults on the ports of gates and then trying to collapse them, we only place those faults on circuit lines that cannot collapse any further. The criterion used in placing a specific stuck-at fault on a line is based on the gate driven by the line. Table 3.1 shows faults to be placed on lines based on the gate the line is driving. As before, fanout is treated as an actual component.

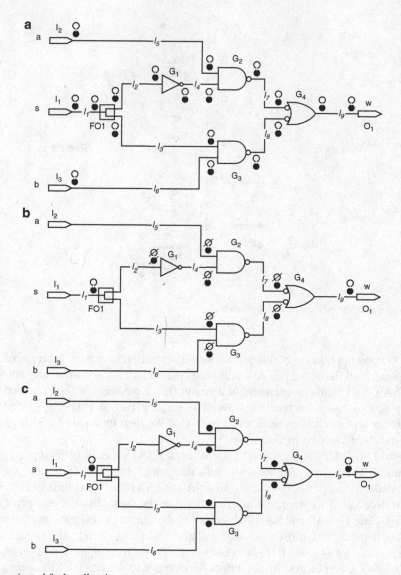

Fig. 3.43 Gate-oriented fault collapsing

An example of this method is shown in Fig. 3.44. Faults shown on the circuit lines were determined by entries of Table 3.1.

As we described, only the faults that cannot be collapsed are placed on the circuit. For example, for G_4 that is a NAND gate, we place stuck-at faults on its inputs. Our algorithm does not decide on faults at the output of a gate. SA0 faults are not placed on input lines of this gate, because these faults are equivalent to SA fault at this gate's output, and an output fault is decided by the type of the gate driven by it. This process is repeated for all gates of the circuit, while deciding on faults of input lines based on entries of Table 3.1. If we are to apply this method to a sequential circuit, all input lines of a flip-flop should be treated as pseudo primary outputs (such as data input and clock signals) and flip-flop outputs should be treated as pseudo primary inputs. Pseudo primary outputs and inputs are treated the same as primary outputs and inputs as shown in Table 3.1.

Table 3.1 Line-oriented fault collapsing

Type of target gate	Put this (these) fault(s) on the gate's input line(s)
ADD, NAND	SA1
OR, NOR	SA0
INV, BUF	None
FANOUT	SA0, SA1
XOR	SA0, SA1
Primary output	SA0, SA1
Primary input	None

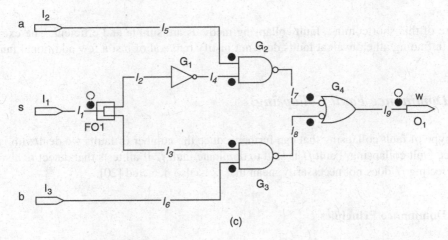

(c)

Fig. 3.44 Line-oriented fault collapsing

3.4.5 Problem with Reconvergent Fanouts

A fanout (divergence point) that reconverges at a gate somewhere in the circuit is called a *reconvergent fanout*. Reconvergent fanouts present problems in most test applications and require special treatment.

Both methods presented above have deficiencies when it comes to reconvergent fanouts. In both methods, a fanout stem takes both SA0 and SA1 faults. This is whether the fanout stems converge at a gate or not. However, if they do converge, there is a possibility that a fault at the convergence point becomes equivalent to a fault at the divergence point (fanout stem). This can be corrected by simulating a section of the circuit that has a reconvergent fanout for faults at the divergence and convergence points. Simulation results will determine if certain faults are to be removed.

As an example for this case, consider the circuit shown in Fig. 3.45. After performing line-oriented fault collapsing, faults placed on circuit lines are as shown in this figure. The SA0 fault at the fanout stem of this circuit produces a 1 on w. At the same time, we also have a SA1 on w. Obviously, these faults are equivalent, and our fault collapsing methods did not detect them. Simulating the area enclosed in the dotted line for faults on l_3 and l_{10} finds the equivalent faults.

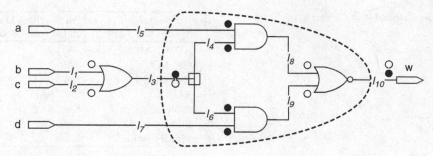

Fig. 3.45 Reconvergent fanout fault problem

In spite of this shortcoming, fault collapsing methods are simple and efficient. The extra work involved in finding all equivalent faults does not justify removal of just a few additional faults.

3.4.6 Dominance Fault Collapsing

Another type of fault collapsing that can further reduce the number of faults we deal with is called dominance fault collapsing. Fault $f1$ is said to dominate fault $f2$ if all tests that detect $f2$ also detect $f1$, but detecting $f1$ does not necessarily mean that $f2$ is also detected [20].

3.4.6.1 Dominance Principles

A simple example for fault dominance is shown in Fig. 3.46. In this figure, a fanout component has a SA1 fault on its stem, s, and a SA1 on its branch a. A test for s:SA1 is to force a 0 into the fanout stem. As a result of this test, if the faulty value propagates to circuit outputs through both fanout branches (a and b), as shown in Fig. 3.46b, then both s:SA1 and a:SA1 have been detected.

In this case, these faults look like they are equivalent. However, if as in Fig. 3.46c, the faulty value of 1 propagates through the lower branch only and is somehow blocked in the upper branch (branch a), then s:SA1 is detected and a:SA1 is not. Since tests for s:SA1 form a superset of a:SA1, s:SA1 dominates a:SA1. We present this dominance case in a larger example. Figure 3.47 shows two identical circuits with one having a SA1 fault on its fanout stem and the other on its branch.

We apply two tests to this circuit: $T_1 = 00X0$ and $T_2 = 10X1$. The good circuit outputs for these tests become $wy = 00$ and $wy = 10$, respectively. When these tests are applied to the faulty circuit with s:SA1 (Fig. 3.47b), circuit outputs become $wy = 10$ and $wy = 11$, respectively. Since for both tests, wy values are different than the good wy values, both tests T_1 and T_2 detect s:SA1. Now we apply these same tests to Fig. 3.47c with branch a:SA1. In this case, wy becomes 10 for T_1, and since this response is different from the good circuit response, T_1 detects a:SA1. However, when T_2 is applied to the circuit with a:SA1, the wy output becomes 10 which is the same as that of the good circuit. This means that any test that detects a:SA1, i.e., T_1, also detects s:SA1, but there are tests that detect s:SA1, i.e., T_2, and not a:SA1. Since tests for s:SA1 form a super set of those for a:SA1, we say that s:SA1 dominates a:SA1. This dominance relation is shown in Fig. 3.48. If we were to choose either T_1 or T_2, since T_1 detects both faults, it is the preferred test. And since this test can be found by considering only a:SA1, dominance fault collapsing removes s:SA1 and only keeps a:SA1.

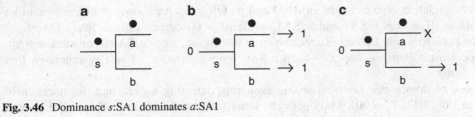

Fig. 3.46 Dominance s:SA1 dominates a:SA1

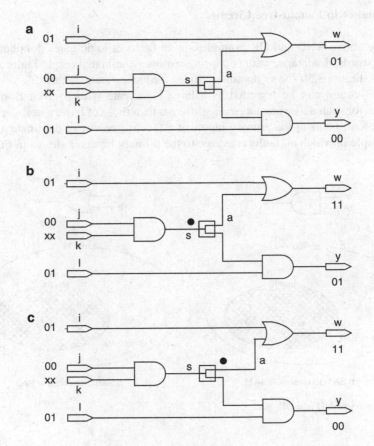

Fig. 3.47 Fault dominance example

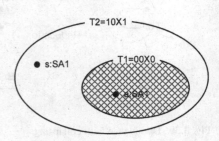

Fig. 3.48 Dominance relation

A dominance relation also exists in an AND and an OR gate. As shown in Fig. 3.49a, in an AND gate, $ab = 10$ detects b:SA1 and w:SA1, while $ab = 00$ detects only w:SA1. Therefore, $ab = 10$ is the preferred test. In the OR gate shown in this figure, since b:SA0 dominates w:SA0, in dominance fault collapsing, we only keep b:SA0 for which $ab = 01$ test is generated that detects both faults.

The purpose of dominance fault collapsing is merely detecting a fault and not necessarily locating it. In Fig. 3.47, $T_1 = 00X0$ produces the same faulty response for either faults shown. This response cannot be used to tell these faults apart.

3.4.6.2 Dominance in Fanout-free Circuits

As shown in the above AND and OR examples, input faults at logic gates dominate those of the output. In other words, dominance fault collapsing removes dominated output faults and only keeps gate input faults. Figure 3.50 shows dominance rules for basic gate structures.

A fanout-free circuit can be regarded as a large logic gate with a given Boolean function. Dominance rules for such a gate cannot be any different than those of basic gates, such as AND and OR gates. Therefore, faults at the primary inputs of a fanout-free circuit dominate all faults in the circuit. An example in which all faults converge to the primary inputs is shown in Fig. 3.51.

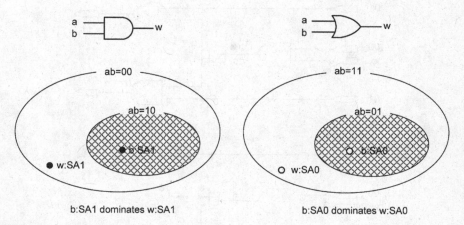

Fig. 3.49 Dominance in AND and OR gates

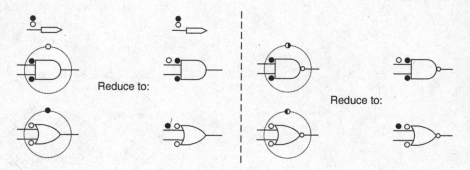

Fig. 3.50 Dominance fault collapsing

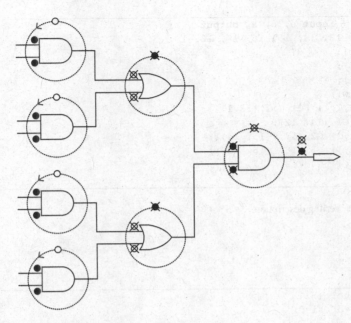

Fig. 3.51 Dominance fault collapsing example

3.5 Fault Collapsing in Verilog

Perhaps one of the most basic steps in any test method, e.g., fault simulation and test generation, is obtaining a fault list. We have shown methods for obtaining and reducing fault lists for such applications. This section shows an automated way of doing this. We show how a Verilog testbench can be developed to perform fault collapsing on the netlist it is testing and then show PLI implementation of PLI task used for this purpose.

3.5.1 *Verilog Testbench for Fault Collapsing*

For applying test methods at the gate-level to a circuit-under-test, a netlist of the circuit is required. The netlist follows a certain format which is known to the testbench and the PLI functions used by the testbench.

Figure 3.52 shows a Verilog module that corresponds to the 2-to-1 multiplexer example that we used in this chapter. Components used in this description correspond to those of Fig. 3.14, and line names and labels used here are the same as the labels in Fig. 3.14.

Figure 3.53 shows a Verilog testbench for *mux2to1* module of Fig. 3.52. This testbench instantiates this circuit using *DUT* instance name. In an **initial** statement, PLI function **$FaultCollapsing** is called with the name of the module for which a fault list is to be generated, and the output file that will contain this list. Following this call, the testbench performs its normal functions of applying tests and other tasks it may have.

At time 0, before any simulation begins, the **$FaultCollapsing** function is called and a collapsed list of faults is generated. Figure 3.54 shows this fault list. Faults generated here for the multiplexer match those obtained manually in Sect. 3.4.4 and shown in Fig. 3.44.

```
module mux2to1 ( input a, b, s, output w );
   wire l1, l2, l3, l4, l5, l6, l7, l8, l9;
   pin I1 (s,l1);
   pin I2 (a,l5);
   pin I3 (b,l6);
   pout O1 (l9,w);
   fanout_n #(2,1,1) FO1 (l1,{l2,l3});
   notg #(1,1) G1 (l4, l2);
   nand_n #(2,1,1) G2 (l7, {l5, l4});
   nand_n #(2,1,1) G3 (l8, {l6, l3});
   nand_n #(2,1,1) G4 (l9, {l7, l8});
endmodule
```

Fig. 3.52 Multiplexer Verilog description

```
module TESTmux2to1 ();
   reg ai=0, bi=0, si=0;
   wire wo;
   mux2to1 DUT (ai, bi, si, wo);

   initial begin
      $Faultcollapsing (TESTmux2to1.DUT, "Mux1Faults.flt");
      repeat (15) #73 {ai, bi, si} = {ai, bi, si} + 1;
   end
endmodule
```

Fig. 3.53 Verilog testbench performing fault collapsing

Fig. 3.54 Multiplexer collapsed faults

```
TESTmux2to1.DUT.O1.in[0] s@1
TESTmux2to1.DUT.O1.in[0] s@0
TESTmux2to1.DUT.FO1.in s@1
TESTmux2to1.DUT.FO1.in s@0
TESTmux2to1.DUT.G2.in[1] s@1
TESTmux2to1.DUT.G2.in[0] s@1
TESTmux2to1.DUT.G3.in[1] s@1
TESTmux2to1.DUT.G3.in[0] s@1
TESTmux2to1.DUT.G4.in[1] s@1
TESTmux2to1.DUT.G4.in[0] s@1
```

For a testbench performing test generation or fault simulation, this fault list will be read by the testbench, and appropriate functions will be performed for such applications. Other testbench functions include fault injection, test data applications, and output response evaluations. PLI functions for some of these tasks will be discussed in the later chapters.

```
module and_n
   #(parameter n = 2, tphl = 1, tplh = 1)(out, in);
   input [n-1:0] in;
   output out;
   wire [n-2:0] mwire;
   genvar i;
   and and_0 (mwire [0], in [0], in [1]);
   generate
      for (i = 1; i <= n-2; i = i+1) begin : AND_N
         and inst (mwire [i], mwire [i-1], in [i+1]);
      end
   endgenerate
   bufif1 #(tplh, tphl) inst(out, mwire [n-2], 1'b1);
endmodule
```

Fig. 3.55 Test primitive AND gate

Gate models known to the **$FaultCollapsing** function and other PLI functions that we use in testing follow a certain format as discussed in Chap. 2.

Figure 3.55 shows an AND gate to use as reference in the PLI function implementation that we discuss next.

3.5.2 PLI Implementation of Fault Collapsing

The **$FaultCollapsing** function that has been implemented in PLI uses the line-oriented fault collapsing method discussed in Sect. 3.4.4.

The top-level PLI C++ function is *FaultCollapsing*. This function is registered as **$FaultCollapsing** that becomes available to Verilog testbench descriptions. The main function of fault collapsing shown in Fig. 3.56 is to open the fault file passed to it, look for sub-modules in the design-under-test, and pass the module handler and *faultFile* pointer to the *traceChildren* function.

The *traceChildren* function shown in Fig. 3.57 takes a module handle of a top-level module in *DUT* and looks for its children. This is a recursive function, and when it reaches a child module that is of a known primitive type, e.g., *and_n* of Fig. 3.55, it passes the primitive type, module handle, and *faultFile* to the *tracePorts* function.

The *tracePorts* function is shown in Fig. 3.58. This function traces ports of the primitive module that is passed to it, and looks for an input vector that contains bits corresponding to the primitive inputs. Once such port is found, it passes the port handle, primitive type, and *faultFile* to *dumpPortBitFaults*.

Recall that inputs of primitives, e.g., *and_n*, are grouped into an *n*-bit vector called *in*. Function *dumpPortBitFaults*, shown in Fig. 3.59, takes port handle of this vector as input, and it looks in this vector for *portbit* handles that correspond to the individual primitive inputs. For every *portbit_handle* of a primitive, depending on the type of the primitive (switch statement in Fig. 3.59), it executes Table 3.2 and writes the corresponding faults and the hierarchical name of the primitive to the fault file that is passed to this function.

The PLI C++ code that was presented here is typical of most PLI functions that we have developed and used in our test applications in this book. Because of the hierarchies involved, this code contains many of the techniques used in our other functions. PLI functions in other chapters will not be explained at this level of detail.

```
int FaultCollapsing ()
{
   handle mod_handle;
   acc_initialize();
   mod_handle = acc_handle_tfarg(1);
   FILE* faultFile =fopen((char*)acc_fetch_tfarg_str(2),"w+");
   traceChildren ( mod_handle, faultFile );
   fclose(faultFile);
   fclose(logFile);
   acc_close();
return 0;
}
```

Fig. 3.56 Top-level PLI fault collapsing function

```
static void traceChildren ( handle mod_handle, FILE* faultFile )
{
   handle child;
   int moduleType = -1;

      moduleType = knownModuleType (acc_fetch_defname ( mod_handle ));

      if(moduleType >= 0) // Trace ports of known primitives
      {
        tracePorts (mod_handle, moduleType, faultFile );
      }
      else

        for ( child = acc_next_child( mod_handle, 0 );child;
        child = acc_next_child( mod_handle, child ))
        {
            traceChildren ( child, faultFile );
        }
}
```

Fig. 3.57 Tracing children of top-level design modules

```
static void tracePorts ( handle mod_handle, int moduleType,FILE* faultFile )
{
   handle port_handle;
   int tmpint;

   for ( port_handle = acc_next_port( mod_handle, 0 );
       port_handle; port_handle = acc_next_port( mod_handle,
       port_handle )
     )
   {
       tmpint = acc_fetch_direction( port_handle );
       if(tmpint == accInput)
       {
          dumpPortBitFaults (port_handle,moduleType,faultFile);
       }
   }
}
```

Fig. 3.58 Tracing primitive ports, looking for input port

```
void dumpPortBitFaults( handle port_handle, int moduleType, FILE*
faultFile )
{
    int tmpint = 0;
    handle portbit_handle;

    for ( portbit_handle = acc_next_bit( port_handle, 0 );
        portbit_handle;portbit_handle=acc_next_bit(port_handle,
        portbit_handle ))
        {
            switch(moduleType)
            {
                    //_NOT_G:
                    //_AND_N
            case 1:
                    fprintf(faultFile,"%s s@1\n",
                    acc_fetch_fullname(portbit_handle));
                    break;
                        //_NAND_N

            case 2:
                    fprintf(faultFile,"%s s@1\n",
                    acc_fetch_fullname(portbit_handle));
                    break;
                        //_OR_N

            case 3:
                    fprintf(faultFile,"%s s@0\n",
                    acc_fetch_fullname(portbit_handle));
                    break;
                        //_NOR_N

            case 4:
                    fprintf(faultFile,"%s s@0\n",
                    acc_fetch_fullname(portbit_handle));
                    break;
                        //_XOR_N

            case 5:
                    fprintf(faultFile,"%s s@1\n",
                    acc_fetch_fullname(portbit_handle));
                    fprintf(faultFile,"%s s@0\n",
                    acc_fetch_fullname(portbit_handle));
                    break; //_XNOR_N

            case 6:
                    fprintf(faultFile,"%s s@1\n",
                    acc_fetch_fullname(portbit_handle));
                    fprintf(faultFile,"%s s@0\n",
                    acc_fetch_fullname(portbit_handle));
```

Fig. 3.59 Dumping primitive hierarchy and its faults

```
                          break;
                              //_FANOUT_N

            case 7:
                          fprintf(faultFile,"%s s@1\n",
                          acc_fetch_fullname(portbit_handle));
                          fprintf(faultFile,"%s s@0\n",
                          acc_fetch_fullname(portbit_handle));
                          break;
                              //_PO

            case 8:
                          fprintf(faultFile,"%s s@1\n",
                          acc_fetch_fullname(portbit_handle));
                          fprintf(faultFile,"%s s@0\n",
                          acc_fetch_fullname(portbit_handle));
                          break;
                              //_DFF

            case 9:

                          if(partialCompare(acc_fetch_name(portbit_handle), "D", 1))
                             {fprintf(faultFile,"%s s@1\n",
                          acc_fetch_fullname(portbit_handle));
                          fprintf(faultFile,"%s s@0\n",
                          acc_fetch_fullname(portbit_handle));}
                          break;
                              //_TFF

            case 10:

                          if(partialCompare(acc_fetch_name(portbit_handle), "T", 1))
                             {fprintf(faultFile,"%s s@1\n",
                           acc_fetch_fullname(portbit_handle));
                           fprintf(faultFile,"%s s@0\n",
                           acc_fetch_fullname(portbit_handle));}
                           break;


                 }
             }
         }
```

Fig. 3.59 (continued)

3.6 Summary

This chapter discussed faults. Perhaps the most important topic to understand in testing is that there are defects and faults at various levels model these defects. The first part of the chapter discussed faults from transistors to systems. Although most test programs and methods are based on gate-level faults, this discussion gives the reader a good understanding of higher level abstractions and prepares him or her for research seeking higher level testing. After the discussion of abstraction levels, the chapter narrowed down fault models to the single stuck-at fault model, as the most practical fault model used in today's electronic testing. Once this was established, several fault collapsing and fault reduction methods that apply to the stuck-at fault model were presented. The last part of the chapter showed an automatic HDL-based implementation for a method for obtaining a collapsed fault list.

References

1. Timoc C, Buehler M, Griswold T, Pina C, Stott F, Hess L (1983) Logical models of physical failures, Proceedings of IEEE international test conference, pp 546–553
2. Hayes JP (1985) Fault modeling. IEEE Des Test Comput, pp 88–95
3. Shen JP, Maly W, Ferguson FJ (1985) Inductive fault analysis of MOS integrated circuits. IEEE Des Test Comput 2(6):13–26
4. Abraham JA, Fuchs WK (1986) Fault and error models for VLSI. Proc IEEE 74(5):639–654
5. Bushnell ML, Agrawal VD (2000) Essentials of electronic testing for digital, memory, and mixed-signal VLSI circuits. Kluwer, Dordecht
6. Wadsack RL (1978) Fault modeling and logic simulation of CMOS and MOS integrated circuits. Bell Syst Tech J 57(5):
7. Jha NK, Gupta S (2003) Testing of digital systems. Cambridge University Press, Cambridge
8. Eldred RD (1959) Test routines based on symbolic logical statements. J ACM 6(1):33–36
9. Galey JM, Norby RE, Roth JP (1961) Techniques for the diagnosis of switching circuit failures. In: Ledley RS (ed) Proceedings of the second annual symposium on switching circuit theory and logical design (Detroit), AIEE, pp 152–160, Oct
10. Malaiya YK, Rajsuman R (1992) Bridging faults and IDDQ testing, Los Alamitos, California. IEEE Computer Society Press, Silver Spring, MD
11. Bossen DC, Hong SJ (1971) Cause-effect analysis for multiple fault detection in combinational networks. IEEE Trans Comput C-20:1252–1257
12. Jha NK (1986) Detecting multiple faults in CMOS circuits. In: Proceedings of the international test conference, pp 514–519
13. Hayes JP, Polian I, Becker B (2006) A model for transient faults in logic circuits. Int'l Workshop on Design and Test, Dubai, UAE, Nov. 2006
14. Agarwal VK, Fung AFS (1981) Multiple fault testing of large circuits by single fault test sets. IEEE Trans Comput C-30(11):855–865
15. Hughes JLA, McCluskey EJ (1986) Multiple stuck-at fault coverage of single stuck-at fault test sets. In Proceedings of the international test conference, pp 368–374
16. Jacob J, Biswas NN (1987) GTBD faults and lower bounds on multiple fault coverage of single fault test sets. In: Proceedings of the international test conference, pp 849–855, Sept
17. Roth JP (1966) Diagnosis of automata failures: a calculus and a method. IBM J Res Develop 10(4):278–291
18. McCluskey EJ, Clegg FW (1971) Fault equivalence in combinational logic networks. IEEE Trans Comput 20(11): 1286–1293
19. Schertz DR, Metze G (1972) A new representation for faults in combinational digital circuits. IEEE Trans Comput C-1(8):858–866
20. Abramovici M, Breuer MA, Friedman AD (1994) Digital systems testing and testable design.IEEE Press, Piscataway, NJ3.6 Summary

Chapter 4
Fault Simulation Applications and Methods

Simulating a faulty model of a circuit is called fault simulation. This process is used by test and design engineers; it is the most used test method, and perhaps is one of the most computationally intensive test applications. This chapter is on fault simulation, and Verilog codes and testbenches are used throughout the chapter for illustrating concepts and showing how fault simulation is used in practice.

In order to address the above-mentioned issues, in the first section, after the introduction of basic concepts and terminologies, we show HDL environments for fault simulation. By this, we try to make it easier for designers to understand fault simulation, and use this process for improving test and testability of their designs during the design phase. We then discuss various applications of fault simulation, and where possible, we show HDL environments for implementing them. In the third section of this chapter, we focus our attention on various techniques for speeding up the fault simulation process. Some of these techniques opt to HDL environments better than others, and for those, we discuss related HDL implementations.

4.1 Fault Simulation

We use fault simulation for test data generation, test set evaluation, circuit testability evaluation, providing information for testers, finding faults in a circuit, diagnostics, and many other applications. In all such applications, a faulty model of the circuit being analyzed is made, and it is simulated for a test vector or a complete test set.

The level at which fault simulation is done depends on the level of the circuit being simulated and the level at which faults are injected in the circuit. Most fault simulations deal with gate-level circuits and stuck-at structural gate-level faults.

This section familiarizes the reader with the gate-level fault simulation process, presents related definitions and terminologies, and shows a complete HDL setup for fault simulation. The HDL codes show the complete process of performing fault simulation on several gate-level circuits using single stuck-at fault models.

4.1.1 Gate-level Fault Simulation

The block diagram of Fig. 4.1 shows components involved in fault simulation. The *Simulation Environment* is where everything happens.

Z. Navabi, *Digital System Test and Testable Design: Using HDL Models and Architectures*,
DOI 10.1007/978-1-4419-7548-5_4, © Springer Science+Business Media, LLC 2011

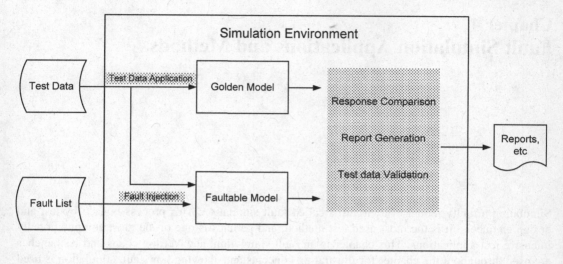

Fig. 4.1 Fault simulation process

In this environment, the good circuit model (*Golden Model*), and the *Faultable Model* are instantiated and a setup for comparing and reporting their responses is provided. The golden model can be a good netlist or a behavioral description of the circuit being simulated. The faultable model is at the gate-level and it is in the netlist form. There are provisions in this netlist for making circuit lines faulty, and thus creating various faulty models.

The inputs to the simulation environments are *Test Data* and *Fault List*. Faults from the fault list input of the Simulation Environment are read and by the use of the *Fault Injection* process injected into the Faultable Model to make various faulty models. Test Data (or test set) is read by the Simulation Environment and applied to the golden and faulty circuit models. For single stuck-at fault model, faults are injected one at a time. With each new fault injection, a new faulty model is created. Each faulty model is compared with the golden model for test vectors in the test set.

As mentioned, the Simulation Environment is also responsible for collecting simulation results and reporting discrepancies and generating other required reports. The exact tasks performed here determine the application for which fault simulation is being performed. One such application may be identifying test vectors in the test set that can detect a certain number of faults. Applications such as this are discussed in Sect. 4.2.

4.1.2 Fault Simulation Requirements

With the above discussions, information that a fault simulation environment requires are:

- Circuit netlist with provisions to become faulty
- Good circuit netlist or behavioral description
- A file containing test data
- A file containing fault list

A fault simulator produces:

- Report files
- Flags and messages

Capabilities that a fault simulator requires for processing the above-mentioned information are as given below.

4.1.2.1   Gate-level Simulation

Since the golden and faulty circuit models are usually available at the gate level, a fault simulation environment should have gate-level simulation capability.

4.1.2.2   Behavioral Simulation

While the faulty model is being simulated at the gate-level for stuck-at gate-level faults, the golden model is simulated for its functionality only. Therefore, a fault simulation environment with behavioral simulation capability can save processing time by simulating the behavioral description of the golden model instead of its gate-level netlist.

4.1.2.3   Reading Data Files

Capability to read text inputs is required in a fault simulation environment for reading test data and list of faults from external data files. Since in many instances, test data and fault lists are generated by different programs, capability to read different formats of data by the fault simulator may be needed. This eliminates the need for extra format conversion programs.

4.1.2.4   Fault Injection Capability

What distinguishes a fault simulation program from a standard logic simulator is the capability of fault injection in a fault simulator. To create a faulty model, a mechanism for injecting faults in the netlist of circuit being simulated is needed. This capability can be internal to the fault simulator or a function added to a standard logic simulator.

4.1.2.5   Writing Report Files

Producing output and report files noting various comparison results and simulation reports is the direct result of a fault simulation run. A fault simulation environment must be capable of writing external data files for creating such reports. As in the case of input files, having formatting capabilities to produce reports readable by other test programs is an added advantage.

Fault simulation can be done by using a dedicated fault simulator. Alternatively, some fault simulation techniques can be implemented within standard logic simulators. This may require extra functions and procedures added to the logic simulator which is usually not as efficient as dedicated fault simulation programs.

4.1.3   An HDL Environment

After a discussion of characteristics of a fault simulation environment, the last section presented two choices of using a dedicated program, or adapting an existing logic simulation environment to perform this task. In this section, we take the latter route. Namely, we use a Verilog based simulation environment and provide utilities in Verilog to satisfy the requirements discussed in Sect. 4.1.2. The advantages of this choice are that we are using a tool that designers are familiar with (a Verilog simulation tool), there are flexibilities for input and output file format conversions, we can use behavioral as well as gate-level circuit descriptions, and last but not least, we have a working environment that concepts of fault simulation can be illustrated with. The issue of inefficiency of this option versus a dedicated program is dealt with in Sect. 4.3.

Figure 4.2 shows the Verilog fault simulation environment that is discussed and used here. The complete environment is a Verilog testbench that performs tasks discussed in Sect. 4.1.2.

We use the familiar 2-to-1 multiplexer to illustrate how data and description files are prepared, and how these tasks are done in Verilog. For reference, Fig. 4.3 shows the above mentioned multiplexer.

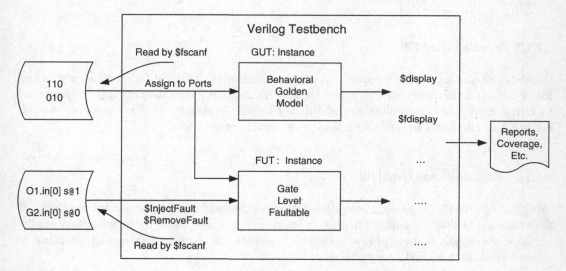

Fig. 4.2 Verilog fault simulation

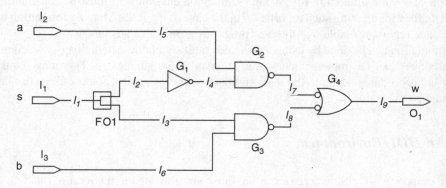

Fig. 4.3 2-to-1 multiplexer

4.1.3.1 Input Files and Information

As mentioned in Sect. 4.1.2, for the preparation of fault simulation, *Faultable Model, Golden Model, Test Data*, and *Fault List* are needed. These are shown here for Verilog fault simulation.

Figure 4.4 shows the gate-level Verilog code of the multiplexer. The formatting is the same as that of Chap. 3 for fault collapsing. The utilities for fault injection use this format to access gate inputs where stuck-at faults are to be injected.

This netlist uses gates from our test primitive library (*component_library.v*), the components of which are shown in Appendix B. These gate models perform the usual gate-level simulations, and are known to various fault simulation functions handling circuit faults. An example of a NAND gate is shown in Fig. 4.5. The input number of which is passed to it via its *n* parameter. Note that the first parameter of all three NAND gates of Fig. 4.4 uses number 2 for the number of inputs. Verilog coding details of the components of this library were discussed in Chapter 2. Appendix F discusses a translation software program (*NetlistGen*) that produces netlists such as that of Fig. 4.4 from behavioral Verilog descriptions. *NetlistGen* appends to commercial synthesis programs for netlist generation.

```
module mux2to1 ( input a, b, s, output w );
   wire 11, 12, 13, 14, 15, 16, 17, 18, 19;
   pin I1 (s,11);
   pin I2 (a,15);
   pin I3 (b,16);
   pout O1 (19,w);

   fanout_n #(2,1,1) FO1 (11,{12,13});
   notg #(1,1) G1 (14, 12);
   nand_n #(2,1,1) G2 (17, {15, 14});
   nand_n #(2,1,1) G3 (18, {16, 13});
   nand_n #(2,1,1) G4 (19, {17, 18});
endmodule
```

Fig. 4.4 Multiplexer faultable netlist

```
module nand_n
   #(parameter n = 2, tphl = 1, tplh = 1)
   (out,in);
   input [n-1:0] in;
   output out;
   wire [n-2:0] mwire;
   genvar i;

   and and_0 (mwire [0], in [0], in [1]);
   generate
      for (i=1; i <= n-2; i=i+1) begin : NAND_N
         and inst (mwire [i], mwire [i-1], in [i+1]);
      end
   endgenerate
   not #(tplh, tphl) inst(out, mwire [n-2]);
endmodule
```

Fig. 4.5 NAND test primitive

The next item in the list of information needed for fault simulation is the golden model. We can, of course, instantiate the same module as that used as the faultable model, and just do not inject faults on its lines. Alternatively, for faster simulation, we can use a behavioral description of the multiplexer. Figure 4.6 shows this description of the multiplexer.

Figure 4.7 shows test data we are using for fault simulating the multiplexer. The data shown here consists of all combinations of data for the three multiplexer inputs. In an actual fault simulation, where real circuits have a large number of inputs, test data is generated by a test generation process or are obtained from a part manufacturer. Note that one of the applications of fault simulation is test generation. This topic is addressed in Sect. 4.2 of this chapter.

The last item in the list of necessary information for fault simulation is the list of faults to create faulty circuit models. This list can be generated by a fault collapsing program or manually. Our Verilog fault simulation environment expects a Verilog hierarchical name starting from the top-level test bench leading to the line where fault is to be injected. For example, stuck-at-0 fault on l_4 input of G_2 of Fig. 4.3 should be described as:

$$\text{Tester.FUT.G2.in[0] s@1}$$

This assumes that our faultable module (Fig. 4.4) is instantiated in *Tester* using faultable-under-test (FUT) instance name.

Instead of manually creating a fault list, or just listing every stuck-at fault of our circuit, we can use the *FaultCollapsing* PLI function of Chap. 3. This function generates the fault list shown in Fig. 4.8.

```
module mux2to1B ( input a, b, s, output w );
   assign w = ~s ? a : b;
endmodule
```

Fig. 4.6 Behavioral description of the multiplexer

```
000
001
010
011
100
101
110
111
```

Fig. 4.7 Multiplexer test data

```
Tester.FUT.O1.in[0] s@1
Tester.FUT.O1.in[0] s@0
Tester.FUT.FO1.in s@1
Tester.FUT.FO1.in s@0
Tester.FUT.G2.in[1] s@1
Tester.FUT.G2.in[0] s@1
Tester.FUT.G3.in[1] s@1
Tester.FUT.G3.in[0] s@1
Tester.FUT.G4.in[1] s@1
Tester.FUT.G4.in[0] s@1
```

Fig. 4.8 Multiplexer faults after fault collapsing

4.1.3.2   Fault Injection

Before we show our testbench performing tasks of Sect. 4.1.2, we need to introduce two new PLI functions for fault injection and fault removal.

PLI task **$InjectFault (wireName, faultValue)** forces its *faultValue* argument on *wireName*. The *wire-name* argument is a Verilog hierarchical name. This name starts with the module name within which **$InjectFault** is called, followed by instance names of module hierarchies leading to the wire-name that the value is being forced on. Names in this hierarchical name are separated by dots.

PLI function **$RemoveFault (wireName)** has a *wireName* argument. When called, the forced value from the wire identified in its argument is removed, and the wire values are determined by values reaching it through normal simulation. Figure 4.9 shows injecting a stuck-at-0 fault on input 1 of G_2 of Fig. 4.3, and removing it after 450 ns. While fault injection and removal are taking place, random test vectors are being applied to the instantiated multiplexer.

4.1.3.3   Performing Fault Simulation

Fault simulation processings listed in Sects. 4.1.2.1 through 4.1.2.5 are executed by the Verilog testbench shown in Fig. 4.10a.

The first part of the testbench declares wires and variables for data application and test control. We discuss these variables when discussing their use. The golden-under-test (*GUT*) and faultable-under-test (*FUT*) modules are instantiated at the beginning of the testbench. *GUT* is the behavioral module of Fig. 4.6 and *FUT* is that shown in Fig. 4.4. These instantiations have the same set of inputs (*ai, bi, si*), and have different output signals (*woG, woF*) to be compared. If delays are to be considered the netlist module, *mux2to1*, should be instantiated for the golden model instead of *mux2to1B*.

The **initial** block shown in Fig. 4.10a is responsible for reading tests and faults, injecting faults in *FUT* and analyzing the outputs. Near the beginning of this statement, the *Mux.flt* file is opened using Verilog **$fopen** system task. This is followed by a **while** *loop* that uses Verilog's **$fscanf** (formatted scan from a file) to read *wireName* and *stuckAtVal* from the fault list. *wireName* is a

```
module Tester ();
   reg ai=0, bi=0, si=0;
   wire woG, woF;
   mux2to1B GUT (ai, bi, si, woG);
   mux2to1B FUT (ai, bi, si, woF);

   initial begin
      #000;$InjectFault("TESTmux2to1.FUT.G2.in[1]", 1'b0);
      #450 $RemoveFault("TESTmux2to1.FUT.G2.in[1]";
      //repeat above for more faults
      #50
      $stop;
   end
   always #50 {ai, bi, si} = $random;
   // compare woG and woF and make decisions
endmodule
```

Fig. 4.9 Fault injection and removal

```
module Tester ();
   reg ai=0, bi=0, si=0;
   wire woG, woF;
   reg detected;
   integer testFile, faultFile, status;
   reg[2:0] testVector;
   reg [8*60:1] wireName;
   reg stuckAtVal;

   mux2to1B GUT (ai, bi, si, woG);
   mux2to1 FUT (ai, bi, si, woF);

   initial begin
      faultFile = $fopen("Mux.flt", "r");

      while( ! $feof(faultFile))begin
         detected = 1'b0;
         status = $fscanf(faultFile,"%s@%b\n",wireName, stuckAtVal);
         $InjectFault ( wireName , stuckAtVal);
         testFile = $fopen("Mux.tst", "r");

         while((!$feof(testFile))&(detected == 0)) begin
            #30;
            status = $fscanf(testFile,"%b\n",
            testVector);
            {a, b, si} = testVector;
            #60;
            if (woG != woF) begin ... end
         end //while eof test
         $RemoveFault(wireName);
         #30;
      end//while eof faults
      $stop;
   end// end of initial
endmodule
```

```
Fault:    Tester.FUT.I4.in[0] SA1 was detected by 000 at 90.
Fault:    Tester.FUT.I4.in[0] SA0 was detected by 011 at 1110.
Fault:    Tester.FUT.FO1.in SA1 was detected by 010 at 1770.
Fault:    Tester.FUT.FO1.in SA0 was detected by 011 at 2610.
Fault:    Tester.FUT.G2.in[1] SA1 was detected by 000 at 3090.
Fault:    Tester.FUT.G2.in[0] SA1 was detected by 101 at 4290.
Fault:    Tester.FUT.G3.in[1] SA1 was detected by 001 at 4680.
Fault:    Tester.FUT.G3.in[0] SA1 was detected by 010 at 5520.
Fault:    Tester.FUT.G4.in[1] SA1 was detected by 100 at 6450.
Fault:    Tester.FUT.G4.in[0] SA1 was detected by 011 at 7110.
```

Fig. 4.10 (a) Testbench performing fault simulation. (b) Fault simulation report

string and *stuckAtVal* is an integer. The **$InjectFault** function, that comes after **$fscanf** injects the *stuckAtVal* value into the *FUT* wire identified by the *wireName* hierarchical name.

Recall that the *FUT* instance name is used in the hierarchical names in the fault list sample of Fig. 4.8. This fault injection causes the *FUT* instance to become a faulty model of the multiplexer.

Now that we have a faulty model, the test file is opened in a similar fashion to the fault list file, and in a **while** loop, test data are read from it. This is the inner **while** loop in Fig. 4.10. As shown, each test vector read from the test file is named as *testVector*. The 30 ns delay at the beginning of this **while** loop is needed for its next iterations. After applying *testVector* to circuit inputs, a 60 ns delay allows the propagation of values through the faulty and golden models. Following this, an **if** statement compares golden and faulty outputs.

Outputs *woG* and *woF* are different when a test has been applied that detects fault injected in *FUT*. Making fault reports, counting the number of detected faults, saving tests that detect a fault, and other functions that fault simulation is being done for go in this **if** block. In our example, we are only reporting the detected fault, the input value, and the time that fault is detected. Sect. 4.2 elaborates on other functions that this environment can perform.

After the **while** loop applying all tests in *Mux.tst* file to a given faulty model, **$RemoveFault** removes the fault, allows a settling time required by the PLI function, and goes into the next iteration of injecting a new fault. Figure 4.10b shows report generated by **$display** task of this testbench.

4.1.4   *Sequential Circuit Fault Simulation*

Sequential circuit fault simulation is different from combinational fault simulation because of the internal states of the circuit and its clocking and resetting requirements.

A resetting is required with every fault that is injected in the circuit. This is due to the fact that the internal flip-flops of the circuit may be holding values from the last fault and last set of inputs, which are completely irrelevant to new fault being injected.

Furthermore, clocking the circuit after the application of a test vector, allows faults hidden in the internal states of the circuit to appear on the circuit primary outputs. Take for example a fault the effect of which reaches circuit flip-flop input, but not any of the circuit's primary outputs. With clocking the circuit, this fault gets another chance to come back into the logic of the circuit, and possibly appears on the circuit primary outputs.

4.1.5   *Fault Dropping*

A term that is often used in relation with fault simulation is *fault dropping*. This term refers to dropping a fault from further processing (or simulation), once the fault is found to be detected. Fault dropping is done for reducing simulation time by skipping extra testings for the fault [1].

See for example the testbench of Fig. 4.10a. There are two loops in this figure, the outer loop is related to fault and the inner one is for tests. Once *detected* becomes 1, the inner loop exits, and the outer loop removes the detected fault and injects a new fault.

Implementations of fault dropping are different. In our example, since the inner loop was for test vector application, by simply exiting the loop, we were able to implement fault dropping. Other fault simulation applications may require applying test vectors to be done in the outer loop, and fault injection in the inner loop. In such cases, detected faults must be marked as such, so when the next test is applied the detected fault is dropped from being processed again.

In some fault simulation applications, we cannot do fault dropping. An example is fault dictionary generation that is discussed in the next section.

4.1.6   *Related Terminologies*

In the texts that follow, we use certain terminologies for explaining methods and procedures. Although definitions may be clear from the context, to eliminate ambiguities, we give a formal definition of fault simulation terminologies here.

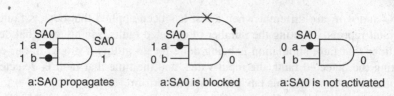

Fig. 4.11 Fault propagation

4.1.6.1 Fault Activation

A fault is activated if the line value reaching the fault is different from its faulty value. A 0 on a line with SA1 fault activates it, and a 1 activates a SA0 fault [2, 3].

4.1.6.2 Fault Propagation

If a fault that has become active appears at an input of a gate and causes the output of the gate to have a different value than its good value, the fault is said to have propagated to the gate output. For example, two 1s at inputs *a* and *b* of a 2-input AND gate (Fig. 4.11) generate a good 1 at the output of the gate. A SA0 fault on *a* causes the output of the gate to become 0, which is different than the good value 1. In this case, the *a:SA0* has propagated to the gate output. Whereas, if *b* is 0, the gate output will be 0 regardless of *a* being faulty or not, in this case *a:SA0* is said to have been blocked [2, 3].

4.1.6.3 Fault Detection

With the above terminology, a fault is detected by a given test, if the fault is activated, and it propagates through all gates leading to at least one of circuit outputs.

4.1.6.4 Fault Blocking

As mentioned above, fault blocking occurs if good value at a gate input forces the gate output to a fixed value, and the faulty input does not play a role in the determination of the gate output. For an AND (OR) gate, a 0 (1) on an input blocks all faults on the other inputs.

For a given test vector, if a fault is blocked and does not propagate to any circuit output, the fault is not detected by that test.

4.2 Fault Simulation Applications

There are many uses of fault simulation programs. A fault simulator can be a stand-alone program, or a program used as a procedure as part of another program such as a test generation program. Usually, fault simulators are run on netlists generated by a synthesis program for obtaining the information for test and diagnosis of manufactured parts.

Perhaps the most well-known application of a fault simulator is the calculation of fault coverage. However, this is not the only use, and applications such as test generation, test refinement, and generation of fault dictionaries are just some of other uses of fault simulators. This section discusses some of the important areas that fault simulation, as a stand-alone program, or part of a larger program can be used in.

4.2.1 Fault Coverage

For a test set, fault coverage (or test coverage) is the ratio of detected faults over total faults in a circuit [4]. Total faults are often those obtained after fault collapsing.

$$\text{For test set } T, \ \ FC = \text{Detected faults/Total faults}$$

For a given test set, fault simulation is used as a stand-alone program to calculate fault coverage to measure how good the test set is. For a test set being developed, fault simulation is used to measure the coverage of an existing test set, to stop generating more tests when enough coverage has been obtained.

Another area of test generation where fault coverage is important is the evaluation of deterministic tests. Such tests are generally generated without regard to circuit and gate delays. Because of this, due to actual gate delays, some tests may not detect faults that they have been targeted for. For this reason, performing fault simulation of gate-level circuits with gate and wire delays gives a realistic measure of faults covered by the generated tests.

4.2.1.1 Fault Coverage Procedure

Figure 4.12 shows the procedure for fault coverage calculation. We start with a given test set and a fault list. We assume a good circuit model and a faultable model are instantiated and being compared (*GUT* and *FUT*). Injecting faults of the fault list into *FUT* makes our faulty circuits.

The procedure consists of two nested loops. The outer loop considers every fault in the fault list, and the inner loop applies tests in the test set to the faulty circuit until one detects the fault. In the inner loop, the outputs of the good circuit and faulty circuits are compared. The two outputs differ when test *t* detects fault *f*. In that case, the number of detected faults is incremented, the rest of tests are skipped, and the next fault is considered (this implements fault dropping).

```
Given Test Set T, n test vectors, t₁:ₙ;
Given Fault List F, m faults, f₁:ₘ;f₀ no fault;
For j in 1 to m loop -- every f in F
    Inject fⱼ;
    For i in 1 to n loop -- every t in T
        While fⱼ is not detected begin
            Simulate faulty circuit;
            Increment DF if fⱼ is detected;
        End while;
    End for;
    Remove fⱼ;
End for;
Record DF, detected fault in F;
Calculate %FC based on m and DF;
```

Fig. 4.12 Fault coverage calculation

When completed, every fault has been examined for detection by the test set. The first test vector of the test set that detects a fault causes the fault to be marked as detected, and the number of detected faults to be incremented. When all faults are considered, fault-coverage is calculated by the ratio of the detected faults, *DF*, over the original number of faults, *NF*.

4.2.1.2   HDL-based Fault Coverage

Figure 4.13 shows Verilog implementation of fault coverage calculation for the 2-to-1 multiplexer. This is a Verilog testbench that instantiates good and faultable models, and with procedural code within an **initial** block implements the procedure of Fig. 4.12.

We are to calculate the fault coverage of test given in "*Mux.tst*" file. Fault list for which coverage is to be calculated is obtained by calling the PLI **$FaultCollapsing** function. As shown in the **initial** block in Fig. 4.13, this function finds faults of *FUT* and dumps them into "*Mux.flt*." This file is closed when fault collapsing function is completed.

The two **while** loops are related to faults and tests are in the body of the **initial** block, in this order. The outer loop reads a fault from the fault list, and injects it into *FUT* by using **$InjectFault**. This repeats for all faults. The inner loop applies tests to inputs of good and faulty models until either a fault is detected or end-of-file of test file is reached. In the former case, *detectedFault* is incremented, and next fault is taken.

At the end of two nested loops, the **$display** task reports the fault coverage. In addition, a report of faults that are detected and the time of their detection can be generated by display tasks in the **if** block that checks faulty and good outputs.

Fault coverage calculation shown here takes advantage of fault dropping. This is implemented by exiting the inner loop once a fault is detected.

4.2.1.3   Sequential Circuit Fault Coverage

Calculation of fault coverage for sequential circuits is only different from combination fault coverage in the application of reset after injection of each fault, and the application of clock after assigning a test vector to the circuit inputs. This topic was discussed in Sect. 4.1.4.

In a testbench that we developed for fault coverage calculation of Residue-5 sequential circuit, an **always** statement was used for generating clock pulses. After a fault was injected, the reset input of the circuit was asserted and held high for several clocks, and then deasserted. In addition, after a new test vector was placed on circuit inputs, the @(**posedge** *clk*) statement caused clocking of test data effects into circuit flip-flops.

This topic is more elaborated when we discuss a testbench for sequential circuit fault dictionary generation.

4.2.2   Fault Simulation in Test Generation

Another important use of fault simulation is in test generation. Test generation is the process of obtaining test vectors for detecting circuit faults [4, 12]. This section gives a general outline of where fault simulation can be useful in test generation. Where possible, some procedures are discussed, however, detailed descriptions and examples for these applications are presented in the chapter on test generation. Some areas that fault simulation can be useful in test generation are test refinement, random test generation, and deterministic test.

```
module Tester ();
    reg ai=0, bi=0, si=0;
    wire woG, woF;
    reg detected;
    integer i;
    integer testFile, faultFile, status;
    real faultCount, detectedFault;
    reg[2:0] testVector;
    reg [8*60:1] wireName;
    reg stuckAtVal;

    mux2to1 GUT (ai, bi, si, woG);
    mux2to1 FUT (ai, bi, si, woF);

    initial begin
        faultCount = 0;
        detectedFault = 0;

        faultFile = $fopen("Mux.flt", "w");
        $FaultCollapsing(Tester.FUT, "Mux.flt");
        $fclose(faultFile);
        faultFile = $fopen("Mux.flt", "r");

        while( !$feof(faultFile))begin
            detected = 1'b0;
            status = $fscanf(faultFile,"%s
        s@%b\n",wireName, stuckAtVal);
            $InjectFault ( wireName , stuckAtVal);
            testFile = $fopen("Mux.tst", "r");

            while((!$feof(testFile))&(detected == 0)) begin
                #30;
                status = $fscanf(testFile,"%b\n", testVector);
                //$display("testVec = %b\n", testVector);
                {ai, bi, si} = testVector;
                #60;
                if (woG != woF) begin
                    detected = 1'b1;
                    detectedFault = detectedFault + 1;
                    $display("Fault:%s SA%b detected by %b at %t.",
                             wireName, stuckAtVal, {ai, bi, si}, $time);
                end//if
            end //while eof test
            $RemoveFaulT(wireName);
            #30;
            faultCount = faultCount+ 1;
        end//while eof faults

        $display("coverage = %f\n", detectedFault/faultCount);
        $stop;
    end// end of initial
endmodule
```

Fig. 4.13 Fault coverage calculation in Verilog

4.2.2.1 Test Refinement

Test efficiency is the number of faults covered by a test vector. For a given test set, fault simulation can be used to identify test vectors that are low in efficiency, or test vectors that detect faults already covered by other test vectors. Based on efficiency and coverage, if it is found that there are tests that do not have a significant contribution to the overall fault coverage, or have no contribution at all, they can be removed from the test set. As such, a test set can be refined for the fewest number of tests and the highest coverage.

```
Start with fault list F;
Repeat for as long as good t's are being found
   Generate random test t;
   For j in 0 to m loop -- every remaining f in F;
      Inject f_j;
         Simulate faulty circuit with t;
         Increment number of faults detected if detected;
      Remove f_j;
   End for;
   Evaluate t by the number of faults it detects;
   Save t in T or discard;
   Decide to repeat finding another t or quit;
End repeat;
```

Fig. 4.14 Fault simulation for random tests

4.2.2.2 Random Test Generation

Random test generation can be regarded as a complement or a replacement for costly automatic test pattern generation algorithms. By means of fault simulation, a randomly generated input vector is examined for detection of faults, and based on this, it is decided whether to keep the vector as a test vector or to drop it. For this purpose, simulation of the new vector is checked for detection of remaining undetected faults [5–9].

Figure 4.14 shows a general outline of the procedure for random test generation. As shown, test t is randomly selected and its efficiency in detecting undetected faults is measured by a fault simulation loop. In this loop, the remaining undetected faults are injected into the circuit-under-test, and a counter is incremented if the random test vector, t, detects the injected fault f. At the end, the value of this counter is used in deciding on keeping or discarding the test.

This procedure only highlights the use of fault simulation in random test generation. Issues such as when to stop generating tests, or what an acceptable test efficiency is, are not discussed here. Such factors, are related to test generation, and are discussed in the chapter on this topic.

4.2.2.3 Fault-oriented Test Generation

Fault-oriented test generation is a process that a certain circuit fault is selected, and a test is generated for it. This test generation scheme is a complex process, and requires many iterations though the circuit for coming up with a test vector that detects the targeted fault. Once the test is found, it is important to identify other faults that are detected by the same test vector. This eliminates the need for repeating the test generation process for faults that are already detected.

Identifying faults detected by a test vector, in addition to the fault that was targeted, is done by fault simulation. As such, a fault-oriented test generation program usually includes fault simulation, as part of its process.

Figure 4.15 shows an outline of a test generation procedure. As shown, when test t is generated, it is applied to the circuit input, and faults that are not detected are injected in the circuit one at a time. Faults in a faulty circuit that produce a different output than the good circuit output are marked as detected. Such faults are removed from the list of faults for which test needs to be made.

```
Start with fault list F;
For j in 0 to m loop -- every remaining f in F;
   Generate test t for fj;
   Mark fj as detected;
   For j in 0 to m loop -- every remaining f in F;
      Inject fj;
         Simulate faulty circuit with t;
         If response is different, mark fj as detected;
      Remove fj;
      Report if fault is detected;
      Remove fault from F if detected ;
   End for;
End for;
```

Fig. 4.15 Fault simulation part of fault-oriented test generation

4.2.3 Fault Dictionary Creation

This section shows how fault simulation can be used for creating a fault dictionary. We discuss what a fault dictionary is, how it is created, and what it is used for. Fault dictionaries for combinational and sequential circuits are discussed, and a Verilog testbench is developed for creating them.

4.2.3.1 Fault Dictionary

Fault dictionaries are data bases in which a correspondence is made between circuit faults and test vectors that detect them. Fault dictionaries are used in hardware testing for fault diagnosis. When a faulty response of a circuit is observed, the site of fault, with a close approximation, can be found using the information included in the fault dictionary.

A fault dictionary database is generated by simulation for a given test set, and results are available in hardware testing. The same test set used in simulation is also used during hardware testing. The database provides anticipated results for every fault for test vectors of the test set. Anticipated results are put in the form of *fault syndromes*. A fault syndrome contains information about test vectors that do and those that do not detect the fault [10].

In its simplest form, a fault dictionary for a combinational circuit is a two-dimensional table (see Fig. 4.16), with its rows designating circuit faults and its columns representing test vectors. Check marks in a row, corresponding to a fault, show test vectors that detect the fault. Contents of a row with 1's replacing check marks and 0's elsewhere are regarded as fault syndromes.

Figure 4.16 shows collapsed list of faults of our 2-to-1 multiplexer, and test vectors that detect these faults. A simple syndrome has been created for each fault. In selection of test vectors, we have selected enough vectors to produce unique syndromes for each fault. Note in Fig. 4.16 that if *v0* and *v1* are removed from the test set, we still have 100% fault coverage, but we will not have unique syndromes, e.g., *G2.in[0] s@1* and *G3.in[1] s@1* will both have 0001 for syndrome. In this case, if this syndrome is obtained during hardware testing and diagnosis, we will be uncertain as to which fault has occurred in the circuit.

Having the same syndrome for multiple faults is called *fault aliasing*. Fault aliasing does not cause loss of fault coverage but causes loss of diagnostic resolution. Since fault dictionaries are mainly used for diagnostic purposes, selection of a good and brief syndrome, especially in sequential circuits, is an important topic.

	V_0 000	V_1 001	V_2 010	V_3 011	V_4 100	V_5 101	Syndrome
Tester.FUT.O1.in[0] s@1	✓	✓	✓			✓	111001
Tester.FUT.O1.in[0] s@0				✓	✓		000110
Tester.FUT.FO1.in s@1			✓		✓		001010
Tester.FUT.FO1.in s@0				✓		✓	000101
Tester.FUT.G2.in[1] s@1	✓		✓				101000
Tester.FUT.G2.in[0] s@1						✓	000001
Tester.FUT.G3.in[1] s@1		✓				✓	010001
Tester.FUT.G3.in[0] s@1			✓				001000
Tester.FUT.G4.in[1] s@1					✓		000010
Tester.FUT.G4.in[0] s@1				✓			000100

Fig. 4.16 Multiplexer fault dictionary

4.2.3.2 Generating a Fault Dictionary

As mentioned, a fault dictionary is generated with the use of fault simulation using faulty and good circuit models. The circuit for which a fault dictionary is to be generated is simulated for every fault and every test vector in a given test set. Figure 4.17 shows a procedure for performing this.

Figure 4.18 shows a Verilog testbench performing fault simulation on *mux2to1* using fault list from "*Mux.flt*" and test vectors from "*Mux.tst*," where the **initial** block shown here implements the procedure of Fig. 4.17. This code generates a fault dictionary similar to that shown in Fig. 4.16.

As in the case of fault coverage calculation of Fig. 4.13, good and faultable circuit models are instantiated and their outputs are compared. The circuit is simulated for every fault in "*Mux. flt*" and every test in "*Mux.tst*." As responses of good and faulty circuit models for a test vector are compared (*woG* and *woF*), a 1 will be inserted in the fault syndrome position corresponding to the test vector if responses are different, otherwise a 0 is placed in the syndrome. When all tests are applied, after the end of the loop corresponding to the test vectors (the inner-loop), circuit fault and its syndrome are written into the "*Mux.dct*" file. When completed, the fault dictionary file, consisting of information in first and last columns of Fig. 4.16 is generated.

The testbench discussed above is very similar to the testbench of Fig. 4.13 for fault coverage calculation. However, important differences exist that need attention. In general, the generation of fault dictionaries is different from coverage analysis in that for fault dictionaries simulation must be done for each fault and each test vector, while coverage analysis skips test vectors after the first one is found to detect a given fault. In other words, fault dropping is done in coverage analysis, but not in fault dictionary generation. This is evident in the inner loops of Figs. 4.13 and 4.18 that correspond to test vectors. In Fig. 4.13 *detected* becomes 1 when a test detects an injected fault. This variable is used as part of the **while** loop to force exiting the loop when such happens. On the other hand, the **while** loop in Fig. 4.18, continues for as long as there are test vectors in *testFile*. The exhaustive looping of the nested **while** loops makes fault dictionary creation much slower than applications, such as fault coverage calculation.

```
Given test T, n test vectors, t_1:n ;
For i in 1 to n loop -- every t in T;
   Simulate good circuit;
   Collect good partial syndrome gs_1:i;
End for;
Report good circuit syndrome gs_T;
For j in 0 to m loop -- every f in F;
   Inject f_j;
   For i in 1 to n loop -- every t in T;
      Simulate faulty circuit j;
      Collect partial syndrome fs_j,1:i;
   End for;
   For fault f_j report syndrome fs_j,T;
   Remove f_j;
End for;
```

Fig. 4.17 Fault dictionary generation procedure

4.2.3.3 Sequential Circuit Fault Dictionary

Because of the required clocking and internal registers of a sequential circuit, generating a syndrome is more involved than in combinational circuits. Furthermore, as in fault coverage analysis, we must reset the golden and faulty circuit models every time a new fault is injected.

In combinational circuits, a fault syndrome is generated just by concatenating 1's and 0's for tests that do and do not detect the fault. This method has a domino-effect in sequential circuits due to the possibility of propagation of the first fault effect into the states of such circuits. This causes the bits of the syndrome corresponding to test vectors after the first vector that detects the fault to become 1, resulting in very few unique syndromes, and increasing the chance of fault aliasing.

A possible alternative solution would be to form the syndrome for a fault by concatenating all faulty circuit output vectors for all tests in a test set. A syndrome made this way is very large and not practical. A better alternative to cascading the faulty circuit responses is to compress them into an acceptable size syndrome. The method of compression and its size can be decided experimentally by fault simulation for a minimum fault aliasing.

A parameterized test response compression function for sequential circuits is shown in Fig. 4.19. The *poly* and *seed* parameters, and the input and output lengths are decided based on the required compression.

This compression module is used in a testbench for creating a fault dictionary for a sequential circuit. This is called a multi-input signature register (MISR) that is used for parallel data compression and signature calculation. MISRs are discussed in later chapters when we discuss Built-in Self-test (BIST).

For compressing a parallel output vector of a sequential circuit over several clocks, the output vector is connected to some inputs of module of Fig. 4.19, and this module is clocked every time the sequential circuit is clocked. When a fault is injected in the sequential circuit, the *misr* module is reset along with the sequential circuit being tested. When an input test data is applied to the sequential circuit, both MISR and the sequential circuit are clocked. The MISR collects and compresses output data as they are produced. When all tests have been applied, the output of the *misr* module provides a syndrome for the injected fault.

Figure 4.20 shows sections of a testbench for our Residue-5 sequential circuit example. The *misr* module is instantiated with a *poly* parameter passed to it. Three bits of the 8-bit input of *misr*

```verilog
module Tester ();
   parameter tstCount = 8;
   reg ai=0, bi=0, si=0;
   wire woG, woF;

   reg detected = 0;
   integer i;
   integer testFile, faultFile, dictionaryFile, status;
   reg[2:0] testVector;
   reg [8*60:1] wireName;
   reg [tstCount-1:0] syndrome;
   reg stuckAtVal;

   mux2to1 GUT (ai, bi, si, woG);
   mux2to1 FUT (ai, bi, si, woF);

   initial begin
      dictionaryFile = $fopen("Mux.dct", "w");
      faultFile = $fopen("Mux.flt", "r");
      while( ! $feof(faultFile)) begin
         i = 0;
         status = $fscanf(faultFile,"%s s@%b\n",wireName, stuckAtVal);
         $InjectFault ( wireName , stuckAtVal);
         testFile = $fopen("Mux.tst", "r");
         detected = 1'b0;
         while((!$feof(testFile))) begin
            #30;
            status = $fscanf(testFile,"%b\n", testVector);
            {ai, bi, si} = testVector;
            #60;
            if (woG != woF) begin
               detected = 1'b1;
               syndrome[i] = 1;
            end else syndrome[i] = 0;
            i = i + 1;
         end //while eof test
         $RemoveFault(wireName);
         $fwrite(dictionaryFile, "%s, %b \n",wireName, syndrome);
         #30;
      end//while eof faults
      $fclose(dictionaryFile);
      $stop;
   end// end of initial
endmodule
```

Fig. 4.18 Fault dictionary for a combinational circuit

```verilog
module misr #(parameter n = 8, poly = 187, seed = 0) (input clk,
   rst, input [n-1:0] d_in, output reg [n-1:0] d_out);

   always @(posedge clk)
      if (rst)
         d_out = seed;
      else
         d_out = d_in ^({n{d_out[0]}} & poly)^ {1'b0, d_out[n-1:1]};
endmodule
```

Fig. 4.19 Module used for syndrome compression

```verilog
module Tester ();
   . . .
   residue5_net GUT (global_reset, clk, reset, in, outG);
   residue5_net FUT (global_reset, clk, reset, in, outF);

   misr #(8,187,0) Signature_Generator (clk,reset,misr_in,misr_out);

   initial begin
     faultFile = $fopen ("Res5.flt", "r");
     dictionaryFile = $fopen("Res5.dct", "w");
     while(! $feof(testFile)) begin
        status = $fscanf(testFile, "%b\n", testVector);
        in = testVector;
        misr_in = {2'b0, outG, 3'b0};
        @(posedge clk); @(negedge clk);
     end
     SigG = misr_out;
     $fwrite (dictionaryFile, "Golden Signature: %b\n", SigG);
     $fclose (testFile);
     f = 0;
     while(! $feof(faultFile)) begin
        @(negedge clk); reset=1; @(negedge clk); reset=0;
        i = 0;
        status = $fscanf(faultFile,"%s s@%b\n",wireName,stuckAtVal);
        $InjectFault (wireName, stuckAtVal);
        testFile = $fopen("Res5.tst", "r");
        while(!$feof(testFile)) begin
           status = $fscanf(testFile, "%b\n", testVector);
           in = testVector;
           misr_in = {2'b0, outF, 3'b0};
           @(posedge clk); @(negedge clk);
           if (outG != outF) begin
              syndrome[i] = 1;
              detected = 1;
              DetectedFaults[f] = 1;
           end else syndrome[i] = 0;
           i = i + 1;
        end // while of test
        SigF = misr_out;
        $RemoveFault(wireName);
        $fwrite(dictionaryFile, "%s, %b \n", wireName, SigF);
        faultCount = faultCount + 1;
        if (SigG != SigF)
           detectedFault = detectedFault + 1;
        f = f + 1;
     end // while of fault
     // Generate reports
     $display("F Coverage: %f/%f = %f", . . .
     $stop;
   end // initial
   always #100 clk = ~clk;
endmodule
```

Fig. 4.20 Partial residue5 testbench, using MISR for syndrome calculation

are connected to the outputs of the faulty instance of the *residue5* module. The output of MISR, that is an 8-bit vector, is the fault syndrome that is written into the fault dictionary along with the fault id (*wireName*). As faulty Residue-5 outputs are produced, they are compressed by the *misr* module, and after all test inputs are applied, the compressed MISR output becomes the fault syndrome.

4.2.3.4 Using Fault Dictionaries

As mentioned, fault dictionaries are primarily used in hardware testing for fault diagnosis. An environment for this purpose, that is, in a way the physical reproduction of the testbench within which a fault dictionary was created, is shown in Fig. 4.21.

A faulty circuit is tested for all test vectors for which the fault dictionary was created. As circuit responses are produced, they are accumulated and compressed in exactly the same fashion as syndromes were created during fault simulation. At the end of a test run, the accumulation of circuit responses is compared with all available syndromes in the fault dictionary. A match in this case, gives a fairly good estimate of the fault and its location in the faulty circuit.

Problems with fault dictionaries include the size of the dictionary database, the complexity of algorithms for syndrome lookup and fault aliasing. The large size of a fault dictionary database can partially be overcome with data compression. A related issue is software algorithms for looking up a syndrome in the compressed data. Such issues are more related to software programing in the testers, and are not in the scope of this book. As for fault aliasing, more compression results in more fault aliasing which can again be resolved with proper compressions.

Another problem that relates to hardware testing is the issue of multiple faults versus single stuck-at faults. As discussed, fault dictionaries are created for single stuck-at faults while in reality multiple faults or faults not directly representable by the stuck-at model can occur in real circuits. This causes the inability of finding the response of a faulty circuit in the syndromes included in the fault dictionary.

Some of the issues discussed above and the aliasing problems that are due to complexity of circuits being tested make fault dictionaries more applicable to printed circuit board (PCB) testing. PCBs are less complex than integrated circuits and have fewer tests and faults. Because of the complexity of integrated circuits, fault dictionary generation, and testing with fault dictionaries do not play an important role in today's microelectronic test technology. In spite of this, we covered this topic in this chapter because it is still an interesting problem, and it is a good application for fault simulation. Understanding this concept, the reader may be triggered thinking of other related ideas that are more applicable to the newer technologies.

4.3 Fault Simulation Technologies

At this point in this chapter, we have discussed fault simulation, we have presented example fault simulations for combinational and sequential circuits, and we have talked about ways in which fault simulation can be used in electronic testing.

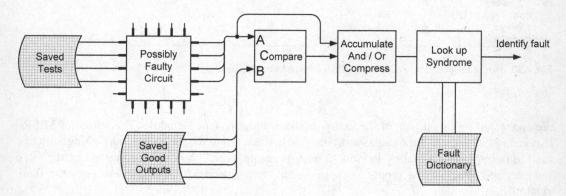

Fig. 4.21 Diagnosis using fault dictionary

An issue that we have not discussed so far is the complexity of fault simulation. The simulation environments in Sects. 4.1 and 4.2 used a very simple simulation flow by injecting a fault and applying all tests to each and every faulty model of the CUT. In real circuits with hundreds of thousands of gates, this method of fault simulation is forbiddingly slow. As the number of gates grows, the number of lines in the circuit grows that result in more faults. Also with the number of gates, the number of test vectors required to test the circuit grows. Taking gates, faults, and test vectors into account, complexity of fault simulation becomes k^3 for a circuit with k gates. More efficient fault simulation approaches than our two nested loops (Sect. 4.1) approach are needed.

This section discusses several fault simulation techniques for performance improvements. Although the complexity remains as k^3, but various fault simulation techniques try to obtain better performances by parallel propagation of test vectors, propagating faults instead of tests, simulating only parts of the circuit, taking advantage of behavioral codes, and other optimization techniques. These techniques are discussed here, and as they are presented, comparisons between different techniques are given. In these presentations, we use the circuit shown in Fig. 4.22 as our test case. Circuit diagram, Verilog netlist, collapsed list of faults, and test vectors with which we examine our circuit are shown here.

We will show fault simulation for four test vectors (t_1, t_2, t_3, t_4), and four circuit faults (f_1, f_2, f_3, f_4), as highlighted in Fig. 4.22. Good circuit simulation of this circuit is shown in Fig. 4.23. Test vectors that are applied to the circuit are shown on the left of the circuit next to the inputs, starting with the first test in the right most columns. These tests are the same as those shown in Fig. 4.22 (t_1, t_2, t_3, t_4). Line values shown correspond to test vectors in the same order, e.g., right most values are for t_1.

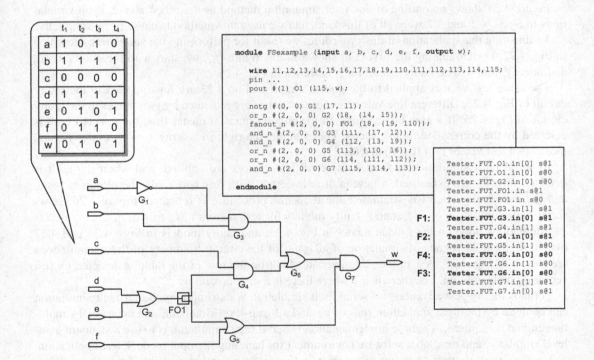

Fig. 4.22 Example circuit for fault simulation

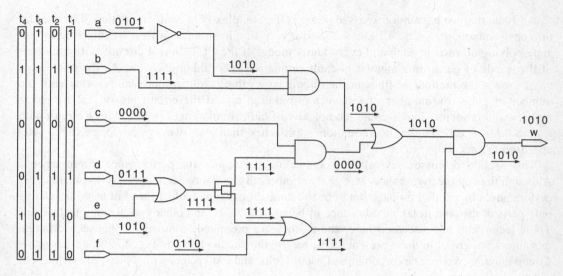

Fig. 4.23 Good circuit simulation of our test case

4.3.1 Serial Fault Simulation

The simplest fault simulation method is serial fault simulation. In this simulation, the circuit being simulated is faulted and a faulty model is obtained. This faulty model is simulated for all test vectors and results are compared for the detection of the injected fault. This process repeats for every circuit fault [2, 10].

Figure 4.24 shows an outline of this fault simulation method as described above. Fault simulations in Sects. 4.1 and 4.2 were all of this kind, and we used this method because of its simplicity.

To illustrate the application of this procedure, we use it for performing the simulation presented in Fig. 4.22. For simulating the first fault shown in this figure (f_1), we start with the good circuit, and inject f_1 into it. The faulty circuit for f_1 is shown in Fig. 4.25.

The same test vectors applied to the golden circuit of Fig. 4.23 are now applied to the faulty circuit of Fig. 4.25. Different line values in these diagrams are indicated by gray boxes in the faulty circuit of Fig. 4.25. If a gray box appears on the circuit output, it means that, the injected fault is detected by the corresponding test vector. For example, the right most gray 1 on w, in this Figure, means that test vector t_1 (110100) detects f_1.

In serial fault simulation, a fault is injected, test vectors are applied, and when the fault is detected or tests are exhausted, a new fault is injected. With this new fault, the circuit is reset, and the new faulty circuit is simulated and the same procedure is repeated. Figure 4.26 shows serial fault simulation of f_2, f_3, and f_4 faulty models for test vectors t_1, t_2, t_3, and t_4.

Simulations done on the golden model in Fig. 4.23, and faulty models in Figs. 4.26 and 4.27 overall perform 20 rounds of simulation of all gates of the circuit. Summary of these simulations are shown in Fig. 4.27. A 1 in the table indicates fault on the row of the table is detected by test vector on table column. I.e., there is a 1 where there is a gray box on w.

Perhaps the biggest advantage of serial fault simulation is its simplicity. Serial fault simulation can be done by multiple simulation runs of a standard gate-level simulator, and no special simulation engine is required. A simple implementation of serial fault simulation is to use a standard gate-level simulator, and develop a software environment for handling its input models, test application, and collection of outputs. A software program in this environment, responsible for fault injection, takes the good circuit netlist, alters it with a fault, and feeds the circuit into the standard simulation

```
Given test T, n test vectors, t_{1:n};
Given fault list F, m faults, f_{1:m};, f_0 no
fault;

For j in 0 to m Loop -- every f in F
    Inject f_j;
    For i in 1 to n Loop - every t in T
        Simulate faulty circuit j;
        Compare results, record results, etc.
    End for;
    Remove f_j;
End for;
```

Fig. 4.24 Serial fault simulation procedure

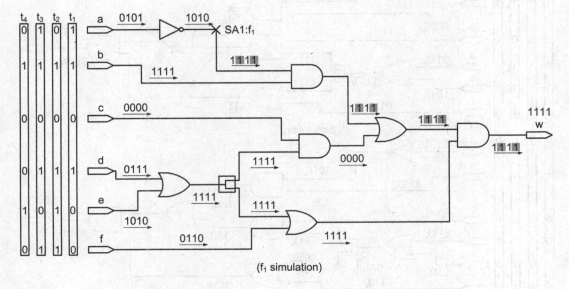

(f_1 simulation)

Fig. 4.25 Serial simulation for f_1

programs. Other programs in this environment can be developed to provide test application, user interfaces, and output collection and report generation.

Serial fault simulation does not require complex data structures, and all input and output information are at the bit-level, which makes handling of this data very easy. This fault simulation takes advantage of fault dropping, i.e., once a fault is detected no more test vectors are applied to the same faulty circuit.

On the other hand, some of the disadvantages of serial fault simulation include its large overhead for preparing faulty circuits and test data for simulation. In addition, the simple data structure of the simulation engine does not allow the use of parallelism, partial simulation of the circuit, or any other optimization technique.

The HDL-based implementation of serial fault simulation in the earlier part of this chapter has the advantage of automated faulty model creation. Furthermore, the Verilog testbench provides a convenient environment for data application, fault handing, and output data processing. This implementation suffices for small circuits, and it has educational value.

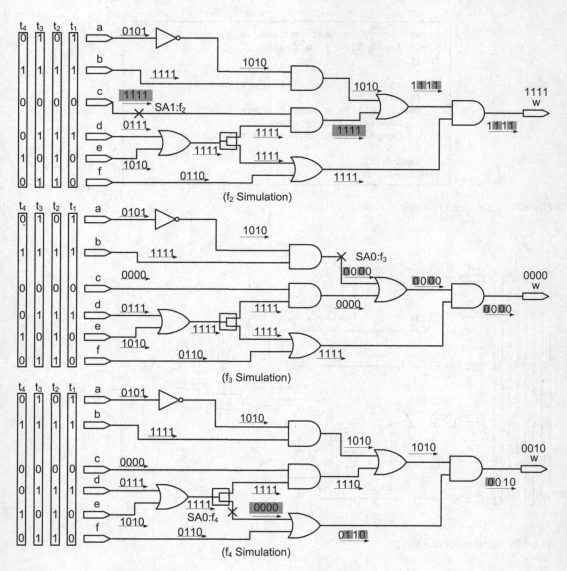

Fig. 4.26 Continuing serial fault simulation of test case of Fig. 4.22

	t_1	t_2	t_3	t_4
Golden	0	0	1	0
Faulty f_1	1	0	1	0
Faulty f_2	1	0	1	0
Faulty f_3	0	1	0	1
Faulty f_4	0	0	0	1

Fig. 4.27 Fault simulation results of Fig. 4.22 test case

4.3.2 Parallel Fault Simulation

A simple extension of serial fault simulation, where faults are handled one at a time, is to process several faults in parallel. This, at least, distributes the overhead of preparing faulty circuits between several faults [11, 12].

In the older simulation software technologies, this parallelism directly translated to packing data corresponding to many faults into a computer word. For example, in a 32-bit machine, circuit line data for 31 faults and one good circuit can be packed into a word. In this case, all logical operations take place at the word level, and, in theory, improves simulation speed by a factor of 32.

The newer simulation engines that have to handle circuit descriptions at different abstraction levels do not have such simple data structures that can directly translate parallel simulations to machine word-length.

Nevertheless, parallel fault simulations can still improve simulation performance. This is partly due to combining data values in data structures of simulator, and partly due to reducing fault handling overhead by loading several faults simultaneously.

4.3.2.1 Parallel Fault Simulation Algorithm

Figure 4.28 shows an outline of the parallel fault simulation algorithm. Like serial fault simulation, there are two nested loops related to faults and tests. The only difference is that the outer-loop reads a group of faults instead of only one. Fault injection, removal, and propagating values happen for all group members in parallel.

Applying this algorithm to the test case of Fig. 4.22 is shown in Fig. 4.29. Recall that we are simulating the circuit for four faults and one golden model. As t_1, t_2, t_3, and t_4 are applied to the circuit, these values travel through five virtual circuits (faulty f_1, f_2, f_3, f_4 and golden model) is parallel. Values on lines of these circuits are packed into a word of the machine running the simulation, or into a data structure of the simulation software. In any case, we are gaining performance by simultaneously handling five values.

4.3.2.2 Verilog Implementation

In Verilog implementation of parallel fault simulation, the testbench reads multiple faults and multiple parallel circuits are activated. Since we cannot access the internal data structure of the

```
Given test T, n test vectors, t_1:n;
Given fault list F, m faults, f_1:m; f_0 no fault;
Divide m faults in g groups of p faults

For jj in 1 to g Loop -- every group of p faults
    Inject f_0; Inject f_(jj-1)xp+1:jj xp faults
    For i in 1 to n Loop - every t in T
        Simulate p faulty circuits;
        Compare results, record results, etc.
    End for;
    Remove f_(jj-1)xp+1:jj xp faults;
End for;
```

Fig. 4.28 Parallel fault simulation algorithms

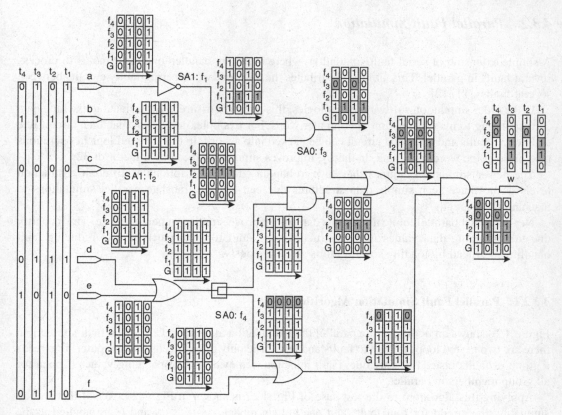

Fig. 4.29 Parallel fault simulation, test case of Fig. 4.22

Verilog simulator, we count on the internal simulator optimizations to combine our multiple line values and give a better performance than the serial fault simulation.

Implementing parallel fault simulation in Verilog requires changes in the fault simulation primitives: for reading multiple faults instead of just a single fault, in the PLI functions for injecting and removing faults, and in the testbench running the simulation. These issues are discussed by the use of examples.

Gate Primitives. For parallel simulation of multiple gates, our gate primitives for fault simulation (*and_n*, *or_n*, etc.) that we discussed in Sect. 4.1.3, have been modified as shown in Fig. 4.30. This figure shows a parameterized AND gate with the same interfaces as gates discussed before. A new parameter, *nf*, has been added here. This parameter is the number of faults, and expands the AND gate into *nf* parallel gates. The input vector of the gate that is shown in Fig. 4.30 is a vector of $n \times nf$ bits, where *n* is the number of inputs. This vector contains input values for *nf* gates each of which have *n* inputs. The output of the gate primitives are *nf* bits in size.

Circuit Netlist. Partial netlist for parallel fault simulation of circuit of Fig. 4.29 is shown in Fig. 4.31. The number of parallel simulations is passed to this netlist from its testbench, and this number is passed to the *nf* parameters of all gates that are instantiated in the netlist.

Fault Injection. Parallel injection of *nf* faults in *nf* parallel circuits requires a different PLI function than the one used for one fault for serial fault simulation. As shown in Fig. 4.32 the new **$ParInjectFault** function, takes the *nf* parameter, *nf* wire names, and *nf* corresponding fault values. The **$ParRemoveFault** works in a similar fashion.

Parallel Testbench. Partial code of a testbench performing parallel fault simulation is shown in Fig. 4.33. This testbench virtually instantiates *nf* parallel faultable models and simultaneously

```
module and_n
  #(parameter n = 2, tphl = 1, tplh = 1, nf = 1)
  (out,in);
  input [(n*nf)-1:0] in;
  output [nf-1:0]out;
  reg [nf-1:0]val;
  integer i;

  always@(in) begin
     val = in[nf-1:0];
     for(i=1; i<n; i=i+1) begin
        val=val & in[(nf*i)+:nf];
     end
  end
  assign out=val;
endmodule
```

Fig. 4.30 *nf* parallel gates

```
module ParSimpleCKT #(parameter nf = 5)
       (input a, b, c, d, e, f, output [rf-1:0] w);
  wire [nf-1:0] l1,l2,l3,l4,l5,l6,l7,l8,l9,l10,l11,l12,l13,l14,l15;
  pin #(nf) I1 (a, l1);
  . . .
  pout #(nf) O1 (l15, w);

  notg #(0, 0, nf) G1 (l7, l1);
  fanout_n #(2, 0, 0, nf) FO1 (l8, {l9, l10});
  and_n #(2, 0, 0, nf) G3 (l11, {l7, l2});
  . . .
  and_n #(2, 0, 0, nf) G7 (l15, {l14, l13});
endmodule
```

Fig. 4.31 Partial netlist for parallel fault simulation

```
module Tester #(parameter nf = 5);
  . . .
  reg [8*60:1] wireName [nf-1:0];
  reg [nf-1:0] stuckAtVal;

  ParMux2to1 #(nf) GUT (ai, bi, si, woG);
  ParMux2to1 #(nf) FUT (ai, bi, si, woF);

  initial begin
     . . .
     $ParInjectFault (nf,
     wireName[0], wireName[1], wireName[2], wireName[3],
     wireName[4],
     stuckAtVal[0], stuckAtVal[1], stuckAtVal[2],stuckAtVal[3],
     stuckAtVal[4] );

     . . .
     $ParRemoveFault(nf,
     wireName[0], wireName[1], wireName[2], wireName[3],
     wireName[4]);
     . . .
  end// end of initial
endmodule
```

Fig. 4.32 Parallel fault injection and removal

```
module TestSimpleCKT ();
   reg a, b, c, d, e, f;
   wire w;
   integer i;

   SimpleCKT inst (a, b, c, d, e, f, w);

   initial begin
      for(i=0;i<10;i=i+1) begin
         #10;
         a=$random+19;
         b=$random+654;
         c=$random-54;
         d=$random*434;
         e=$random+76;
         f=$random*5546;
         #20;
      end
      $stop;
   end
endmodule
```

Fig. 4.33 Testbench for parallel fault simulation

injects *nf* faults in these circuits. As test vectors are read, the same test vector is applied to all *nf* faulty models. Finally, report generation of faulty responses is done for *nf* responses at a time.

4.3.2.3 Comparing Parallel Fault Simulation

In theory, parallel fault simulation of *nf* faulty models is *nf* times faster than serial fault simulation. Although this is the biggest advantage of this fault simulation method, in reality, this never happens. This fault simulation method has relatively low fault injections overhead that help its performance over serial fault simulation. As in serial fault simulation, processing of values is still at the bit-level, which makes input data and response handling relatively easy.

A disadvantage of parallel fault simulation is the requirement of a special simulation engine that can handle parallel simulation of bit values. Even with such an engine, we are still at a disadvantage because fault dropping cannot be done, and all faults must be simulated for all test vectors. Recall in serial fault simulation that as soon as a fault was detected, no more test vectors were applied to the faulty circuit. However, in parallel fault simulation, since we have many parallel faulty circuits, we have to continue with all tests to give a chance to all faults to be detected.

Another important disadvantage of parallel fault simulation is its large data structure, requiring large memory in the host machine.

Our Verilog based implementation of parallel fault simulation does not fit into the classical definition of this method. Obviously, we are still using the same simulation engine that we use for standard simulations. In a way, we are making parallel simulations of multiple gates possible, by having gate models that expand themselves into *nf* virtual parallel gates. Our Verilog implementation heavily depends on the simulator's optimizations for performance improvement. Performance gains we experienced were no more than 5X over serial fault simulation.

4.3.3 Concurrent Fault Simulation

Parallel fault simulation requires multiple complete parallel circuits. For all test vectors, all such circuits are simulated. This results in large memories and long simulation time for simulating every line of every circuit.

Concurrent fault simulation is based on the observation that, not all faults propagate to all parts of the circuit being simulated. Therefore, there is no need to repeat parts of the circuit that simulate the same for all faults, and thus saving memory. Furthermore, repeating only gates for faults that propagate to them, leaves us with fewer gates to simulate, and thus saving processing time [13, 14].

Concurrent fault simulation has the ability to consider all faults at the same time. This simulation method is the most popular fault simulation in commercial tools.

4.3.3.1 Concurrent Fault Simulation Algorithm

Concurrent fault simulation expands only those parts of the circuit that have a different fault effect than the rest of the circuit. In other words, all faults considered, only those gates to which faults propagate are duplicated. Duplication of gates happens as many times as there are faults that have propagated to the gate output. Each duplicated gate must be identified by the fault that has caused the duplication.

In concurrent fault simulation, all faults are considered at the same time, and test vectors are applied one at a time. Figure 4.34 shows a general outline of this fault simulation method.

The algorithm begins with a list of test vectors and a fault list. As shown, virtual injection of all faults happens before individual test vectors are applied to the circuit.

A **for** loop in this algorithm handles the application of test vectors. As shown, for a given test vector, only gates to which circuit fault propagate are duplicated.

We use our test case of Fig. 4.22 with four faults and four test vectors as an example for concurrent fault simulation. Figure 4.35 shows how circuit gates are duplicated when $t_4 = 010010$ is applied. Recall that all faults are considered at the same time.

When $abcdef = 010010$ reaches a gate with a designated fault, and if the output value of the gate activates the fault (i.e., 1 for SA0, 0 for SA1), then the gate is duplicated. The duplication of gates continues for as long as the fault propagates. If a fault propagates in part of a circuit, and then it is blocked (by a 0 on another input of AND, or a 1 on another input of an OR gate), duplication of gates ceases.

Gates in Fig. 4.35 show some of these cases. The good value of G_3 AND gate is 1, but, since f_3 puts a SA0 fault on this output, a new gate with 0 output is created. This output becomes the input

```
Given test T, n test vectors, t_{1:n};
Given fault list F, m faults, f_{1:m}; f_0 no fault;

Consider all faults for concurrent injection
For i in 1 to n Loop - every t in T
    Propagate t_i;
    If due to fault f_j a gate output is faulty
        Duplicate gate with faulty output
    End if;
End for;
```

Fig. 4.34 Concurrent fault simulation

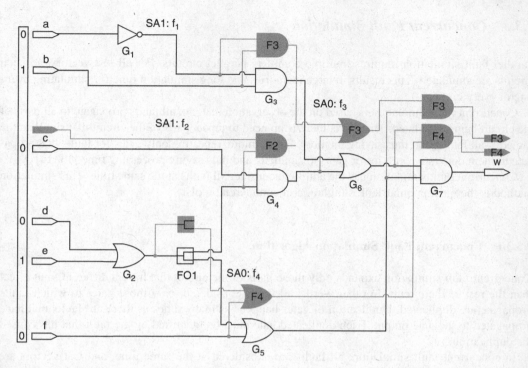

Fig. 4.35 Duplication of gates in concurrent fault

of G_6 OR gate, the other input of which is 0, enabling propagation effect of f_3 to the output of the OR gate. Therefore, a faulty version of G_6 identified by F_3 is also created. Propagation of this fault continues to the circuit output causing the duplication of all gates on its path.

On the other hand, f_4 at the output of *FO1* propagates through G_5 and G_7, causing the duplication of G_5, and the triplication of G_7, which has already been duplicated due to the propagation of f_3.

The f_1 fault in Fig. 4.35 is never activated, and no extra gates are created for it. On the other hand, f_2 is activated by a 0 on the c input, and propagates to the output of G_4. This causes the duplication of G_4 for this fault. However, f_2 at G_4 is blocked by the upper input of G_6 being 1, and no further duplication occurs.

4.3.3.2 Implementing Concurrent Fault Simulation

Duplication of gates in concurrent fault simulation gives a good graphical view of this algorithm, but the actual implementation may be different than just duplicating gates. The implementation of this approach in standard software environments can be done by dynamic linked lists for gate outputs.

Figure 4.36 shows the circuit of Fig. 4.35 for t_1, t_2, t_3, and t_4 tests. Instead of expanding gates that have faulty outputs, using a linked list, we expand output values to which a fault propagates. As Fig. 4.35 identifies duplicated gates by the fault id that causes their duplication. Similarly, in Fig. 4.36, faulty output values are identified by fault ids that reach the gates. See for example, for t_4, the link list at the circuit output is expanded to include f_3 and f_4. This is consistent with the expansion of gates in Fig. 4.35.

Linked lists shrink back to just the golden model values before a new test is to be applied. Our linked list suggestion is just a possible implementation of concurrent fault simulation.

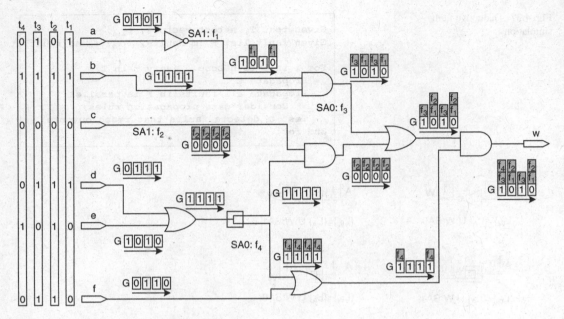

Fig. 4.36 Using linked list for faulty outputs

4.3.3.3 Comparing Concurrent Fault Simulation

The biggest advantage of concurrent fault simulation is smaller memory usage in comparison with parallel fault simulation, due to duplicating gates only when a fault propagates to them. Furthermore, we have fewer gates to propagate data and thus simulation takes less processing time. On the other hand, in concurrent fault simulation we can insert all the faults simultaneously in the circuit and provide a better situation for further implicit optimization in the simulation environment.

4.3.4 Deductive Fault Simulation

In serial, parallel, and concurrent fault simulations, faulty values propagate through lines of circuit being simulated. Alternatively, in deductive fault simulation, list of activated faults propagate through circuit lines [15]. We present the algorithm and an example of deductive fault simulation in this section.

4.3.4.1 Deductive Fault Simulation Algorithm

Figure 4.37 shows an outline of the deductive fault simulation algorithm. As in other algorithms discussed, we start with a test set and a fault list. A **for** loop in this algorithm applies test vectors from the test set to the circuit inputs. As a result of a test vector being applied, good line values are calculated. Then, all line faults are identified when line values are different than those of the good values. The faults identified as such form fault lists that travel from their sites toward circuit outputs. Fault lists collect other faults in their paths as they make this journey. Formation of fault lists is governed by rules discussed below [16].

Fig. 4.37 Deductive fault simulation

```
Given test T, n test vectors, t₁:ₙ;
Given fault list F, m faults, f₁:ₘ; f₀ no fault;

For i in 1 to n Loop - every t in T
   Propagate tᵢ;
   Propagate all faults in F in parallel;
      Consider gate propagation rules;
   Test tᵢ detects faults that reach outputs;
End for;
```

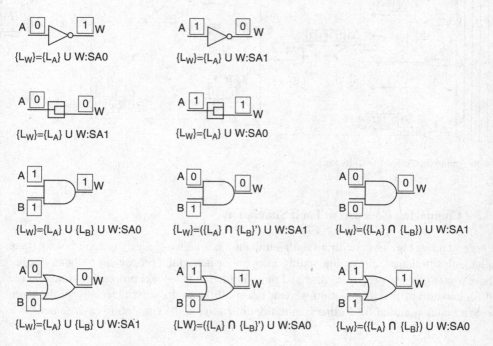

Fig. 4.38 Fault list propagation values

4.3.4.2 Gate Fault List Propagation

Figure 4.38 shows fault list propagations for an inverter, fanout, AND, and OR gates. Propagation of a fault list to the output of a gate depends on good values on gate's inputs and output. Therefore, Fig. 4.38 shows list propagations for all input output combinations. We discuss some of these rules below.

Starting with the inverter, with a 0 input and a good 1 output, suppose that fault list $\{L_A\}$ reaches the A input. This fault list that has been formed by the collection of faults prior to reaching A propagates to the inverter output. In addition, if the inverter output has a SA0 fault, it is also identified (because of value 1 on W), and appends to this fault list. Therefore, fault list at the output of an inverter is that of the input, union with the fault that is stuck-at complement of the good output value.

Propagation of faults through fanout follows the same rules as that of the inverter. The third row in Fig. 4.38 shows fault propagation rules for AND gates. An AND with 1s at the inputs and a good 1 at the output propagates fault lists at both inputs, and adds its output SA0 to the list. This is justified because if a fault list reaches input A, because we are dealing with single stuck-at faults, we are sure that B remains 1. Therefore, the fault list on A propagates to W. The same is true for B.

The next case of AND gate is when input A is 0 and B is 1. In this case, the fault list on A propagates to the output for as long as B remains 1. However, if a fault in L_A also appears in L_B because of the fault, B becomes 0 and does not allow that same fault in L_A to travel to the gate's output. Therefore, for an AND gate with $A = 0$ and $B = 1$, faults on A that do not appear on B propagate to the output of the gate (i.e., intersection of L_A with complement of L_B). Output SA1 is also added to this list.

The last case for an AND gate is the case of two 0's on its inputs. In this case, if a fault appears on A, this fault only propagates to the gate's output if input B is 1. Because of single stuck-at model, the only way for B to become 1 is if the same fault also appears on B. Therefore, the intersection of fault lists on A and B appears on W. The output SA1 is also added to this list.

Fault propagation rules for an OR gate are similar to those of an AND gate. These rules, based on good circuit values, are shown in the last row of Fig. 4.38.

4.3.4.3 Deductive Fault Simulation Example

Based on the procedure shown in Fig. 4.37 and rules of Fig. 4.38, we show deductive fault simulation for $t_1 = 010010$ of the example of Fig. 4.22. This is done for all faults of this example circuit, and not just those exercised for other fault simulation methods.

Figure 4.39 shows our example circuit, test vector t_1, good line values for this test vector, and all circuit faults that we are interested in. Circuit lines are labeled to identify their corresponding faults. With the given good values, we show the propagation of fault lists, and those that reach the circuit output are detected by t_1.

In the left column in Fig. 4.40 the initial list of faults on circuit lines are shown. This list is the list of faults we are interested to check for the detection by t_1. As collection of faults and formation of fault lists begin from the primary inputs, starting lists of activated faults on lines immediately driven by the primary inputs are shown in Fig. 4.40b (on the right).

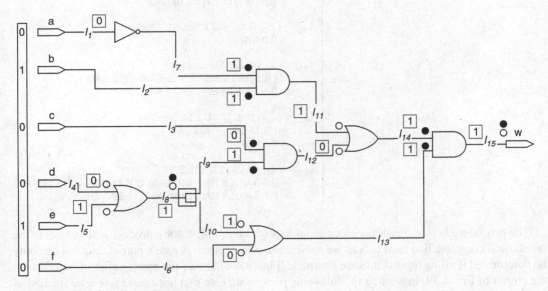

Fig. 4.39 Starting point for deductive fault list propagation

Fig. 4.40 (a) Initial line
faults, (b) Faults at primary
inputs

a

Initial Fault List:

{L_{l2}} = L2:SA1
{L_{l3}} = L3:SA1
{L_{l4}} = L4:SA0
{L_{l5}} = L5:SA0
{L_{l6}} = L6:SA0
{L_{l7}} = L7:SA1
{L_{l8}} = {L8:SA0, L8:SA1}
{L_{l9}} = L9:SA1
{L_{l10}} = L10:SA0
{L_{l11}} = L11:SA0
{L_{l12}} = L12:SA0
{L_{l13}} = L13:SA1
{L_{l14}} = L14:SA1
{L_{l5}} = {L15:SA0, L15:SA1}

b

Activated Faults:

{L_{l1}} = Ø
{L_{l2}} = Ø
{L_{l3}} = L3:SA1
{L_{l4}} = Ø
{L_{l5}} = L5:SA0
{L_{l6}} = Ø

Fig. 4.41 Fault list
propagation

Propagating Faults:

L_{l7} = L_{l1} U L7:SA0
     = Ø

L_{l11} = (L_{l7} U L_{l2}) U L_{l11}:SA0
      = {L_{l11}:SA0}

L_{l8} = (L_{l5} ∩ L_{l4}') U L_{l8}:SA0
     = ({L5:SA0} − {Ø}) U L8:SA0
     = {L5:SA0, L8:SA0}

L_{l9} = L_{l8} U L9:SA0
     = {L5:SA0, L8:SA0}

L_{l10} = L_{l8} U L10:SA0
      = {L5:SA0, L8:SA0, L10:SA0}

L_{l12} = (L_{l3} ∩ L_{l9}') U L12:SA1
      = {L3:SA1}

L_{l13} = (L_{l10} ∩ L_{l6}') U L13:SA0
      = ({L5:SA0, L8:SA0, L10:SA0} − {L6:SA0})
      = {L5:SA0, L8:SA0, L10:SA0}

L_{l14} = (L_{l11} ∩ L_{l12}') U L14:SA0
      = ({L11:SA0} − {L3:SA1})
      = {L11:SA0}

L_{l15} = (L_{l13} U L_{l14}) U L15:SA0
      = {L5:SA0, L8:SA0, L10:SA0} U {L11:SA0}) U L15:SA0
      = {L5:SA0, L8:SA0, L10:SA0, L11:SA0, L15:SA0}

The procedure is to start with primary input lists (those in Fig. 4.40b), process gates closet to the inputs, and complete line fault lists as we move toward the output. A gate's output fault list can only be determined if all its inputs' lists are complete. Figure 4.41 shows the propagation of fault lists in the circuit of Fig. 4.39. In reading the following paragraph, note that uppercase L is used for List of faults. This is subscripted by the line that the list belongs to. A lines is identified by a lowercase l, and the number that follows it is the line number. Line numbers are subscripted except when used as list (L) subscripts. L_{17} designates the fault List that propagates to line 7.

The fault list at the output of inverter on line l_7 is calculated first. As shown in Fig. 4.38, the list at the inverter output is the union of the list on its input and stuck-at complement of its output. As shown in Fig. 4.40b, L_{11} is empty. Also L_7:$SA0$, that is the fault at the inverter output, is not in our initial fault list (Fig. 4.40a). Therefore, L_{17} is empty. Next, fault list on line l_{11} is calculated. Fault lists at the inputs of the AND gate driving this line are already known (L_{17} in Fig. 4.41, and L_{12} in Fig. 4.40b). Good line values of this gate are also calculated already (Fig. 4.39). Using this information, propagation rule is determined from Fig. 4.38 (first AND gate from left in this figure). Following this procedure for all circuit gates, the list that reaches line l_{15} (Line 15) is calculated as shown in Fig. 4.41. This list indicates that $t_1 = 010010$ detects l_5:$SA0$, l_8:$SA0$, l_{10}:$SA0$, l_{11}:$SA0$, and l_{15}:$SA0$.

4.3.5 Comparison of Deductive Fault Simulation

The least we can say about the deductive fault simulation is that it is an interesting algorithm. It has fewer computations than parallel and serial fault simulations, and can take advantage of optimized list processing engines. Some of the disadvantages of this method are large lists to propagate, requiring a special simulation engine, and a complex data structure.

In addition to the above software and implementation issues, a disadvantage of this method that is particular to fault simulation is the inability of performing fault dropping. Since all faults are processed simultaneously, none can be treated separate from others.

4.3.6 Critical Path Tracing Fault Simulation

A fault simulation with a different approach than those discussed in the previous sections is critical path tracing (CPT). The methods discussed so far were fault-oriented, which means that faults are injected and tests are applied to check for their detection. In CPT, for a given test vector, the circuit is traced from outputs to inputs, and faults that can be detected are identified [17–20].

4.3.6.1 Basic CPT Implementation

We present definitions regarding critical paths before discussing the implementation of CPT fault simulation.

Critical Lines. With a set of logic values at a gate's input and output, an input line is critical if the output is critical and toggling the input toggles the output value. Figure 4.42a shows several critical and noncritical gate inputs. Lines driving primary outputs are always critical.

Critical Path. Critical path is a continuous sequence of critical lines starting from a line in a circuit leading to a primary output, from a critical path.

Critical Path Faults. Stuck-at faults along a critical path, with fault values that are complement of those of the critical lines of the path, can be detected by the test vector causing the critical path.

Figure 4.42b shows an outline of the CPT fault generation algorithm. Note here that no faults are injected, and detected faults are deduced from critical paths of the circuit.

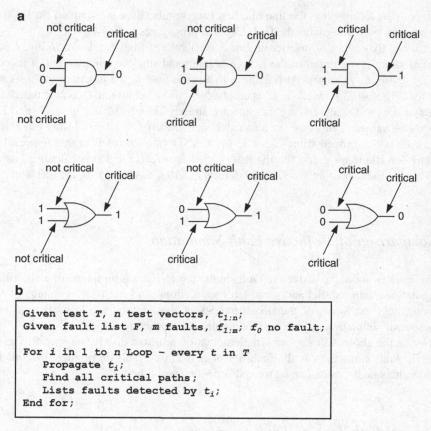

Fig. 4.42 (**a**) Critical values. (**b**) CPT algorithm

4.3.6.2 Reconvergent Fanouts in CPT

Finding critical paths is simple, and the toggling rule mentioned above determines critical lines for basic logic gates. The problem with CPT implementation is when we reach a reconvergent fanout. In this case, a critical branch or a noncritical branch does not determine the criticality of stem.

Take, for example, partial circuits shown in Fig. 4.43. We show critical lines by bold line segments. In Fig. 4.43a, branches of the fanout element are both critical, and the stem is also critical. We can tell that the fanout stem is critical because toggling s line value toggles the reconvergent point, w. In Fig. 4.43b, the upper branch of fanout is critical, but its stem is not. This can again be verified by toggling value of s and observing that w does not change. In Fig. 4.43c, none of the branches of the reconvergent fanout are critical, but its stem is. Verify this by changing s from 0 to 1 and seeing that w also changes from 0 to 1.

In the above examples, since the effect of a fanout stem reaches a gate at the reconvergent point from several paths, considering each path independently does not decisively determine criticality of the fanout stem. The simplest solution, in this case, is to isolate the reconvergent part of the circuit, and apply the toggling rule from divergence point to the convergence point, e.g., line s to line w in our examples.

4.3.6.3 CPT Example

We use the example of Fig. 4.22 to find faults detected by test vector t_1. Recall that this test vector was also used in the deductive fault simulation in Fig. 4.39.

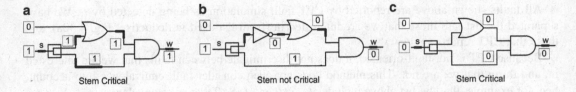

Stem Critical   Stem not Critical   Stem Critical

Fig. 4.43 Reconvergent fanouts in CPT

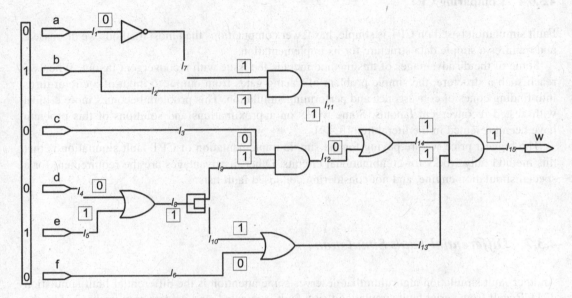

Fig. 4.44 CPT fault simulation example

We use Fig. 4.44 for illustrating CPT fault simulation. As shown here, $t_1 = 010010$ is applied to the circuit inputs, and the circuit is simulated determining all circuit line values. We start with the line driving the w output (line l_{15}) that according to definition is a critical line. Since lines l_{13} and l_{14} are 1, and toggling their values toggles line l_{15} output value (see Fig. 4.42a), both l_{13} and l_{14} are critical lines.

We continue with l_{14} going backward, and postpone l_{13} for now. Since l_{14} is critical (according to Fig. 4.42a), with good line values as shown in Fig. 4.44, line l_{11} is critical. Going backward from l_{11}, the AND gate case is the third AND gate in Fig. 4.42a, therefore, both l_7 and l_2 are critical. Because of l_7, l_1 is also critical.

Now that we have reached the primary inputs and there are no more lines to analyze, we go back to the AND gate, the l_{13} input of which we postponed. Going back from l_{13}, l_{10} is critical. Line l_{10} is a fanout branch, and we cannot decide if its stem is critical by just considering l_{10}. As we suggested before, to determine if l_8 is a critical line, we change its value from 1 to 0, and we see that l_{15} also changes from 1 to 0, therefore l_8 is a critical line. Because of this, lines l_5 is also critical. Considering complete paths, path formed by lines l_5–l_8–l_{15} is a critical path.

Now that all critical paths in the example of Fig. 4.44 are known, for t_1, generating a list of faults detected by t_1 is a simple matter. Detected stuck-at faults are those on critical paths with the stuck-at value being the complement of the line value. According to this, detected faults are those shown below:

$$\{l_5\text{:SA0}, l_8\text{:SA0}, l_{10}\text{:SA0}, l_{11}\text{:SA0}, l_{15}\text{:SA0}\} \quad \{l_{13}\text{:SA0}, l_2\text{:SA0}, l_1\text{:SA1}, l_7\text{:SA0}, l_{14}\text{:SA0}\}$$

All faults shown above are reported by CPT fault simulation as being detected by t_1. We have separated the list into those that we are interested in (from list used in deductive, Fig. 4.40a), and those that CPT reports anyway.

Because CPT is not fault-oriented, it does not discriminate between faults that we are interested in, and those that we are not. This method does not even consider fault equivalence into account. See, for example, that the list above includes l_{11}:SA0 and l_2:SA0 that are equivalent.

4.3.6.4 Comparing CPT

Fault simulation based on CPT is simple, has fewer computations than most methods we discussed, and requires a simple data structure for its implementation.

Some of the disadvantages of this method include the issue with reconvergent fanout. When we reach such a structure, the simple problem of tracing gates from output to input of a circuit turns into finding cones of convergence and performing simulation. This problem becomes more critical with nested reconvergent fanouts. Some works on approximations and solutions of this problem have been presented in the literature [18, 19].

The biggest price we are paying for the simple implementation of CPT fault simulation is that this method only works for combinational circuits. Other disadvantages are the requirement for a special simulation engine, and not considering collapsed fault lists.

4.3.7 *Differential Fault Simulation*

Another fault simulation algorithm that deserves some attention is the differential fault simulation [21]. Recall from serial fault simulation that a fault is injected, and its detection is checked by the application of all test vectors. When simulating large logic circuits, a changing bit from one test vector to another creates many events in its path from circuit inputs to the outputs. This has to repeat for every test vector and every fault of the circuit being analyzed.

On the other hand, if we keep a test vector at the circuit inputs and just inject and remove faults, far fewer events occur in the circuit. Two reasons for this are:

1. The process of removing a single stuck-at fault and injecting another only creates two events at the site of the faults. Whereas, the difference between bit values of two consecutive test vectors is usually much greater.
2. Faults affect somewhere in the circuit between inputs and outputs, and closer they are to the outputs, cause fewer events. Whereas, events caused by changing circuit inputs have to travel a longer distance to reach the outputs.

Based on the above, a simple implementation of differential fault simulation is to change the order of fault injection and test application loops. Figure 4.45 shows an outline of this simulation method. As shown and as compared with serial fault simulation algorithm of Fig. 4.24, only the loop orders have changed. The outer loop applies a test vector, and by the inner loop injecting and removing faults, the test vector is checked for the detection of all circuit faults. Detected faults are marked, and the outer loop applies the next test vector.

Unlike in serial fault simulation that fault dropping is done implicitly, in the differential fault simulation fault dropping is done by explicitly making detected fault as such.

```
Given test T, n test vectors, t_{1:n};
Given fault list F, m faults, f_{1:m};, f_0 no fault;

For i in 1 to n Loop - every t in T
   For j in 0 to m Loop -- every f in F
      Inject f_j;
         Simulate only those parts affected;
         Compare results, record results, etc.
      Remove f_j;
   End for;
End for;
```

Fig. 4.45 Differential fault simulation

4.4 Summary

Fault simulation is used in every aspect of digital system testing. The designers use it to evaluate their design, and the test engineers use it to generate test vectors. Designers and test engineers use it to help them with design testability and insertion of design for test (DFT) or Built-in Self-test hardware. IP core designers and core integrators also use it for the evaluation of test vectors and test integration.

Because of this wide use of fault simulation and because of its complexity, many algorithms have been developed for the implementation of fault simulation. Like any other book on digital system testing, this chapter discussed the applications of fault simulation and presented the various algorithms for it. However, a difference in the way we presented fault simulation was the incorporation of hardware description languages. On the one side, since HDLs and HDL-based tools are integrated parts of most today's design environments, an HDL-oriented fault simulation helps designers incorporate this important process in their design cycle. On the other side, HDLs helped the presentation of fault simulation techniques and applications in a concise and unambiguous fashion. Such is helpful in bringing HDLs into the arena of test engineers.

References

1. Jha NK, Gupta S (2003) Testing of digital systems. Cambridge University Press, Cambridge
2. Bushnell ML, Agrawal VD (2000) Essentials of electronic testing for digital, memory, and mixed-signal VLSI circuits, Kluwer, Dordrecht
3. Roth JP (1966) Diagnosis of automata failures: a calculus and a method. IBM J Res Dev 10(4):278–291
4. Agrawal VD, Agrawal P (1972) An automatic test generation system for illiac IV logic boards. IEEE Trans Comput C-21(9):1015–1017
5. Agrawal P, Agrawal VD (1975) Probabilistic analysis of random test generation method for irredundant combinational logic networks. IEEE Trans Comput C-24(7):691–695
6. Agrawal P, Agrawal VD (1976) On monte carlo testing of logic tree networks. IEEE Trans Comput C-25(6):664–667
7. Agrawal VD (1978) When to Use Random Testing. IEEE Trans Comput C-27(11):1054–1055
8. Parker KP, McCluskey EJ (1975) Probabilistic treatment of general combinational networks. IEEE Trans Comput C-24(6):668–670
9. Eichelberger EB, Lindbloom E (1983) Random-pattern coverage enhancement and diagnosis for LSSD logic self-test. IBM J Res Dev 27(3):265–272

10. Miczo A (2003) Digital logic testing and simulation, 2nd edn. Wiley Interscience, New York
11. Seshu S (1965) On an improved diagnosis program. IEEE Trans Electron Comput EC-14(1):76–79
12. Seshu S, Freeman DN (1962) The diagnosis of asynchronous sequential switching systems. IRE Trans Electron Comput EC-11
13. Ulrich EG, Agrawal VD, Arabian JH (1994) Concurrent and comparative discrete event simulation. Kluwer, Boston, MA
14. Ulrich EG, Baker T (1974) Concurrent simulation of nearly identical digital networks. Computer 7:39–44
15. Armstrong DB (1972) A deductive method for simulating faults in logic circuits, IEEE Trans Comput C-21(5):464–471
16. Menon PR, Chappell SG (1978) Deductive fault simulation with functional blocks. IEEE Trans Comput C-27(8):689–695
17. Abramovici M, Breuer MA, Friedman AD (1994) Digital systems testing and testable design. IEEE Press, Piscataway, NJ Revised printing
18. Abramovici M, Menon PR, Miller DT (1984) Critical path tracing: an alternative to fault simulation. IEEE Des Test Comput 1(1):83–93
19. Menon PR, Levendel YH, Abramovici M (1988) Critical path tracing in sequential circuits. In: Proceeding of the international conference on computer-aided design pp. 162–165
20. Menon PR, Levendel YH, Abramovici M (1991) SCRIPT: a critical path tracing algorithm for synchronous sequential circuits. IEEE Trans Comput-Aided Des 10(6):738–747
21. Cheng W-T, Yu M-L (1990) Differential fault simulation for sequential circuits. J Electron Test Theory Appl 1(1):7–13

Chapter 5
Test Pattern Generation Methods and Algorithms

Test vectors are generated for post manufacturing test of a digital system. Because of the complexity of digital systems, the size of necessary tests, and test quality factors, automatic methods are used for generation of test patterns. This process is referred to as automatic test pattern generation (ATPG). For a circuit under test (CUT), test pattern generation must be due to the testing of the circuit as thoroughly as possible, and in the shortest possible time.

ATPG is done by the utilization of programs, methods, and algorithms; all of which use some forms of circuit and fault models. Often, ATPG refers to test generation from a netlist model of CUT using the stuck-at fault model.

This chapter discusses the basics of ATPG methods and shows how test generation programs fit in the overall test cycle. We focus on random test generation (RTG) methods for combinational and sequential circuits. In the sections that follow, we discuss the basics of test generation, testability measures, RTG methods, and test methods that incorporate random and deterministic methods.

5.1 Test Generation Basics

In this section, basic methods of test generation using several small examples are presented. The purpose is to familiarize the reader with the basics of test generation procedures, the terminology, categorization of ATPG algorithms, and the role of various utilities facilitating test generation.

5.11 Boolean Difference

Test vectors are generated to detect faults. A test vector is an input vector that creates different outputs for faulty and good circuits. Test generation is finding such input vectors. Functionally, we can find such inputs by finding the Boolean difference of a good circuit model and its faulty model, and then finding input vectors that satisfy this difference [1, 2, 3].

For Boolean difference, we start with good circuit model, and modify it according to the fault that we are seeking the test for. The XOR of the good circuit and faulty circuit is the equation for the Boolean difference. Test(s) for the given fault are the input vectors that make the Boolean difference 1.

As an example, consider the circuit in Fig. 5.1. This is the multiplexer circuit we used in the previous chapters, with an extra input and an extra output. Fault for which the Boolean difference is to be calculated is shown in this figure.

Z. Navabi, *Digital System Test and Testable Design: Using HDL Models and Architectures*, DOI 10.1007/978-1-4419-7548-5_5, © Springer Science+Business Media, LLC 2011

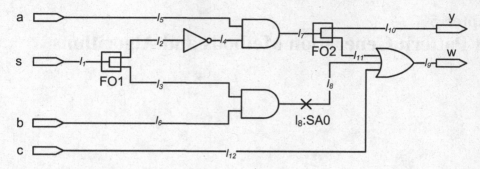

Fig. 5.1 Example circuit for Boolean difference

The circuit shown here has two outputs as described below. The g subscript for y and w mean that they are "good" circuit's outputs.

$$y_g = a.\overline{s}$$

$$w_g = a.\overline{s} + b.s + c$$

The fault we are considering in this example is l_8:SA0. Forcing l_8 to 0, the faulty outputs become:

$$y_{f8} = a.\overline{s}$$

$$w_{f8} = a.\overline{s} + c$$

Boolean difference for the two outputs become:

$$BD(y_g, y_{f8}) = y_g \oplus y_{f8}$$
$$= (a.\overline{s}) \oplus (a.\overline{s})$$
$$= 0$$
$$BD(w_g, w_{f8}) = w_g \oplus w_{f8}$$
$$= (a.\overline{s} + b.s + c) \oplus (a.\overline{s} + c)$$
$$= \overline{(a.\overline{s})}(b.s)(\overline{c})$$
$$= b.s.\overline{c}$$

Tests that detect fault shown in Fig. 5.1 are all input combinations that make Boolean differences of any of the outputs 1. Obviously, no such combination exists for the y output. On the other hand, the following is true for the w output:

$$BD(w_g, w_{f8}) = b.s.\overline{c}$$

Expression $b.s.\overline{c}$ become 1 for $b = 1$, $s = 1$, and $c = 0$. Therefore, tests that detect fault shown in Fig. 5.1 are:

$$absc = 0110$$

and

$$absc = 1110$$

Defects of all types can be processed similarly, as long as they can be modeled by Boolean functions. Faults that make Boolean differences of all outputs equal to 0 are undetectable.

The Boolean difference calculation can be regarded as the ultimate solution for generating tests for given circuit faults. This method is exact, covers all fault types, and is complete in that it finds all tests for fault being considered. However, this problem cannot be solved in nonpolynomial time complexity.

Another solution for test generation that has the same problem as the Boolean difference is to apply all possible input combinations to the faulty circuit model and search for those that produce a different output than the good circuit. This has to be repeated for every fault. This solution for the test generation problem is an exhaustive search one, and like the Boolean difference, is not practical for large circuits.

Test generation techniques and algorithms that we discuss in this chapter try solving this problem using heuristics to limit the search space. The rest of this section covers some basics for these algorithms and definitions.

5.1.2 Test Generation Process

We present test generation techniques that try to simplify exhaustive solutions discussed above. For this, we use the same example we used above. A deterministic and a random method is presented to optimize the search of test vectors.

5.1.2.1 Deterministic Search

For generating a test for l_8:$SA0$, we started at the site of the fault line l_8. Suppose this fault exists in an actual circuit and we are trying to make it show itself at a primary output. For this purpose, first we have to use values of the inputs at the circuit to drive a value into l_8 that is different than the faulty value. Since the fault we are trying to detect is a stuck-at-0 fault, we have to drive a 1 into this line. This requires s and b to be 1. Next, we need to provide input conditions to propagate the effect of this fault to one of the circuit outputs. For this purpose, l_{11} and l_{12} must both be 0 to propagate the effect of l_8:$SA0$ to l_9 and then to w. Line l_{11} is already 0 since requiring $s = 1$ causes l_4 and thus l_{11} to become 0. The other condition for propagating l_8:$SA0$ is $l_{12} = 0$. This can be accomplished by $c = 0$. This analysis results in generation of test $asbc = $ X110 for l_8:$SA0$. The value of a is X, which means that either 0 or 1 is acceptable.

5.1.2.2 Random Search

An alternative to the deterministic search discussed above, is to use random test vectors and check for faults they can detect. Such a solution can result in the detection of a good number of faults with very few random tests.

5.1.2.3 Methods and Algorithms

In the above discussions, we casually discussed the ways of reaching test vectors for detecting faults, faster than exhaustive or complete solutions. These casual methods are the basis of more formal algorithms that are used for test generation.

Other factors than just detecting faults that must be considered in test generation include the reduction of test vectors, reducing test time for testing the physical device, and the detection of multiple faults by the same test vectors. We deal these issues and the problem of fault detection in the test generation solutions that we present in this chapter.

5.1.3   Fault and Tests

As mentioned above, there are several ways that test vectors can be generated for a circuit. Some search for a test using circuit topology, some use a functional model of the circuit, and some use a mix of both. This section categorizes various ways that tests can be generated.

5.1.3.1   Fault-oriented Test Generation

Considering a fault in a circuit, and then looking for a test that can detect the fault is fault-oriented test generation. This method of test generation is most appropriate when there are few faults remaining in the circuit to be detected.

5.1.3.2   Fault Independent Test Generation

In fault independent test generation, tests are generated independent of faults. In one case, a test is generated and then evaluated for faults it can detect, and in another scenario, tracing a circuit results in applying test vectors and faults that they can detect. In either case, specific faults cannot be targeted. Using fault independent test generation is most appropriate when there are still many faults left in the circuit to be detected. In this case, a random test or a deterministic test has a good chance of detecting a good number of faults.

5.1.3.3   Random Test Generation

RTG selects test vectors in random [4–8]. This is most efficient at the beginning of a test generation session when there are many undetected faults in the circuit.

Often RTG programs are complemented with evaluation procedures for a better selection of test vectors. Some RTG programs target specific areas of a CUT that has faults that are hard to detect. Decisions about the number of random tests that can be useful in detecting faults, and expected the number of faults to detect can be made based on how many hard-to-detect faults are in the circuit, and how hard it is to detect them. Section 5.2 discusses controllability and observability that will be helpful in making some of these decisions.

5.1.3.4   Unspecified Inputs

In fault-oriented test generation, certain inputs do not play a role in detecting a fault and their values can be either 0 or 1. This was the case in generating a test for l_8:SA0 in Fig. 5.1. We use 'X' for indicating values for such inputs. X's are also possible in some RTG programs that use a subset of the inputs to target specific areas of a circuit.

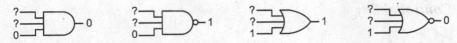

Fig. 5.2 Basic gate control values

Because X input values can take either a 0 or a 1 value, they are useful in merging several test vectors into one and thus reducing the number of test vectors and saving circuit test time. This issue is addressed in the last section of Chap. 6 on test compaction.

5.1.4 Terminologies and Definitions

In the presentations that follow, we use terminologies that are brief and concise, and help understanding of the materials. Unambiguous definition of such terms is important that is described in this section.

Circuit Under Test. As before, CUT, MUT, GUT, and FUT are used for Circuit, Model, Good circuit, and Faultable circuit that are being tested. We continue using these labels for models for which test is being generated.

Stuck-at Models. Unless otherwise specified, methods and algorithms in this chapter apply to the single stuck-at fault model.

Control Value. A 0 input of an AND gate makes the output 0 regardless of values of all other AND inputs. This value is called control value. For an OR gate, a 1 is its control value. Figure 5.2 shows control values for four basic gates. A control value on an input of a gate blocks propagation of faults from other inputs.

Inversion Value. Inversion value of a gate is 0 if no inversion is done, and it is 1 otherwise. For an AND gate the inversion value is 0, while the inversion value of a NAND is 1. Inversion in a path is 0 if there are even number of inversions, and is 1 if there are odd number of gates with 1 inversion values. A control value on an input of a gate generates the same value on the output if the gate has inversion 0, and the complement of the control value if the inversion is 1. For example, the control value on an input of a NOR gate (1) generates the complement of this value (0) on its output.

Test Efficiency. Efficiency of a test vector is measured with the number of faults it detects. The required number of faults to detect for a test vector to be regarded as efficient depends on many factors. Some of these factors are which test generation method we are using, the remaining undetected faults, difficulty of detecting remaining faults, and where we are in the test generation process, i.e., just starting or near the end.

5.2 Controllability and Observability

In the fault-oriented test generation process, circuit input values must be adjusted to enable the detection of faults. Often in this process, we face situations that we have to select between several inputs and/or internal nodes of a circuit. There are also places where decisions have to be made in fault-independent test generation. The base of most of these decisions is how hard or easy it is to put *control* values on some internal lines from circuit input, or *observe* internal line values on circuit outputs. These are controllability and observability issues that we discuss in this section [9, 10].

5.2.1 Controllability

Controlability is defined as a measure of difficulty of setting a circuit line to a certain value. Primary circuit inputs are the most controllable. Consider, for example, circuit shown in Fig. 5.3. For finding a test for l_7:$SA1$, we first need to have a 0 on this line. This being the AND gate control value, requires either input of the gate to be 0. We choose input a since it is more directly controled and achieves the designated l_7 value.

5.2.2 Observability

Observability is defined as a measure of difficulty of observing the value change of a line on a primary output. Primary circuit outputs are the most observable.

Figure 5.3 shows that the effect of l_7:$SA1$ can most easily be seen from the path shown to output y of the circuit.

Controllability and observability examples presented above were very simple and could be decided by inspection. However, for larger circuits and for lines and gates deep inside a circuit, calculation of controllability and observability are not as simple and more systematic methods are needed.

5.2.3 Probability-based Controllability and Observability

A measure of controllability for a line in a circuit is: given a random test vector at the primary inputs of the circuit, how probable it is for the line value to become 1[11]. We refer to this probability measure as P1(*line*). Similarly, we have P0(l). Probability of a primary input receiving a 1 when a random data is applied to the circuit is 0.5, thus P1(pi) = P0(pi) = 0.5.

A measure of observability for a line in a circuit is: given a random test vector at the primary inputs of the circuit, how probable it is for the effect of the line value to propagate to a primary output[11]. A line value propagates to an output if toggling the line value toggles the output.

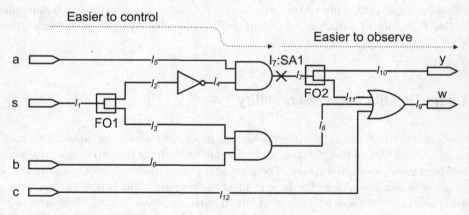

Fig. 5.3 Controllability and observability

Table 5.1 Probability-based controllability and observability parameters

Controllability		Observability
With 1	With 0	
P1(e)	P0(e)	PB(e)
P1(pi) = 0.5	P0(pi) = 0.5	PB(po) = 1

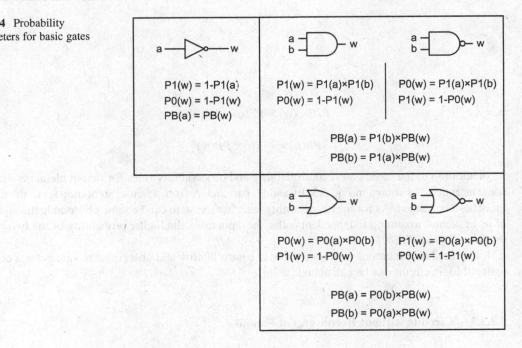

Fig. 5.4 Probability parameters for basic gates

We refer to this probability as PB(*line*). B is used here for observability instead of O, so it does not get confused with 0 (zero). Probability of a primary output being observed is always 1; thus PB(po) = 1.

Table 5.1 shows definitions for probability-based controllability and observability parameters. Equations 5.1 and 5.2 show these parameters. Figure 5.4 Shows probability-based controllability and observability parameters for five basic logic gates [12, 13]. We are assuming independent probabilities, thus no reconvergent fanouts. Also, these are for two-input gates and can easily be extended to gates with higher input counts.

As shown here, for the output of a two-input AND gate to become 1, both inputs must be 1, thus multiplying the two probabilities. Similarly, probability of a 0 on an OR gate output is calculated by multiplying 0 probabilities of its inputs. Observation values of the inputs of a gate depend on that of its output. In order to be able to observe value on input *a* of an AND gate on its output, its *b* input must be 1 and its output must be observable. Therefore, PB(*a*) is calculated by multiplying the probability of *b* being 1 and observability value of *w*.

Probability-based controllability and observability values for AND, NAND, OR, and NOR gates can be calculated using Eq. 5.1 and Eq. 5.2. These equations use *c* for control value, and *i* for inversion value. The expressions immediately following "p" becomes 1 or 0 depending on *i* and *c* values. E.g., for *i* = 0 and *c* = 1 in Eq. 5.1, the left-hand side of this equation becomes P0.

Fig. 5.5 Probability for independent fanout paths

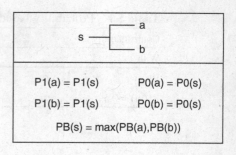

$$P1(a) = P1(s) \qquad P0(a) = P0(s)$$

$$P1(b) = P1(s) \qquad P0(b) = P0(s)$$

$$PB(s) = max(PB(a), PB(b))$$

$$Pi \overline{\oplus c}(w) = P\overline{c}(a) \times P\overline{c}(b) \tag{5.1}$$

$$PB(a) = P\overline{c}(b) \times PB(w) \tag{5.2}$$

In addition to the basic gates, controllability and observability rules for fanout elements deserve some attention. As shown in Fig. 5.5, P1 and P0 parameters from a fanout stem transfer to the fanout branches unchanged. As for the observability, because the stem can become observable through any of its branches, assuming independent paths, the stem takes the higher probability of the two paths, thus $max(PB(a), PB(b))$.

Using the basic relations mentioned above, controllability and observability values for a combinational logic circuit can be calculated.

5.2.3.1   Circuits without Reconvergent Fanout

Figure 5.6a shows a fanout-free circuit for whose lines probability-based controllability and observability values are to be calculated. The numbers in curly brackets are: {P1, PB}.

Initially, P1 values for lines connected to the primary input are determined and placed in curly brackets. Primary input P1 values for a random test vector are 0.5. Then, based on the rules of Fig. 5.4, P1 parameters for the inverter, AND, and OR gates are calculated. This calculation begins with primary inputs; moves toward the primary outputs, and gates closer to the primary inputs are processed first. When calculating P1 values for a gate, P1 values for all its inputs must be known.

As an example, consider the OR gate. P1 parameters for the inputs of this gate are 0.25 and 0.5. This makes P0 input parameters 0.75 and 0.5, respectively, which results in 0.375 for the P0 of the output, i.e., P1 = 0.625. These are shown in Fig. 5.6b.

Calculation of PB values starts with circuit primary output, and move toward the primary inputs, primary output probabilities are 1. For an OR gate, PB of an input is calculated by P0 of other inputs and PB of the output, according to Eq. 5.2. The value 0.75 for PB on the lower input of the OR gate is calculated by PB of 1 on its output and P0 of 0.75 on its upper input. Calculation of PB values for the inputs of the AND gate are done in a similar fashion. Figure 5.6c shows P1 and PB values of all lines of our example circuit.

Calculation of P1 and PB values for circuits with fanout and without reconvergent fanouts is done in the same way. In this case, in addition to rules of Fig. 5.4, those of Fig. 5.5 also apply.

Fig. 5.6 P1 and PB for a fanout-free circuit

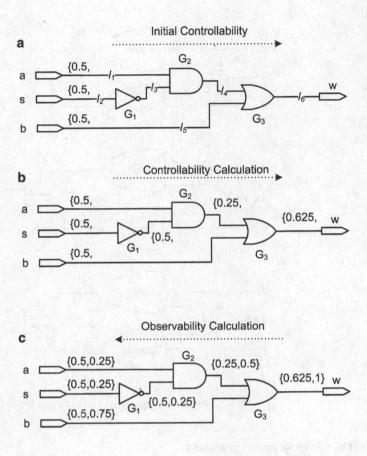

5.2.3.2 Reconvergent Fanouts

Calculation of P1 and PB values for lines of a circuit with one more reconvergent fanouts becomes more difficult than that mentioned above. In such a case, at the divergence point (fanout), controllability figures from fanout stems propagate to branches as shown in Fig. 5.7. However, at the convergence point, the probabilities coming into the inputs of the gate are not independent, and rules of Fig. 5.4 or Eq. 5.1 are no longer valid.

For reconvergent fanouts (see Fig. 5.7), the circuit must be split at the fanouts and controllability parameters for two circuits, one with stem at 1 and another with stem at 0 must be calculated. Controllability values at the convergent point are calculated by conditional probabilities.

The example we consider for this case is shown in Fig. 5.8. Figure 5.8a shows all P1 and PB figures that can be calculated independent of reconvergence at G_7. The missing parameters in this figure are the P1 parameter of the output of G_7 and PB values that are related to this controllability. The output of G_7 is line l_{21} that is not shown here for legibility of the diagram.

Figure 5.8b shows controllability parameters that are based on controllability of the fanout stem (output of G_2) assuming value 0. Note that G_2 drives this fanout that converges at G_7. Likewise, Fig. 5.8c assumes that same line has a 1 value and shows the corresponding controllability values in the rest of the circuit. Based on conditional probabilities, P1 for the output of G_7 is calculated in Fig. 5.8d and shown on the corresponding line. Note that this is the controllability that was missing

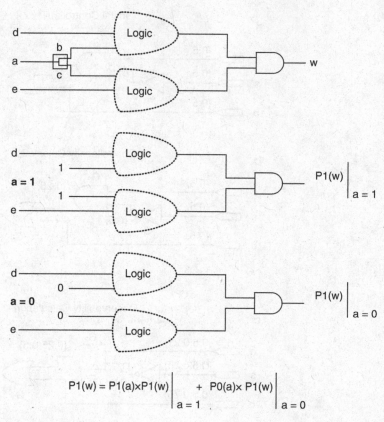

Fig. 5.7 P1 for reconvergent fanout

from Fig. 5.8a. As shown, $P1(l_{21})$ is 0.5: Had we not considered the conditional probabilities, and calculated $P1(l_{21})$ from Fig. 5.8a, 0.464 would result, which we leave for the reader to verify.

5.2.3.3 Detection Probability

We define *detection probability* as follows. Given a random test vector at the primary inputs of a circuit, the difficulty of detecting a fault in the circuit is called detection probability (DP). An obvious (though incorrect, but a good estimate for our purpose) value for DP is obtained by multiplying controllability and observability probabilities [12, 14, 15]. This is justified by the fact that if a line can be controlled with the complement of a fault value, and the line is observed, then the fault is detectable. Thus,

$$DP_{(l:SAv)} = P\overline{v}(l) \times PB(l) \qquad\qquad (5.3)$$

In Eq. 5.3, for a SAv on line l, controllability value of line l is multiplied by observability of line l. In Fig. 5.8d, DP for $l_{11}:SA0$ (lower input of G_4) is calculated by multiplying 0.75 by 0.33, which is 0.24. This means that if we randomly select a test vector and apply it to the inputs of the circuit, the chance of $l_{11}:SA0$ being detected by that test vector is 0.24.

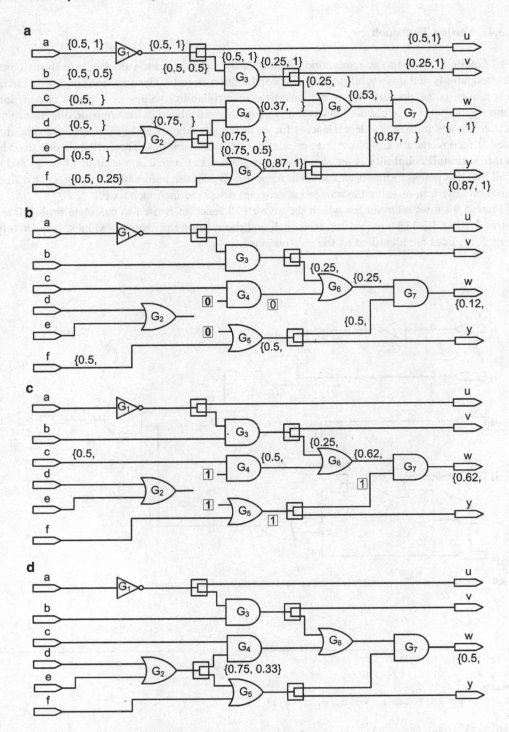

Fig. 5.8 Controllability and observability figure for reconvergent fanouts

Inaccuracy of this method of calculating detection probabilities is that the probabilities that we are multiplying are not necessarily independent. However, for our purposes, where other inaccuracies in RTG are more significant, this is an acceptable estimate.

5.2.3.4 Verilog Testbench

A PLI function referred to as *combinational probability* is developed to calculate probability based on controllability and observability values for combination circuits. This function starts with circuit levels closer to the primary inputs and calculates controllability values as it traces circuit lines toward the outputs of the circuit. Observability values are calculated in the opposite direction. This function ignores probability dependencies for reconvergent fanouts and calculates controllability values at reconvergent outputs by just considering input probabilities. The netlist format used by **$CombinationalProbability** is the same as that explained in Chap. 2 and used in Chaps. 3 and 4. Recall that this netlist format is explained in Appendix B, and can automatically be generated from behavioral descriptions using the translation program discussed in Appendix F.

Figure 5.9 shows a circuit for which the above PLI function is used to calculate probabilities. Figures 5.10 and 5.11 show Verilog testbench and the resulting probability values, respectively. In Fig. 5.11, lines are identified by the gate outputs.

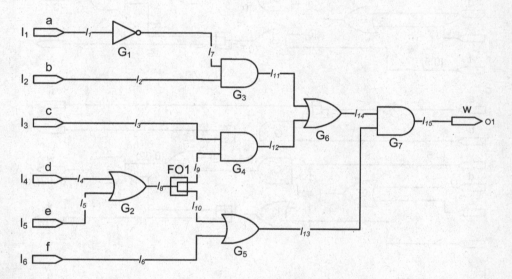

Fig. 5.9 Circuit diagram for probability based on controllability and observability

```
module SimpleCKTPROBABILITY();
   reg a, b, c, d, e, f;
   wire w;
   SimpleCKT INST (a, b, c, d, e, f, w);

   initial begin
      $CombinationalProbability (SimpleCKTPROBABILITY.INST, "CP.txt");
   end
endmodule
```

Fig. 5.10 Testbench producing probability-based values

```
        GateName                    GateType      Level      P0       P1        PB
SimpleCKTPROBABILITY.INST.I1          pin           0       0.50     0.50      0.27
SimpleCKTPROBABILITY.INST.I2          pin           0       0.50     0.50      0.27
SimpleCKTPROBABILITY.INST.I3          pin           0       0.50     0.50      0.49
SimpleCKTPROBABILITY.INST.I4          pin           0       0.50     0.50      0.16
SimpleCKTPROBABILITY.INST.I5          pin           0       0.50     0.50      0.16
SimpleCKTPROBABILITY.INST.I6          pin           0       0.50     0.50      0.13
SimpleCKTPROBABILITY.INST.G1          notg          1       0.50     0.50      0.27
SimpleCKTPROBABILITY.INST.G2          or_n          1       0.25     0.75      0.32
SimpleCKTPROBABILITY.INST.FO1         fanout_n      2       0.25     0.75      0.32
SimpleCKTPROBABILITY.INST.G3          and_n         2       0.75     0.25      0.54
SimpleCKTPROBABILITY.INST.G4          and_n         3       0.62     0.37      0.65
SimpleCKTPROBABILITY.INST.G5          or_n          3       0.12     0.87      0.53
SimpleCKTPROBABILITY.INST.G6          or_n          4       0.46     0.53      0.87
SimpleCKTPROBABILITY.INST.G7          and_n         5       0.53     0.46      1.00
SimpleCKTPROBABILITY.INST.O1          pout          6       0.53     0.46      1.00

                          Muximum Level = 6
                      Number of Gates = 15
```

Fig. 5.11 Probability of controllability and observability

5.2.4 SCOAP Controllability and Observability

In deterministic or random test generation, controllability and observability measures are used for simplifying the related algorithms. However, complexity in calculation of these parameters defeats the main purpose for which they are used. The fanout problem, and the method of calculating probability based on controllability and observability parameters for circuits with reconvergent fanouts is too complex for the methods to be useful for test generation.

Sandia Controllability/Observability Analysis Program (SCOAP) [16] is a testability measure, the complexity of which grows only linearly with the size of the circuit. SCOAP is based on the topology of the circuit, is a static analysis, and does have some inaccuracies due to reconvergent fanouts. Nevertheless, it is easy to calculate and provide a good estimate for test generation programs, as well as design for test techniques.

SCOAP defines a set of parameters for combinational and sequential controllability and observability measures. The combinational parameters have to do with the number of lines that need to be set for controlling and observing a line. The sequential parameters are related to the number of clocks it takes for controlling and observing a line.

In SCOAP parameters, lower values mean more controllable and observable, and lines that are more difficult to control and observe have higher SCOAP parameter values.

5.2.4.1 SCOAP Combinational Parameters

SCOAP defines a set of parameters for combinational controllability and observability, and another set for sequential. The combinational parameters are:

1. $CC0(l)$: Combinational 0-controllability of line l. Relates to the number of lines from the primary inputs that have to be traced to put a 0 on line l.
2. $CC1(l)$: Combinational 1-controllability of line l. Relates to the number of lines from the primary inputs that have to be traced to put a 1 on line l.

3. CB(*l*): Combinational observability of line *l*. Relates to the number of lines that have to be traced to observe value of line *l* on a primary output.

Primary input combinational controllability values are 1 (most controllable) and primary output combinational observability values are 0 (most observable).

Figure 5.12 shows combinational SCOAP parameters for basic logic gates. For the inverter, controllability of the output is 1 added to its input. This means that as more lines are traced to reach a certain line, it becomes more difficult to control the line (higher CC numbers). For an AND gate, you can put a 0, i.e., CC0(*w*), on its output by putting a 0 on either of its inputs. So, we are using the minimum 0-controllability values of the inputs, and add a 1 to it for 0 controllability of *w*.

Again, for an AND gate, since the output becomes 1 if both inputs are 1, the difficulty of controlling *w* with a 1 is the sum of difficulties of all inputs becoming 1, plus a 1 to consider the number of lines traced. This justifies the equation for CC1(*w*) shown in Fig. 5.12. Controllability figure for other gate structures shown in this figure are calculated similar to the AND gate.

As shown in Fig. 5.12, observability of an AND gate input CB(*a*) is the sum of the gate's output observability CB(*w*) and 1-controllability of all other inputs, plus a 1 for the number of lines traced. This is justified by the fact that an AND gate input is only observed on its output if all other inputs assume noncontrol (1) values.

SCOAP combinational controllability and observability parameters for AND, NAND, OR, and NOR gates can be calculated using Eq. 5.4 and Eq. 5.5. These equations use *c* for control value, and *i* for inversion value.

$$CC_{i \oplus c}(w) = (\overline{i \oplus c}) \times \min(CC_c(a), CC_c(b)) + (i \oplus c) \times (CC_{\bar{c}}(a) + CC_{\bar{c}}(b)) + 1 \qquad (5.4)$$

$$CB(a) = CB(w) + CC_{\bar{c}}(b) + 1 \qquad (5.5)$$

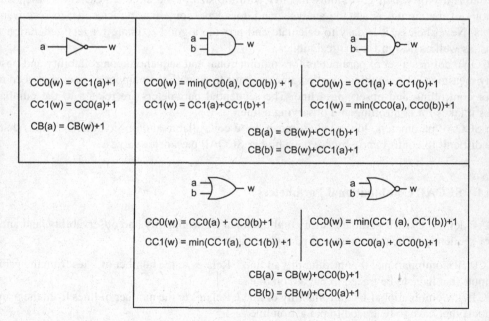

Fig. 5.12 Combinational SCOAP parameters

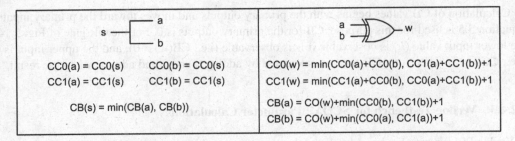

Fig. 5.13 Fanout and XOR SCOAP parameters

Fig. 5.14 SCOAP parameter calculations

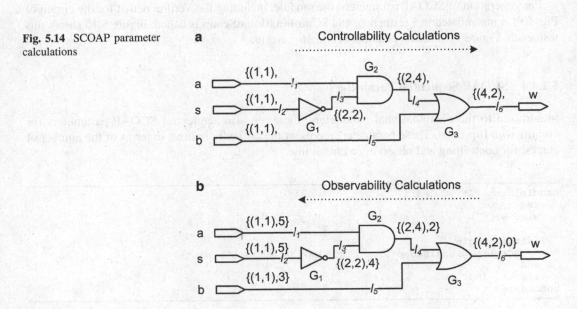

In addition to the basic gates, SCOAP parameters for fanouts and XOR gates are shown in Fig. 5.13. As shown, controllability of fanout branches is the same as that of the stem. A fanout stem can be observed through any of its branches, thus its observability is the minimum of all its branches. Figure 5.13 also shows SCOAP parameters for an XOR gate. The Boolean equation of an XOR justifies calculation of its output controllability and input observabilities.

5.2.4.2 SCOAP Combinational Examples

Figure 5.14a shows a simple circuit for which SCOAP parameters are to be calculated. The numbers in curly brackets are $\{(CC0,CC1),CB\}$. Initially, CC0 and CC1 values for lines connected to primary inputs are determined and put in a set of parenthesis in curly brackets. As discussed before, primary input CC0 and CC1 values are 1, i.e., most controllable. After the determination of controllability values for the primary inputs, SCOAP calculations for the rest of the circuit for logic levels closer to the primary inputs are performed. Therefore, CC0 and CC1 values for l_3 are calculated. G_2 is processed next, since controllability parameters for all its inputs are now known. This procedure continues forward until primary circuit outputs are reached. Controllability values are calculated based on expressions in Fig. 5.12 or Eq. 5.4.

Calculation of CB values begins with the primary outputs and moves toward the primary inputs. Equation 5.5 is used for this purpose. CB for the primary outputs is 0. For the OR gate of Fig. 5.14, the lower input value (l_5) is observable if w is observable (i.e., CB(w) = 0), and the upper input is 0 (i.e., CC0 = 2). Therefore, CB for l_5 is calculated by adding 0 and 2 and adding a 1 to the result.

5.2.4.3 Verilog Testbench for SCOAP Parameter Calculations

A Verilog PLI function has been developed that generates combinational SCOAP parameters of netlists that use the format discussed in Chap. 2, and used in fault simulation applications in Chaps. 3 and 4. We use this PLI function for generating SCOAP parameters for the circuit of Fig. 5.9.

For generation of SCOAP parameters, the module, including the Verilog netlist for the circuit of Fig. 5.9, is instantiated in a testbench, and **$CombinationalScoap** is called. Figure 5.15 shows this testbench. Figure 5.16 shows contents of file *Mux.scoap*.

5.2.4.4 SCOAP Sequential Parameters

In addition to the combinational parameters, there are also sequential SCOAP parameters for circuits with flip-flops. These parameters represent the difficulty in time, in terms of the number of clocks, for controlling and observing a circuit line.

```
module SimpleCKTSCOAP();
   reg a, b, s;
   wire w;

   initial begin
      $CombinationalScoap(SimpleCKTSCOAP.INST, "Mux.scoap");
   end
   mux2to1 INST (a, b, s, w);
endmodule
```

Fig. 5.15 PLI generating SCOAP parameters

GateName	GateType	Level	CC0	CC1	CB
SimpleCKTSCOAP.INST.I1	pin	0	1	1	10
SimpleCKTSCOAP.INST.I2	pin	0	1	1	10
SimpleCKTSCOAP.INST.I3	pin	0	1	1	10
SimpleCKTSCOAP.INST.I4	pin	0	1	1	11
SimpleCKTSCOAP.INST.I5	pin	0	1	1	11
SimpleCKTSCOAP.INST.I6	pin	0	1	1	11
SimpleCKTSCOAP.INST.G1	notg	1	2	2	9
SimpleCKTSCOAP.INST.G2	or_n	1	3	2	9
SimpleCKTSCOAP.INST.FO1	fanout_n	2	3	2	9
SimpleCKTSCOAP.INST.G3	and_n	2	2	4	7
SimpleCKTSCOAP.INST.G4	and_n	3	2	4	7
SimpleCKTSCOAP.INST.G5	or_n	3	5	2	7
SimpleCKTSCOAP.INST.G6	or_n	4	5	5	4
SimpleCKTSCOAP.INST.G7	and_n	5	6	8	1
SimpleCKTSCOAP.INST.O1	pout	6	7	9	0

Maximum Level = 6
Number of Gates = 15

Fig. 5.16 SCOAP parameters generated by PLI function

The sequential parameters are:

1. SC0(l): Sequential 0-controllability of line l. Relates to the number of clocks it takes logic values at primary inputs to put a 0 on line l.
2. SC1(l): Sequential 1-controllability of line l. Relates to the number of clocks it takes logic values at primary inputs to put a 1 on line l.
3. SB(l): Sequential observability of line l. Relates to the number of clocks it takes to propagate value of line l on a primary output.

Primary input sequential controllability values are 0, and primary output sequential observability values are 0.

A simplified set of rules for calculating SCOAP parameters in sequential circuits across a flip-flop are shown in Fig. 5.17. These rules assume flip-flops with a clock and no other set or resetting mechanism, and no clock gating. These are simplification of rules in [17] which give a comprehensive treatment to this topic.

In addition to flip-flop rules, sequential SCOAP parameters are also considered across basic gates. For this, rules are the same as those of Fig. 5.12, without the addition of 1. For example, sequential 0-controllability of an AND gate output becomes:

$$SC0(w) = \min(SC0(a), SC0(b))$$

We present a simple example here to illustrate how combinational and sequential SCOAP parameters are calculated in a circuit with flip-flops. The more complex case of circuits with feedback will not be treated here. Figure 5.18 shows an example with combination (in curly brackets), and sequential (in square brackets [(SC0, SC1), SB]), SCOAP parameters. The process begins with setting

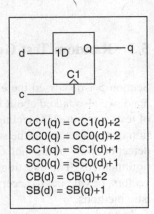

$$CC1(q) = CC1(d)+2$$
$$CC0(q) = CC0(d)+2$$
$$SC1(q) = SC1(d)+1$$
$$SC0(q) = SC0(d)+1$$
$$CB(d) = CB(q)+2$$
$$SB(d) = SB(q)+1$$

Fig. 5.17 Simplified SCOAP for sequential circuits

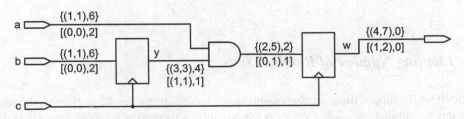

Fig. 5.18 SCOAP parameters in a sequential circuit

primary input controllability and primary output observability values. We move forward with controllability values from circuit inputs to outputs and observability values calculated while traveling circuit lines in the opposite direction.

Combinational controllability values for the AND gate (those in {}) in this figure are calculated as before according to Fig. 5.12. Sequential controllability values (those in) are calculated similarly, except for the addition of 1. Combinational and sequential parameters across the flip-flops are calculated according to rules shown in Fig. 5.17.

Figure 5.18 also shows observability values for going from w to the AND gate output (across flip-flop, going from its output to input). Rules of Fig. 5.17 are used for this purpose. Then, we turn our attention to the AND gate inputs. For calculating combinational observability values of these inputs, the AND gate inputs rules of Fig. 5.12 are used, while for the sequential observability, the same rules without consideration of +1 are used.

5.2.5 Distances Based

A very simple but efficient rule for finding how controllable or how observable circuit lines are is to consider their distances (or logic levels) from primary inputs (for controllability) or primary outputs (for observability).

Consider for example, l_7:$SA1$ in circuit of Fig. 5.1. Observing this fault on y is easier since there are fewer logic gates between the site of fault l_7 and y than between l_7 and w. For controllability, consider l_8:$SA1$. This requires a 0 on l_8. Because of the AND gate b is closest to l_8 than the other inputs, it is easier to set b to 0 than trying to achieve this by setting s. Note that there is only one gate between b and l_8, while considering the fanout as a structure, the distance between s and l_8 becomes 2.

5.3 Random Test Generation

Section 5.1 discussed the exhaustive search for test vectors that detect circuit faults. On the other hand, we also talked about RTG as a way of reducing the number of tests. For reducing the number of tests, various RTG techniques try to set an upper bound, set criteria of selection, or direct formation of test vectors, in the search for a good test set. With a relatively low effort, an RTG is able to detect 60–80% of the faults. The remaining faults are usually detected by fault-oriented test generation methods. The questions that are important to answer are when to stop producing random test vectors, or whether to accept a randomly selected test vector, and when to switch to more deterministic methods.

Instead of selecting all possible test vectors, or exhaustively trying every test and looking for good ones, there are several ways of reducing the number of tests by the application of some random techniques. Two such methods are discussed in detail here. Other possibilities are mentioned in an outline form.

5.3.1 Limiting Number of Random Tests

For a circuit with n inputs, there are $2n$ possible input combinations. For large circuits, the selection of every input combination as a test vector is not possible. A method of selecting fewer than $2n$ tests based on circuit fault properties is presented here.

The number of random tests to use in a test set for testing a circuit with N inputs can be based on how confident we want to be that the test set detects the fault that is least likely to be detected. We define C as the level of confidence, and $Df\text{-}_{hard}$ as the probability that the hardest fault in the circuit (least likely) is detected with one random test.

We use the following definitions to calculate the required number of tests based on $Df\text{-}_{hard}$ and C [18].

N: Required number of tests

C: Confidence level

$Df\text{-}_{hard,\,1}$: The same as $Df\text{-}_{hard}$. This is the probability of detecting the hardest fault with one random test. This parameter is the probability that after the application of only one random test the hardest to detect fault is detected.

$Df\text{-}_{hard}$, N: This parameter is the probability that after the application of N random tests at least one test detects the hardest to detect fault.

The probability that the hardest to detect fault is not detected by a randomly selected test vector is: $1 - Df\text{-}_{hard,\,1}$. Thus, the probability that N tests do not detect this fault is: $(1 - Df\text{-}_{hard,1})^N$. Based on this, the probability that at least one of these N tests detect this fault is 1 minus this figure. By the above definitions, this is the same as $Df\text{-}_{hard}$, N. Therefore, Eq. 5.6 results.

$$D_{f\text{-hard},N} = 1 - (1 - D_{f\text{-hard},1})^N \tag{5.6}$$

We want to apply enough test vectors such that after the application of all tests (N), the probability of detecting the hardest to detect fault by at least one of these tests ($Df\text{-}_{hard}$, N) is greater than C level of confidence. Thus, Eq. 5.7 results.

$$1 - (1 - D_{f\text{-hard},1})^N \geq C \tag{5.7}$$

To solve this equation for N, we move the part with N exponent to the right hand side, and get the ln, of both sides of the resulting expression. Equation 5.8 results.

$$ln(1 - C) \geq N ln(1 - D_{f\text{-hard},1}) \tag{5.8}$$

The smallest value of N for the level of confidence, C, results as follows:

$$N = ln(1 - C) / ln(1 - D_{f\text{-hard},1}) \tag{5.9}$$

For small values of $Df\text{-}_{hard,1}$, which is always the case, Eq. 5.10 holds. Take for example 0.01 for the $Df\text{-}_{hard,1}$. The right hand side of Eq. 5.10 becomes -0.01005.

For small $Df\text{-}_{hard}$:

$$ln(1 - D_{f\text{-hard},1}) = -D_{f\text{-hard},1} \tag{5.10}$$

Using Eq. 5.10 in Eq. 5.9, the smallest value of N results, as shown in Eq. 5.11. We use $Df\text{-}_{hard}$ instead of $Df\text{-}_{hard,1}$ in this equation and discussions that follow.

$$N = -ln(1 - C) / D_{f\text{-hard}} \tag{5.11}$$

Equation 5.11 sets the minimum number of tests needed (N) for a fault with $Df\text{-}_{hard}$ probability of detection to be detected, with C level of confidence. For example, to be 98% sure ($C = 0.98$) that a fault with 0.01 probability of detection ($Df\text{-}_{hard} = 0.01$) is detected, we need at least 400 random test vectors ($N = 400$).

5.3.1.1   Estimating Hardest Detection

For a circuit with n inputs, applying all input combinations will detect the hardest to detect fault. So, a conservative estimate for $Df_{\text{-hard}}$ is as shown in (5.12).

$$D_{f\text{-hard}} \geq (1 / 2^n) \tag{5.12}$$

Since all input combinations are considered, if a fault is detectable, we are sure that it will be detected with at least one of the test vectors. Thus, a fault cannot be harder to detect than the right-hand side of Eq. 5.12. In most cases, this estimate results in far more test vectors than needed.

For a circuit with several outputs, some faults can propagate to more than one output, and have a better probability of detection. Furthermore, not all primary inputs may be needed to propagate certain faults to circuit outputs. To obtain an estimate for the hardest to detect probability, such circuits may be partitioned into fanin cones, and $Df_{\text{-hard}}$ can be estimated based on the number of inputs of the core with most number of inputs. This is because, for a fault in a cone to be detected, it is only necessary to apply test vectors to the inputs of the cone, and observe its effect on the output of the cone.

Consider for example, the circuit shown in Fig. 5.19. Based on the number of inputs, and according to Eq. 5.12, the probability of the hardest to detect fault is estimated as $1/2^8$. For a confidence level of 0.98, according to Eq. 5.11, the required number of tests to detect the hardest fault becomes 1,001 tests. Obviously, is too conservative and is even higher than trying to detect circuit faults exhaustively.

On the other hand, we can partition the circuit of Fig. 5.19 into three fanin cones, each of which feeding one of circuit primary outputs. Calculation of $Df_{\text{-hard}}$ based on this partitioning results in a more realistic estimation for this parameter. As shown in this diagram, there are two cones with four inputs, and one with three. The worst case situation for a fault to be detected is if it is in one of the cones with four inputs. Therefore, the probability to detect the hardest fault is $Df_{\text{-hard}} = 1/16$. For a confidence level of 98%, the number of random tests for testing the circuit of Fig. 5.19 is $N = 63$. Although, this may still be a conservative figure for this simple circuit, it is 16X reduction from the case that considered all circuit inputs.

5.3.1.2   Detection Probability

Detection probability (DP) figures obtained by considering line controllability and observability figures and discussed in Sect. 5.2 may be used for finding random number of tests to satisfy a certain constraint.

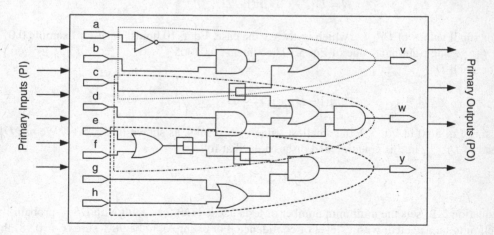

Fig. 5.19 Fanin cones for hard to detect faults

For example, we can search all circuit lines, find the lowest value of DP, and use it in Eq. 5.11 to get an estimate for N.

Although, DP as discussed in Sect. 5.2.3.3 is not accurate, and multiplying controllability by observability is just a rough estimate, using it in heuristics does not cause a serious problem. Furthermore, calculating controllability values without considering reconvergent fanouts, still produces DP figures that can safely be used for estimating the required number of random tests. Potentially, SCOAP parameters can also be used for estimating the number of random tests.

5.3.2 Combinational Circuit RTG

A test set generated by random selection of test vectors is evaluated by fault simulation. Instead of selecting a certain number of tests and evaluating the complete test set at the end, an RTG process can evaluate test vectors before selecting them. Although this form of test generation may not fit into the definition of pure random, but since tests that are to be evaluated are picked in random, we refer to this method as RTG.

RTG methods vary in the way that a random test is selected or discarded. The selection criterion is based on the improvement in fault coverage that each test vector offers. This coverage is usually measured by the number of undetected faults that a new random test detects. The selection criterion considers that fault coverage grows exponentially with the number of test vectors. As shown in Fig. 5.20, in the early stages of RTG process, new random tests detect a good number of faults, and coverage increases rapidly. As the number of selected tests increases, new random tests detect fewer undetected faults.

We present several categories of RTG algorithms that are different in mechanisms they use to follow this exponential curve, and the method they use for exiting the RTG process. Usually, after an RTG algorithm exits, fault-oriented TG programs (that we discuss in Chap. 6) come into action to detect the remaining faults.

5.3.2.1 Fixed Expected Coverage per Test

A simple RTG algorithm that relies on the exponential coverage growth only for exiting the test generation process uses a Fixed Expected Coverage per test (FECpt) vector. This algorithm shown in Fig. 5.21, obtains a random test and performs fault simulation for all undetected faults. As faults are detected by a test vector, they are counted and dropped from the original fault list. The fault count (*fCount*) is a measure of coverage improvement by the random test. This figure is compared

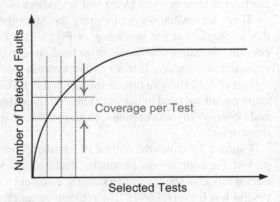

Fig. 5.20 Exponential growth of detected faults

Fig. 5.21 Fixed Expected
Coverage per test

```
Algorithm:FECPT;
Copy F into F';
Use a fixed efCount;
While desired coverage has not reached
    Obtain a random test, t;
    Reset fCount;
    For every f in F';
        Inject f;
        Apply t;
            Simulate faulty circuit;
        Remove f;
        If f is detected
            Increment fCount;
            Drop f from F;
            Increment coverage;
        End if;
    End for;
    If fCount >= efCount
        Keep t;
    End if;
End while;
```

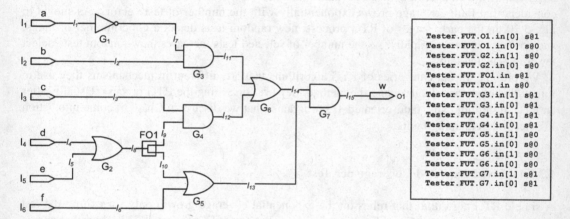

```
Tester.FUT.O1.in[0] s@1
Tester.FUT.O1.in[0] s@0
Tester.FUT.G2.in[1] s@0
Tester.FUT.G2.in[0] s@0
Tester.FUT.FO1.in s@1
Tester.FUT.FO1.in s@0
Tester.FUT.G3.in[1] s@1
Tester.FUT.G3.in[0] s@1
Tester.FUT.G4.in[1] s@1
Tester.FUT.G4.in[0] s@1
Tester.FUT.G5.in[1] s@0
Tester.FUT.G5.in[0] s@0
Tester.FUT.G6.in[1] s@0
Tester.FUT.G6.in[0] s@0
Tester.FUT.G7.in[1] s@1
Tester.FUT.G7.in[0] s@1
```

Fig. 5.22 RTG example circuit

with a fixed expectation fault count (*efCount*), and if it meets this expected number of detected faults, it is kept as a test in the test set, otherwise it is discarded.

There are several ways of exiting this algorithm. One way is to set a low desired fault coverage that is checked at the beginning of Fig. 5.21. Other mechanisms not shown here include one for counting the number of consecutive random tests that do not satisfy (they are unsuccessful) the fixed expectation coverage. If this number exceeds a certain limit, the RTG process exits.

Figure 5.22 shows a circuit for which this RTG method is used in a Verilog testbench. This is the same circuit we used in our fault simulation methods in Chap. 4 and in Verilog-based controllability and observability calculations, earlier in this chapter. After fault collapsing, there are 16 faults in this circuit.

Figure 5.23 shows the outline of a testbench implementing the FECpt RTG algorithm. The upper part of the code shows parameter declarations, variable, and instantiation of good and faultable circuit models. The parameters are the number of faults, expected fault count (*efCount*), and unsuccessful test limit (*utLimit*). The *efCount* parameter is defined in the algorithm of Fig. 5.21. *utLimit*

```
module Tester(); //CRTG_SimpleCKT_Fixed

    parameter numOfFaults = 16;
    parameter efCount = 2;
    parameter utLimit = 200;
    parameter desiredCoverage = 99;
    reg a, b, c, d, e, f;
    wire woG, woF;
    reg [8*50:1] wireName;
    reg stuckAtVal;
    reg [5:0] testVector;
    reg [1:numOfFaults] detectedListCT, detectedListAT;
    integer faultFile, testFile, status;
    integer uTests, detectedFaultsCT, faultIndex;
    integer coverage, detectedFaultsAT;
    SimpleCKT GUT (a, b, c, d, e, f, woG);
    SimpleCKT FUT (a, b, c, d, e, f, woF);

    initial begin

        uTests=0; coverage = 0;
        faultIndex = 1; detectedListAT = 0; detectedFaultsAT = 0;
        testFile = $fopen("CRTG_SimpleCKT_Fixed.tst", "w");
        #10;
        while(coverage < desiredCoverage && uTests < utLimit) begin
            detectedFaultsCT = 0; detectedListCT = 0;
            testVector = $random($time);
            uTests = uTests + 1;
            faultIndex = 1;
            #10;

            while(!$feof(faultFile)) begin // Fault loop
                . . .
            end //end of while(!$feof(faultFile))

            if(detectedFaultsCT >= efCount) begin
                . . .
            end

        #10;
        end //end of while of Coverage
        $display("Number of Random Vectors Generated: %d", uTests);
        $display("Coverage : %d", coverage);
    end //end of initial
endmodule
```

Fig. 5.23 Fixed expected coverage outline

not shown in Fig. 5.21, is the maximum number of consecutive random tests that do not satisfy *efCount*. This parameter provides an exit condition for the RTG algorithm.

Another exit condition for the algorithm is the *desiredCoverage* parameter. The **while** loop in Fig. 5.23 checks both of these conditions.

Declarations in Fig. 5.23 are self-explanatory. Strings *CT* and *AT* used with some of the variable names are for Current-Test and All-Tests, respectively.

In the outer **while** loop shown in this figure, a random test is generated, and using another **while** loop inside it, the test vector is checked for detecting all circuit faults (Fig. 5.24). The number of faults that the current test (*CT*) detects are calculated in this inner **while** loop. Following this loop, an **if** statement checks if the detected faults satisfy *efCount*. This part is shown in Fig. 5.25, which is elaborated in the following sections.

Figure 5.24 shows the **while** loop that finds faults detected by random test *testVector* (obtained randomly in Fig. 5.23). We keep track of detected faults by their fault index (*faultIndex*) that

```verilog
module Tester();//CRTG_SimpleCKT_Fixed
    . . .
    initial begin
        . . .
        while(coverage < desiredCoverage && uTests < utLimit) begin
            . . .
            faultFile = $fopen("SimpleCKT_Fixed.flt", "r");

            while(!$feof(faultFile)) begin //Fault loop
                status = $fscanf(faultFile,
                                 "%s s@%b\n", wireName, stuckAtVal);

                if(detectedListAT[faultIndex]==0)begin //fault dropping

                    $InjectFault(wireName, stuckAtVal);
                    {a, b, c, d, e, f} = testVector;
                    #60;
                    if(woG != woF) begin
                        detectedFaultsCT = detectedFaultsCT + 1;
                        detectedListCT[faultIndex] = 1'b1;
                    end //end of if
                    $RemoveFault(wireName);
                    #20;
                end //if !detected

                faultIndex = faultIndex + 1;
            end //end of while(!$feof(faultFile))

            $fclose(faultFile);
            . . . //details in Fig.5.25

        end //end of while of Coverage
        $display("Number of Random Vectors Generated: %d", uTests);
        $display("Coverage : %d", coverage);
    end //end of initial
endmodule
```

Fig. 5.24 Finding faults detects by a random test, implementing FECpt

corresponds to the line number where a fault is located in the fault list file. The *detectedListAT* array keeps track of faults detected so far by all tests that have been selected. On the other hand, *detectedListCT* records faults detected by the current test.

The **if** statement immediately in the body of the inner **while** loop of Fig. 5.24 checks the *detectedListAT* location of the fault that is being handled to see if it has previously been detected. A fault is only injected if it has not been detected by any of the random tests that have been selected thus far. This **if** statement implements fault dropping. In its body, the fault being handled is injected, and it is checked for being detected by *testVector*. If detection occurs, (i.e., *woG* is different than *woF*) *detectedFaultCT* is incremented, and *detectedListCT* location corresponding to the fault is marked.

This **while** loop repeats for all faults for the same random test vector. The **if** statement that follows this loop comes after the **$fclose** statement in Fig. 5.24, and is shown in Fig. 5.25. This statement decides whether this random test is kept or discarded. As shown in Fig. 5.25, if the number of detected faults by the CT is satisfactory, detected faults lists by the CT will be transferred to *detectedListAT*, which is for all tests. Following that, the new coverage is calculated and selection of *testVector* is reported in the *testFile*.

The calculated coverage is one way that the FECpt algorithm uses for exiting. Running this algorithm on the circuit of Fig. 5.21 results in trying 200 random tests, selecting five tests and getting 87% fault coverage. A modified version of this testbench without fault dropping results in trying 102 random tests, selecting tests with 100% fault coverage.

```
module Tester();//CRTG_SimpleCKT_Fixed
    . . .
    initial begin

        while(coverage < desiredCoverage && uTests < utLimit) begin
            . . .
            if(detectedFaultsCT >= efCount) begin
                for(faultIndex=1;faultIndex<=numOfFaults;
                                 faultIndex=faultIndex+1)
                    if((detectedListCT[faultIndex] == 1'b1)) begin
                        detectedListAT[faultIndex] = 1'b1;
                        detectedFaultsAT = detectedFaultsAT + 1;
                    end
                coverage = 100 * detectedFaultsAT / numOfFaults;
                $fdisplay(testFile, "%b", testVector);
            end

        end //end of while of Coverage
        $display("Number of Random Vectors Generated: %d", uTests);
        $display("Coverage : %d", coverage);
    end //end of initial
endmodule
```

Fig. 5.25 Deciding on keeping or discarding a test

Fig. 5.26 Adjustable
expected coverage

```
Algorithm: AECPT
Copy F into F';
Start with an initial efCount;
While desired coverage has not reached
    Obtain a random test, t;
    Reset fCount;
    For every f in F';
        Inject f;
        Apply t;
            Simulate faulty circuit;
        Remove f;
        If f is detected
            Increment fCount;
            Drop f from F;
            Increment coverage;
        End if;
    End for;
    If fCount >= efCount
        Keep t;
        Adjust efCount, based on coverage;
    End if;
End while;
```

5.3.2.2 Adjustable Expected Coverage per Test

In a different RTG algorithm, instead of using a fixed *efCount*, this parameter is adjusted as more faults are detected. This method tries to adjust the expected number of faults that are detected with the exponential growth of fault coverage (Fig. 5.20).

Figure 5.26 shows the Adjustable Expected Coverage per test (AECpt) algorithm. Basically, this is the same as FECpt algorithm of Fig. 5.21 except the last part (the **if** part) that modifies *efCount* according to the obtained fault coverage at that point in the TG process.

```verilog
module Tester();//CRTG_SimpleCKT_Ajdustable
   parameter numOfFaults = 16;
   parameter initialExpFCount = 2;
   parameter utLimit = 200;
   parameter desiredCoverage = 99;
   reg a, b, c, d, e, f;
   wire woG, woF;
   reg [8*50:1] wireName;
   reg stuckAtVal;
   reg [5:0] testVector;
   reg [1:numOfFaults] detectedListCT, detectedListAT;
   integer faultFile, testFile, status;
   integer uTests, detectedFaultsCT, expFCountCT, faultIndex;
   integer tmp, newDiscovered, coverage, detectedFaultsAT;
   SimpleCKT GUT (a, b, c, d, e, f, woG);
   SimpleCKT FUT (a, b, c, d, e, f, woF);

   initial begin

      uTests = 0; coverage = 0;
      faultIndex = 1; detectedListAT = 0;
      expFCountCT = initialExpFCount;
      testFile = $fopen("CRTG_SimpleCKT_Adjustable.tst", "w");
      #10;

      while(coverage < desiredCoverage && uTests < utLimit) begin
         detectedFaultsCT = 0; detectedListCT = 0; detectedFaultsAT =0;
         testVector = $random($time);
         uTests = uTests + 1;
         faultIndex = 1;
         #10;
         NewDiscovered = 0;

         while(!$feof(faultFile)) begin // Fault Injection loop
            . . .
         end//end of while(!feof(faultFile))

         if(detectedFaultsCT < expFCountCT)
            . . .

         if(detectedFaultsCT>=expFCountCT && (NewDiscovered>0)) begin
            . . .
         end
         #10;
      end //end of while of Coverage
      $display("Number of Random Vectors Generated: %d", uTests);
      $display("Coverage : %d", coverage);
   end //end of initial
endmodule
```

Fig. 5.27 Adjustable expected coverage outline

Figure 5.27 shows the outline of a Verilog testbench implementing this algorithm. Parameter names and declarations in Fig. 5.27 are similar to those of the FECpt testbench of Fig. 5.23. Two differences are *initialExpFCount* and *newDiscovered*. The former is used for expected count that gets adjusted, and the latter counts faults that are detected for the first time. This variable is needed because we are not doing fault dropping in this implementation.

The **while** loop in Fig. 5.27 has the same exit condition as that of Fig. 5.23. Inside this loop, there is a loop for finding faults that are detected by the current random tests, and two **if** statements.

```
module Tester();//CRTG_SimpleCKT_Adjustable

   initial begin
      . . .
      testFile = $fopen("CRTG_SimpleCKT_Adjustable.tst", "w");

      while(coverage < desiredCoverage && uTests < utLimit) begin
         detectedFaultsCT=0; detectedListCT=0;
         detectedFaultsAT=0;

         testVector = $random($time);
         uTests = uTests + 1;
         faultIndex = 1;
         #10;
         newDiscovered = 0;
         faultFile = $fopen("SimpleCKT_Adjustable.flt", "r");
         while(!$feof(faultFile))begin// Fault Injection loop
            status=$fscanf(faultFile,"%s s@%b\n",wireName,stuckAtVal);
            $InjectFault(wireName, stuckAtVal);
            {a, b, c, d, e, f} = testVector;
            #60;
            if(woG != woF) begin
               detectedFaultsCT = detectedFaultsCT + 1;
               detectedListCT[faultIndex] = 1'b1;
               if(detectedListAT[faultIndex] == 0)
               newDiscovered = newDiscovered + 1;
            end //end of if
            $RemoveFault(wireName);
            #20;
            faultIndex = faultIndex + 1;
         end //end of while(!$feof(faultFile))
         $fclose(faultFile);
         . . . //details in Fig.5.21
      end //end of while of Coverage
      $display("Number of Random Vectors Generated: %d", uTests);
      $display("Coverage : %d", coverage);
   end //end of initial
endmodule
```

Fig. 5.28 Finding faults detecting by a random test, implementing AECpt

Figure 5.28 shows the fault injection **while** loop. Differences between this and Fig. 5.26 exist, due to the fact that we are not doing fault dropping here. Namely, *detectedListAT* is not checked, and the fault is injected unconditionally. Instead, *newDiscovered* is incremented when a fault is detected for the first time. This variable is later used in deciding on keeping or discarding the random test.

Figure 5.29 shows the rest of the Verilog testbench implementing AECpt algorithm. This part comes after the **$fclose** statement that is after the fault **while** loop in Fig. 5.28. The first **if** statement reduces the *efCount* if the number of detected faults by the CT (*detectedFaultsCT*) is less than the expected value. If *detectedFaultsCT* is larger than the expected value, the new expected value is raised by a factor influenced by *detectedFaultsCT*. More elaborate methods of adjusting our expectation are implementable in this testbench.

The testbenches for FECpt and AECpt presented the implementation of these algorithms in Verilog. Parameters used and conditions for various decision making provide flexibilities for other implementations. These should be adjusted according to the circuit for which tests are being generated.

```verilog
module Tester(); //CRTG_SimpleCKT_Adjustable
    . . .
    initial begin

        uTests = 0; coverage = 0;
        faultIndex = 1; detectedListAT = 0;
        expFCountCT = initialExpFCount;
        while(coverage < desiredCoverage && uTests < utLimit) begin
            detectedFaultsCT = 0; detectedListCT = 0; detectedFaultsAT = 0;
            faultFile = $fopen("SimpleCKT_Adjustable.flt", "r");
            uTests = uTests + 1;
            faultIndex = 1;
            #10;
            newDiscovered = 0;

            while(!$feof(faultFile))begin
                . . .
            end//end of while(!$feof(faultFile))
            . . .
            detectedFaultsAT = 0;
            if(detectedFaultsCT < expFCountCT)
                tmp = expFCountCT / 2;
            else tmp = (detectedFaultsCT + expFCountCT) / 2;
            expFCountCT = tmp;

            if(detectedFaultsCT >= expFCountCT && (NewDiscovered > 0))
            begin
                detectedFaultsAT = 0;

                for(faultIndex=1; faultIndex<=numOfFaults;
                                  faultIndex=faultIndex+1)
                    if((detectedListAT[faultIndex]==1) ||
                        (detectedListCT[faultIndex]==1))
                    begin
                        detectedListAT[faultIndex] = 1'b1;
                        detectedFaultsAT = detectedFaultsAT + 1;
                    end
                coverage = 100 * detectedFaultsAT / numOfFaults;
                $fdisplay(testFile, "%b", testVector);
            end
            #10;
        end //end of while of Coverage
        $display("Number of Random Vectors Generated: %d", uTests);
        $display("Coverage : %d", coverage);
    end //end of initial
endmodule
```

Fig. 5.29 Adjusting expectation, keeping or discarding a test

5.3.2.3 Precalculated Expected Coverage per Test

Instead of adjusting expected coverage per test, a coverage expectation graph can be calculated before the actual test generation begins. The expectation graph tries to predict the actual exponential growth of fault coverage (Fig. 5.20). Several Verilog implementations of this method are available in the software that accompanies this book. The testbenches have been tested for standard ISCAS benchmarks. They are too lengthy to be included in the text.

Based on the methods discussed above, other algorithm can be devised. For example, in addition to adjusting coverage per test, the overall designed fault coverage can also be modified as tests are being generated.

5.3.3 Sequential Circuit RTG

As fault simulation discussed in Chap. 4, sequential circuit test generation is also influenced by register structures. As we discussed in Chap. 4, after the application of a new test vector, fault effects in sequential circuits may take several clocks to propagate to circuit primary outputs. Because of this, we propose an RTG method for sequential circuits, in which both test vectors and the number of clocks are selected in random. As in fault simulation, resetting a sequential circuit is necessary between faults.

A testbench for sequential circuit test generation is presented here. We use this testbench for generating tests for the Residue-5 circuit that we used for fault simulation in Chap. 4. This circuit has two primary inputs, three primary outputs, and three flip-flops. There are 108 stuck-at faults in the circuit. The block diagram of this circuit is shown in Fig. 5.30.

Figure 5.31 shows parts of the Verilog testbench for generating tests for this circuit. This testbench implements the same algorithm (FECpt) as that of Fig. 5.23. Minor differences between the two testbenches are due to the fact that the latter is for a sequential circuit and handles clocking and resetting.

For brevity, sections of the code in Fig. 5.31, that are the same as the testbench of Fig. 5.23 are not shown. Parameter declarations, including the number of faults in the circuit, expected count, desired coverage, and the limit of the number of unsuccessful tests are shown near the top of this testbench. Next, the code shows the instantiation of the golden and faultable *residue5* Verilog codes. Near the end of this testbench, an **always** statement produces a free running clock, by complementing *clk* every 25 time units.

The first **while** loop shown in Fig. 5.31 provides exit conditions for the algorithm. Immediately following this statement, a random vector is produced, part of which is given to *clockCount* and the rest to *testVector*. These two segments are used for testing faulty models of Residue-5 in the **while** loop, the body of which is blanked in this figure.

Figure 5.32 shows the body of the inner **while** loop of the testbench. As shown, after a fault is read, the circuit is reset by asserting the reset signal, waiting for a clock pulse, and then deasserting the reset. Waiting for a clock pulse is done by flow control statement @(**posedge** *clk*) immediately followed by @(**negedge** *clk*). After injecting the fault (by **$InjectFault**), the test vector portion of the random data (*testVector*) is assigned to the circuit inputs (*in*). The clock count portion is used in a **repeat** statement that waits for *clockCount* clock pulses, effectively clocking the Residue-5 circuit many times. The next statement in this testbench checks the faulty and good circuit outputs to see if the injected fault has been detected. The rest of this testbench is no different than that of Fig. 5.23.

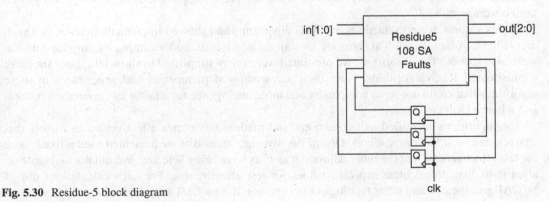

Fig. 5.30 Residue-5 block diagram

```
module Tester ();
   parameter numOfFaults = 108.0;
   parameter efCountCT = 2;
   parameter desiredCoverage = 99;
   parameter utLimit = 140;
      . . .
   residue5_net GUT(global_reset, clk, reset,in, woG);
   residue5_net FUT(global_reset, clk, reset,in, woF);

   initial begin
         . . .
      testFile = $fopen("Res5.tst", "w");

      while((coverage < desiredCoverage) && (uTests < utLimit))begin
         {clockCount, testVector} = $random();
         $display("TV=%b, CC=%d, uTest=%d",
                  testVector, clockCount, uTests);
         uTests = uTests + 1;
            . . .
         while( ! $feof(faultFile)) begin
            . . .
         end //end of while ( ! $feof(faultFile))

         if(detectedFaultCT >= efCountCT) begin
            . . .
         end

         coverage = 100 * (detectedFaultAT/numOfFaults);

         $display("fault coverage = %f\n", coverage );

      end //while coverage $fclose (DetectionFile);

      $stop;
   end //end of initial

   always #25 clk = ~clk;
endmodule
```

Fig. 5.31 Testbench generation tests for a sequential circuit

When a fault is detected, the random test data, and the number of clocks that the test data has been applied are recorded. Figure 5.33 shows test generated for the Residue-5 circuit. The testbench generates a total of 139 vectors, of which only seven are selected. Total test time (adding clock counts in Fig. 5.33) is 45 clock cycles. The seven test vectors applied over 45 clock cycles yield a fault coverage of 82.4%.

In this section, we presented several RTG algorithms and showed implementations of two such algorithms. Adaptation of algorithms to sequential circuits and example of implementation were also shown. The algorithms we presented were fairly simplified to show templates for more sophisticated RTG algorithms. Nevertheless, we showed parameters and procedures in these algorithms that could use more complexity and more intelligence for a faster test generation process and a better fault coverage.

The algorithms identified exit conditions, and random test vector effectiveness as factors that characterize such RTG procedures. One of the two algorithms that we presented used a fixed value for test effectiveness, and the other adjusted it as tests were being selected. We did not elaborate on algorithms that precalculate expected values for test effectiveness. For such calculations use of SCOAP parameters, and other topological information from a CUT may be useful.

```
module Tester ();
    . . .
    initial begin

        testFile = $fopen("Res5.tst", "w");
        while((coverage < desiredCoverage) && (uTests < utLimit))begin
            . . .
            faultFile = $fopen("Res5.flt", "r");

            while( ! $feof(faultFile)) begin
                status=$fscanf(faultFile,"%s s@%b\n",wireName,stuckAtVal);
                faultIndex = faultIndex + 1;

                if(detectedListAT[faultIndex] == 0) begin
                    reset = 1; global_reset = 1;
                    @(posedge clk) @(negedge clk);
                    reset = 0; global_reset = 0;
                    $InjectFault ( wireName, stuckAtVal);
                    in = testVector;

                    repeat(clockCount) @(posedge clk) @(negedge clk);

                    if (woG != woF) begin
                        detectedFaultCT = detectedFaultCT + 1;
                        detectedListCT[faultIndex] = 1;
                        $display("Fault:%s SA%b detected by %b by %d pulse
                        at %t.",wireName,stuckAtVal,in,clockCount,$time);
                    end //if

                    $RemoveFault(wireName);
                end //if !detected
            end //end of while ( ! $feof(faultFile))

            $fclose(faultFile);
            if(detectedFaultCT >= expFCountCT)begin
                . . .
            end
            coverage = 100  * (detectedFaultAT/numOfFaults);
            $display("fault coverage = %f\n", coverage );

        end //while coverage
        $fclose (DetectionFile);
        $stop;
    end //end of initial
    always #25 clk = ~clk;
endmodule
```

Fig. 5.32 Fault injection, input application, and clocking

Fig. 5.33 Test data for Residue-5

Test	Clock Count	
00 ,	1001	9(decimal)
01 ,	0010	2
11 ,	1000	8
01 ,	0011	3
01 ,	1001	9
10 ,	0100	4
10 ,	1010	10

Test generation methods in Sect. 5.3.1 are regarded as fault independent random test generation while Sects. 5.3.2 and 5.3.3 presented random fault-oriented test generation methods.

5.4 Summary

This chapter showed where test generation fits in electronic test cycle. We started with controllability and observability methods that are needed for many test generation methods and programs. We then discussed complete procedures and programs for generation of test. For the specific test generation engines, we focused on RTG and showed several methods for sequential and combinational test generators. Incorporating these engines in Verilog testbenches provides a practical set of tools and environments for test generation and test evaluation.

As discussed, there is a limit as to amount of fault coverage we can get with random test generation. The next chapter discusses deterministic algorithms for test generation that together with random methods presented here, provide complete test generation solutions.

References

1. Sellers FF, Hsiao MY, Bearnson LW (1968) Analyzing errors with the Boolean difference. IEEE Trans Comput C-17(7):676–683.
2. Akers SB (1959) On a theory of Boolean functions. J. SIAM 7.
3. Sellers FF Jr., Hsiao M-Y, Bearnson LW (1968) Error detecting logic for digital computers. New York: McGraw-Hill.
4. Agrawal P, Agrawal VD (1975) Probabilistic analysis of random test generation method for Irredundant Combinational Logic Networks. IEEE Trans Comput C-24(7):691–695.
5. Agrawal P, Agrawal VD (1976) On Monte Carlo testing of logic tree networks. IEEE Trans Comput C-25(6):664–667.
6. Agrawal VD (1978) When to use random testing. IEEE Trans Comput C-27(11):1054–1055.
7. Parker KP, McCluskey EJ (1975) Probabilistic treatment of general combinational networks. IEEE Trans Comput C-24(6):668–670.
8. Eichelberger EB, Lindbloom E (1983) Random-pattern coverage enhancement and diagnosis for LSSD logic self-test. IBM J Res Dev 27(3):265–272.
9. Goldstein LH (1979) Controllability/observability analysis of digital circuits. IEEE Trans Circuits Syst CAS-26(9):685–693.
10. Rutman RA (1972) Fault-detection test generation for sequential logic by Heuristic Tree Search. Technical Report TP-72-11-4, U. of Southern California, Dept. of EESystems, Los Angeles, California.
11. Agrawal VD, Mercer MR (1982) Testability measures – what do they tell us? in Proceedings of the International Test Conference. 391–396 Nov. 1982
12. Savir J (1983) Good controllability and good observability do not guarantee good testability. IEEE Trans Comput, C-32:1198–1200.
13. Jain SK, Agrawal VD (1985) Statistical fault analysis. IEEE Design Test Comput 2(1):38–44.
14. Brglez F (1984) On testability analysis of combinational networks. in Proceedings of the International Symposium on Circuits and Systems, 221–225 May 1984
15. Grason J (1979) **TMEAS** – A testability measurement. Program. Proceedings 16th Design Automation Conference 156–161 June, 1979
16. Goldstein LH, Thigpen EL (1980) SCOAP: Sandia controllability/observability analysis program, in proceedings of the 17th Design Automation Conference, 190–196 June 1980
17. Bushnell ML, Agrawal VD (2000) Essentials of electronic testing for digital, memory, and mixed-signal VLSI circuits, Kluwer Academic Publishers
18. Wang L-T, Wu C-W, Wen X VLSI Test Principles and Architectures: Design for Testability, Morgan Kaufmann, July 2006.

Chapter 6
Deterministic Test Generation Algorithms

The previous chapter provided an understanding of test generation and showed where and how test generation is used in digital system testing.

The random test generation that was discussed gave us some tools and methods to work with. Perhaps an important benefit of random test methods of the previous chapter was their implementations in Verilog testbenches, which enabled us to have HDL environments for design and test. However, random testing alone often comes short of providing a necessary fault coverage and must be complemented with deterministic methods.

This chapter discusses test generation algorithms that are used for deterministic test generation. We show how such methods complement random test generation in a complete system. In presentation of these algorithms, test methods discussed in the previous three chapters such as fault collapsing, fault simulation, and testability evaluation will be used.

In presentation of the algorithms, we try to show the basics and avoid getting into implementation details. The algorithms are at the gate level and require intensive C/C++ programming for their implementations. On the other hand, there are already many implementations of test generation algorithms as stand-alone programs or as parts of complete test suites. Our presentation here tries to motivate the reader to look beyond the gate level and C/C++ programming, and look for ways test generation can be done at higher levels, and perhaps be incorporated in hardware description languages.

In the sections that follow, we discuss deterministic test generation algorithms for combinational and sequential circuits. The last section discusses algorithms for compaction of test data that is a post test generation process for reducing the number of test vectors.

6.1 Deterministic Test Generation Methods

The previous section discussed random test generation, where no analysis of the circuit takes place for generation of an initial test vector. This section discusses deterministic test generation, in which test vectors are resulted from some analysis performed on the CUT. As with RTG, deterministic test generation may be fault-oriented or fault-independent.

In a deterministic fault-oriented test generation procedure, a specific fault in the circuit is targeted, and a test vector is generated to detect that fault. Usually, this method of test generation is accompanied by fault simulation. Once a test is found to detect a given fault, fault simulation is performed to find other faults this test vector can detect. This is done, because generating deterministic fault-oriented test vectors require an involved process, and before repeating this for another fault in the circuit, all faults that can be taken care of by an already computed test vector are marked as detected.

Z. Navabi, *Digital System Test and Testable Design: Using HDL Models and Architectures*,
DOI 10.1007/978-1-4419-7548-5_6, © Springer Science+Business Media, LLC 2011

Fault-oriented test generation deals with single stuck-at fault model, i.e., it generates tests assuming there is only one fault in the circuit, and that is of stuck-at-value type. Although this assumption and model may appear as too limiting, a test set that detects a good percentage of single stuck-at faults also results in a good coverage of multiple faults of stuck-at and other fault types.

Most deterministic test generation methods are fault-oriented, and the subsections that follow discuss various algorithms and methods for this purpose. Before getting into specific algorithms, the section that follows illustrates the use of deterministic fault-oriented test generation in a complete test generation environments.

6.1.1 *Two-phase Test Generation*

Because of the high cost of deterministic fault-oriented test generations and the number of faults that have to be dealt with, such test generation methods are usually used along with and as complimentary to random and functional test generation. Random test generation methods usually yield 60–80% fault coverage and can be generated much faster than deterministic methods. We regard random or functional as the first phase of test generation, and deterministic as the second phase of a complete test generation scheme.

Figure 6.1 shows a two-phase test generation procedure that involves random and deterministic methods.

As shown in this figure, the TG process begins with random test generation. This phase continues until the random test generation procedure exits, either because an acceptable coverage has been reached or because too many random vectors with unacceptable coverage have been tried. The algorithms discussed in Sect. 5.3 of Chap. 5, Figs. 5.24 and 5.28 are the type of algorithms we use in this first phase of TG. After exiting the RTG phase, Fig. 6.1 shows a **for** loop for fault-oriented deterministic test generation. In this loop, faults have not been detected by the RTG

```
Copy F into F';
While desired coverage has not reached
    While more faults are being detected
        Perform random test generation;
    End while;
End while;
For every f in F' (remaining faults)
    Insert f;
        Perform deterministic test generation, t;
    Mark f as detected, drop from F';
    Add t to T;
    For every g in F' (remaining faults)
        Inject g;
        Apply t;
            Simulate faulty circuit;
        Remove g;
        If g is detected by t
            Mark g as detected, drop from F';
        End if;
    End for;
End for;
```

Fig. 6.1 Two-phase random, deterministic test generation

while loop are targeted. Referring back to the exponential growth of detected faults, discussed in the previous chapter (Fig. 5.20), the second phase targets the last 20–40% faults that have not been detected yet.

After injection of a fault and finding test t for detecting that specific fault, a fault simulation loop (inner **for** loop in Fig. 6.1) check t to see which undetected faults, in addition to the fault that was originally targeted, it detects. This fault simulation phase causes fewer deterministic fault-oriented test generation passes.

6.1.2 Fault-oriented TG Basics

By use of several test cases and a small circuit, we present basics of test generation procedures and introduce terms and definitions that will be used in formal TG algorithms. Test generation tries to set values on circuit lines that can be controlled by a tester (primary inputs), and propagate fault to lines that can be observed by the tester (primary outputs), such that value at the primary output of the good circuit and faulty circuit are different.

6.1.2.1 Basic TG Procedure

Test generation for a fault like the SA0 fault in Fig. 6.2 begins with setting all inputs to X and propagating this value through all circuit lines to reach the primary outputs. The value X is unknown or do not care. In a test vector, X means that the corresponding input position can take either a 0 or a 1 value.

Test generation for l_7:$SA0$ begins with trying to force a 1 on this line to make the fault show itself. This is achieved by a 1 on the inverter output, and thus a 0 on a primary input. Therefore, the starting value of X on a changes to 0. Next, we try to propagate the fault effect to the circuit primary output. For now, we use the 1/0 notation for *good/faulty* line values, which means l_7 is 1/0. Effect of fault on line l_7 can only get closer to the circuit's primary output if it passes through G_3 which requires l_2 to become 1. In this case, the good/faulty value on l_{11} becomes 1/0. To have a 1 on l_2, input b must become 1.

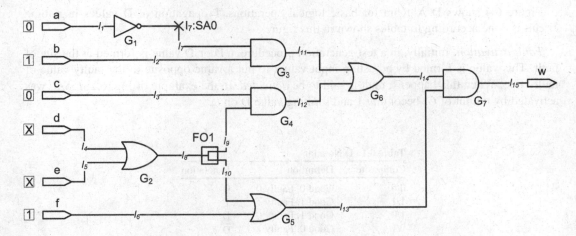

Fig. 6.2 A simple case of test generation

In order to propagate fault effect that is now on l_{11} closer to the circuit's primary output, we have to have a 0 on l_{12}. This makes l_{14} take 1/0 value and in order to achieve l_{12} of 0, either l_3 or l_9 must be 0. We choose l_3 and make this line 0, by just setting input c to 0.

We need to take one more step to make the fault effect get to the circuit's w output and that is to set l_{13} to 1 to propagate value on l_{14}, that is now 1/0, to l_{15} which is the circuit's primary output. As for l_{13}, this line can be set to 1 by setting either l_{10} or l_6 to 1. Since it is easier to set l_6 to 1, we take this route, which requires input f to become 1. At this point, we have input values that cause the effect of l_7:$SA0$ to appear on circuit primary output. Thus, test is made, $abcdef$ = 010XX1. For this input value, good circuit output is 1 and the circuit having the l_7:$SA0$ fault generates a 0 output, thus 1/0.

The X values in the test vector (inputs d and e) can be set to 0 or 1 and have no effect on detection of the fault.

In this example, in each step, we pushed the fault effect one gate closer to the circuit primary output and tried to adjust input values to let this propagation happen. Alternatively, we could find a complete path to the output and make input requirements to satisfy this propagation. In either case, generation of a test vector involves propagation of a fault effect to line in the circuit that can be observed, and adjustment of values of lines in the circuit that let this propagation happen. Several observations can be made by reviewing steps that we took for generating a test in the above example. Definitions and notations are presented as we take a more formal look at what we did above.

Discrepancy values (D). In order to show that a fault effect was propagated, we use 1/0 values to designate good and faulty values. A more formal representation is D for discrepancy, which replaces the composite 1/0 value. Similarly, $\overline{D}$ represents 0/1 which is a good 0 value and faulty 1. The D-notation used in test generation is shown in Table 6.1.

Fault detection. A fault is detected when a D or $\overline{D}$ reaches to a circuit output.

D Algebra. Since detection of faults is known by D or $\overline{D}$ reaching to a primary output, it is important to be able to calculate how D values propagate through individual logic gates. For this purpose, D Algebra is used for logic manipulation of D values. Figure 6.3 shows D Algebra for the basic AND operation. As shown here, to obtain the result of a logical operation whose operands are of the D type (Table 6.1), the operands are turned into their composite form, and the logical operation is performed on bit-by-bit basis, i.e., good values with good values, and faulty ones are operated together. The resulting composite values are then turned into D values according to Table 6.1.

Figure 6.4 shows D Algebra for basic logical operations. Propagation of D values in a logic circuit is done according to tables shown in this figure.

Fault activation. Initially, in a test generation procedure, a D or $\overline{D}$ value is formed at the site of fault. This value is formed by adjusting input values to put a value opposite to the faulty value of faulty line. When this happens, fault is said to be *activated*. In the example of Fig. 6.2, l_7:$SA0$ was activated by output of G_1 becoming 1 and forming value D on l_7.

Table 6.1 D Notation

Composite	Definition	D notation
0/0	Good 0, Faulty 0	0
1/1	Good 1, Faulty 1	1
1/0	Good 1, Faulty 0	D
0/1	Good 0, Faulty 1	$\overline{D}$
X	Don't care	X

Fig. 6.3 AND function D Algebra

Operation	Expanded	Result
0 . 0 = 0/0 . 0/0 = 0/0 =		**0**
0 . 1 = 0/0 . 1/1 = 0/0 =		**0**
0 . D = 0/0 . 1/0 = 0/0 =		**0**
0 . $\bar{D}$ = 0/0 . 0/1 = 0/0 =		**0**
1 . 0 = 1/1 . 0/0 = 0/0 =		**0**
1 . 1 = 1/1 . 1/1 = 1/1 =		**1**
1 . D = 1/1 . 1/0 = 1/0 =		**D**
1 . $\bar{D}$ = 1/1 . 0/1 = 0/1 =		**$\bar{D}$**
D . 0 = 1/0 . 0/0 = 0/0 =		**0**
D . 1 = 1/0 . 1/1 = 1/0 =		**D**
D . D = 1/0 . 1/0 = 1/0 =		**D**
D . $\bar{D}$ = 1/0 . 0/1 = 0/0 =		**0**
$\bar{D}$. 0 = 0/1 . 0/0 = 0/0 =		**0**
$\bar{D}$. 1 = 0/1 . 1/1 = 0/1 =		**$\bar{D}$**
$\bar{D}$. D = 0/1 . 1/0 = 0/0 =		**0**
$\bar{D}$. $\bar{D}$ = 0/1 . 0/1 = 0/1 =		**$\bar{D}$**

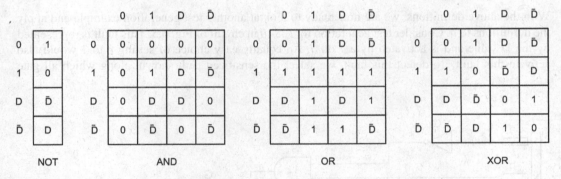

Fig. 6.4 D Algebra for basic logical functions

Fault propagation. *Fault propagation* occurs when a D or $\bar{D}$ value advances toward circuit primary outputs. In our example, circuit propagation of D from line l_7 to l_{11} or propagation of D from l_{11} to l_{14} is regarded as fault propagation.

Sensitized line. A line in a circuit to which a fault has propagated is called a *sensitized line*. Line l_{11} is a sensitized line.

Sensitized path. A path between site of fault and lines in the circuit that contain a continuous sequence of sensitized lines is a *sensitized path*. A fault is detected if there is a sensitized path between site of fault and a primary output. Lines l_7, l_{11}, l_{14}, and l_{15} form a sensitized path between site of l_7:SA0 and w primary output.

Justification. The process of adjusting circuit input values to activate a fault or to facilitate propagation of a fault toward an output is called input *justification*. In Fig. 6.2, activating l_7:SA0 requires a to become 0. Therefore, the good value 1 on l_7 is justified by a 0 on a. Also propagation of D on l_{11} to l_{14} requires l_{12} to become 0. This requirement is justified by a 0 on c.

Implication. In justification, a primary input is set to justify a required value on a circuit line. In circuits with fanout, setting the primary input may affect other parts of the circuit that were originally not intended. The process of propagating a primary input value to all parts of the circuit that it affects is called *implication*.

Making choices. In test generation process, often there are several choices for achieving the same goal. Usually, we select the easiest choice and only go back to other choices if our initial selection fails. In the example of Fig. 6.2, there were several places that we had to make a choice between several possibly good solutions. Take, for example, propagation of D from l_{11} to l_{14} that required l_{12} to be 0. This goal could be achieved by setting l_3 or l_9 to 0. We chose l_3 since it was closer to a primary input and could be set more easily. On the other hand, selecting l_9 to be 0, would be justified by $d = e = 0$. Through implication, this input combination would put a 0 on l_{10}, which would limit our choices for propagating D from l_{14} to l_{15}.

Back tracking. Where choices exist, we make a selection based on what seems to be the best for propagation of values. However, through implication, a choice, which may seem the best at first, may block propagation of fault values elsewhere in the circuit. In such cases, we return to places where choices exist, reset all circuit values to their values before the first choice was made, and make a different choice. This process is referred to as *back tracking*.

6.1.2.2   A More Formal Approach to TG

With the above definitions, we are now ready to look at another test generation example and apply the definitions to it. Consider test generation for l_8:SA0 in circuit of Fig. 6.5. This fault is represented by the D value and is activated by a 1 on d. An equally easy choice of setting e to 1 would also activate this fault. To detect this fault, we select the sensitized path shown, along which all gate

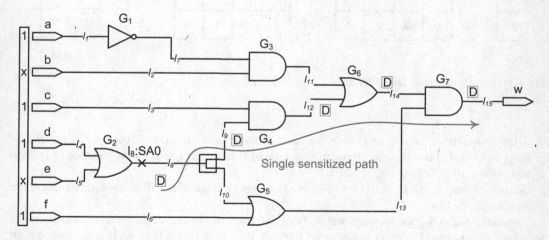

Fig. 6.5 Single sensitized path

outputs leading to w are D. This path sensitization requires justification of 1, 0, and 1 on lines l_3, l_{11}, and l_{13}, respectively. Setting primary input c to 1 justifies a 1 on l_3. Setting a to 1 justifies a 0 on l_{11}, and finally, setting f to 1 justifies a 1 on l_{13}.

After propagations and justifications, a single sensitized path from site of l_8:SA0 to w primary output is created. Input values $abcdef$ = 1X11X1 justify necessary line values for propagation of fault. Thus, a test is found for l_8:SA0.

6.1.2.3 Multiple Sensitized Paths

Example in Fig. 6.5 used a single sensitized path to propagate a D value to the circuit primary output. We show another example here and apply the same test generation procedure to it.

Figure 6.6 shows a simple circuit with a and b inputs and w output. The circuit has three reconvergent fanouts. The following paragraph discusses the procedure we use for generating a test for l_6:SA1.

Activation of l_6:SA1 is done by a 0 on l_6, causing value $\overline{D}$ on this line. This value is justified by $l_3 = 1$ which results in $a = 1$. Implication of $a = 1$ goes forward and sets l_5 to 1.

Now that fault activation is done, we try to find a sensitized path from l_6 to w. Actually, $\overline{D}$ already appears on l_8 and l_9 because of the fanout, and going forward from l_8 or l_9, we choose the upper path shown by an arrow in Fig. 6.6. Along this path, D and $\overline{D}$ values are shown. Propagation of D to the output of G_2 is achieved by $l_5 = 1$, which is already implied by $a = 1$. The last step in propagation is the propagation of D from the output of G_2 to the output of G_5, which requires a 0 on l_{12} input of this gate. Trying to justify this requirement, we have to set both inputs of G_4 to 0. We start with the upper input (l_{10}). For justifying a 0 on l_{10}, a 1 is required on l_9. Unfortunately, this line already has a $\overline{D}$ value, and l_{10} cannot be 0. This means that the requirement of l_{12} having a 0 cannot be satisfied, and we cannot find a test for l_6:SA1.

From the above analysis, it appears that there is no test to detect l_6:SA1. On the other hand, had we propagated D from site of fault (l_6) in two directions through two sensitized paths, the above conflict would not have happened. Figure 6.7 depicts this alternative solution for generating test for l_6:SA1.

Generating test for l_6:SA1 starts with fault activation by justifying a good 0 on l_6. This requires $a = 1$ that propagates to circuit lines as shown. The next step after fault activation is D value propagation, for which we select both paths shown in this figure. These paths require $l_5 = 1$ and $l_7 = 0$ for $\overline{D}$ to propagate in two directions. These requirements are to be realized during the justification phase. However, the first requirement ($l_5 = 1$) is already justified by the implication that occurred when we set a to 1. The other requirement ($l_7 = 0$) can be satisfied by setting b to 0. Implication of 0 on b causes no conflict and is thus accepted. On the output side, the two D values are NORed

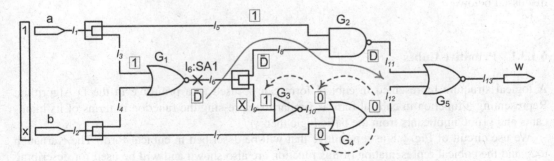

Fig. 6.6 Circuit with reconvergent fanout, trying single sensitized path

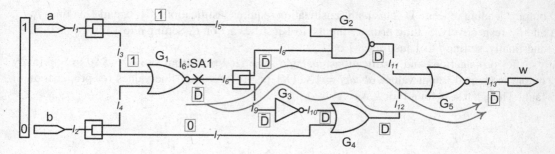

Fig. 6.7 XOR with reconvergent fanout, trying multiple paths

together and create a $\overline{D}$ on w. Thus, test vector $ab = 10$ activates and propagates l_6:$SA1$, and it is considered the test vector for this fault.

The above example illustrated that due to reconvergent fanouts, certain faults require multiple sensitized paths to be detected, and some test generation algorithms handle such situations. The D-Algorithm presented next is one such algorithm.

6.1.3   The D-Algorithm

A formal mechanism putting together procedures and definitions discussed above is the D-Algorithm. The D-Algorithm [1, 2] is a tabular, procedural method for generating tests for combinational circuits.

The D-Algorithm incorporates propagation, justification, and implication and uses a table of circuit line values for representation of results of these processes. This algorithm implements multiple sensitized paths, where needed to reach an answer. Cubical representation of logic functions are used for representing them. To represent line values, the D-Algorithm uses test cube. Initially, all values in a test cube are X. Starting with an all-X test cube, the D-Algorithm injects a fault by introducing a D into the test cube. A series of justifications and propagations in the circuit progressively change Xs in the test cube to 1, 0, D, and $\overline{D}$. When completed, the test cube contains circuit input values for detection of the fault, all circuit line values, and circuit primary output values.

This algorithm works with the concept of cubical representations for logical structures and processes cubes representing logical functions and error representations from various lists. Basic logic primitives are handled in D-Algorithm, but any logical function that can be represented in cubical form can be used as a basic structure in the D-Algorithm. Cubes and corresponding lists are discussed below.

6.1.3.1   Primitive Cubes

A logical structure represented in cubical form can be used as a primitive in the D-Algorithm. Representing a function in cubical form is similar to expressing the function in terms of its implicants and prime implicants from the basic logic theory.

We use circuit of Fig. 6.8 as a primitive that will be described in cubical form. The Karnaugh map and the cubical representation of this function are also shown and will be used for describing cubes for this function.

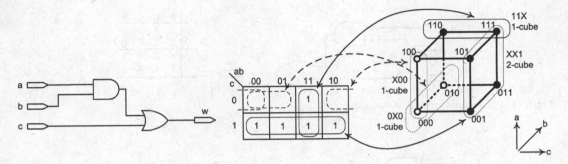

Fig. 6.8 A D-Algorithm primitive

Fig. 6.9 AND–OR logic
0-point and 1-point vertexes

	a	b	c	w
	0	0	0	0
P_0	0	1	0	0
	1	0	0	0
	0	0	1	1
P_1	0	1	1	1
	1	0	1	1
	1	1	0	1
	1	1	1	1

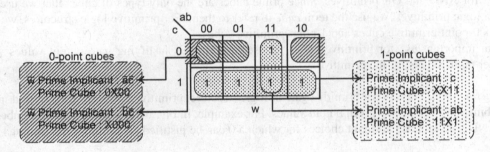

Fig. 6.10 Extracting prime cubes

An input combination and its corresponding output is called a *vertex* in cubical notation. Since $a = 0$, $b = 1$, and $c = 1$ are input combinations for which w is 1, the pattern 0111 is a vertex of logic structure shown in Fig. 6.8. Input combinations for which a primitive output is 1 are called 1-point vertexes, and those for which the output is 0 are called *0-point* vertexes. Figure 6.9 shows 0- and 1-point vertexes of primitive of Fig. 6.8.

A vertex is also called a *0-cube*, a term that we only use in this discussion. We will not use this term later to avoid confusion with 0-point and 1-point cubes. A *0-cube* has all variables of a function. Two *0-cubes* combine to form a *1-cube*. In general, two *n-cubes* combine to form an *(n + 1)-cube*. An n-cube that cannot be combined with another cube to form an upper level cube is a prime cube (see formation of 1-cubes and 2-cubes in Fig. 6.8). Prime cubes of a logical function are its prime implicants. In Fig. 6.8, XX11 and 11X1 are prime cubes of the primitive structure shown. Extracting primitive cubes for the logic structure of Fig. 6.8 by use of Karnough maps is shown in Fig. 6.10.

Prime cubes for which the output of the primitive is 1 are called 1-point prime cubes, and those for which the output is 0 are called 0-point prime cubes. 0-point prime cubes are like prime implicants for the complement of a logic function.

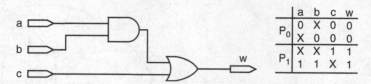

	a	b	c	w
P_0	0	X	0	0
	X	0	0	0
P_1	X	X	1	1
	1	1	X	1

Fig. 6.11 Prime cubes for AND–OR logic

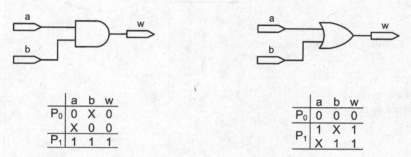

	a	b	w
P_0	0	X	0
	X	0	0
P_1	1	1	1

	a	b	w
P_0	0	0	0
P_1	1	X	1
	X	1	1

Fig. 6.12 Primitive AND and OR cubes

Figure 6.11 shows the logic structure of Fig. 6.8 and its prime cubes. Figure 6.12 shows prime cubes for AND and OR primitives. Since prime cubes are the only types of cubes that we use for given logic primitives, we use the term *cube* to refer to them. For primitive logic structures, we use terms 0-point primitive cubes and 1-point primitive cubes.

An important use for primitive cubes in test generation is justifying gate output values with proper input values for the primitive.

Justification. For a given primitive gate or logic structure, primitive cubes are used to find input combinations that justify given output values. For example, in Fig. 6.12, 0-point primitive cubes of an AND gate indicate all input choices for which a 0 can be justified [1] on the gate's output.

6.1.3.2 Propagation D-Cubes

A propagation D-cube for a logic primitive indicates how D or $\overline{D}$ values on a given input propagate to the primitive's output. Propagation D cubes can be formed by intersection of 0-point and 1-point cubes. Figure 6.13 shows the AND–OR primitive, for which primitive cubes were shown in Fig. 6.11. In Fig. 6.13, we are showing how intersection of primitive cubes is used to form propagation cubes.

Starting with a 0-point primitive cube (e.g., P_0:0X00) and intersecting it with a 1-point primitive cube (e.g., P_1:XX11) forms a propagation cube with $\overline{D}$ output. Similarly, starting with a 1-point primitive cube (e.g., P_1:11X1) and intersecting it with a 0-point primitive cube (e.g., P_0:X000) forms a propagate cube with D output. Figure 6.13 shows the rest of the propagation cubes that are formed by intersecting P_0, P_1 and P_1, P_0 members.

Propagation cubes shown in this figure have only the D or $\overline{D}$ on their inputs. Propagation cubes with multiple D or $\overline{D}$ on the input side can be formed by intersecting 0-point and 1-point vertexes. In the algorithms that follow, most of the time, propagation cubes with only one input D or $\overline{D}$ are used.

Following the same procedure for AND and OR logic gates, propagation cubes for these gates can be obtained. Figure 6.14 shows the corresponding tables.

Fig. 6.13 Forming propagation D-cubes

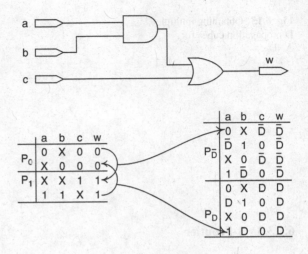

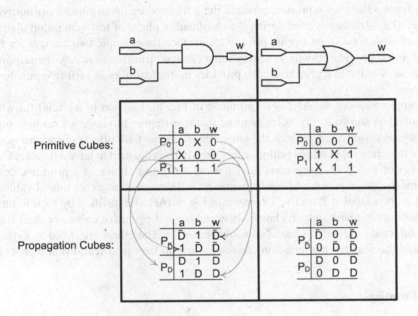

Primitive Cubes:

Propagation Cubes:

Fig. 6.14 Basic gates primitive and propagation cubes

Figure 6.15 shows the complete list of propagation D-cubes for an AND primitive, including those with multiple D or $\overline{D}$ values on their input. These cubes are obtained from AND vertexes, as shown. An important use of propagation cubes in test is for propagating faults to circuit outputs.

Propagation. In test generation algorithms, D values at primitive gate inputs indicate that the fault effect (error) has reached the gate. Propagation D-cubes tell how a D value on a gate input propagates to output (D or $\overline{D}$), and which values other gate inputs must have to allow the propagation (1 or 0). Propagation D-cubes with multiple D or $\overline{D}$ values (e.g., Fig. 6.15) are used in multi-path sensitization test generation methods.

Fig. 6.15 Obtaining multiple
D propagation cubes for
AND

	a	b	w
P_0	0	0	0
	0	1	0
	1	0	0
P_1	1	1	1

	a	b	w
P_0	$\bar{D}$	$\bar{D}$	$\bar{D}$
	$\bar{D}$	1	$\bar{D}$
	1	$\bar{D}$	$\bar{D}$
P_1	D	D	D
	D	1	D
	1	D	D

6.1.3.3 J-Frontier

The J-frontier is a list of gates in a circuit whose outputs are known (0 or 1), but not justified by their inputs. Input values for a primitive gate in the J-frontier are determined by primitive cubes of the primitive. The J-frontier is used during the justification phase of test generation algorithms.

Different choices for a 0- or 1-point primitive cube (see for example two choices for P_0 and two for P_1 in Fig. 6.13) provide options in setting inputs of the primitive to justify the required output. Furthermore, an X value in a primitive cube provides more flexibility in setting inputs for the same output value.

As an example consider the AND–OR primitive of Fig. 6.11 as part of a circuit for which test is being generated. As shown in Fig. 6.16, suppose that it is required to have a 1 on the output of this primitive; perhaps to activate a fault on the output or propagate fault effects elsewhere in the circuit. Also suppose that the c input of this primitive has already been decided to have a 0 value. Considering primitive cubes of Fig. 6.11, these conditions narrow down our choice of a primitive cube to only 11X1. This means that inputs a and b must be set to 1. These conditions set output values for other primitives (e.g., box labeled Primitive Driving a in Fig. 6.16) to be justified by their inputs.

In the justification phase, where choices between several primitive cubes, or their input values exist, one is selected randomly or based on some heuristics. The others are saved in a stack to come back to, should the selected combination create conflicts in later justifications and propagations.

6.1.3.4 D-Frontier

The D-frontier [1] consists of a list of gates in a circuit with at least a D or $\bar{D}$ on one of the inputs that has not propagated to the gate's output. Propagation of input D value is determined by propagation D-cubes. D-frontier is used in test generation during the propagation phase.

For a primitive gate, different propagation cubes that propagate the same input, or X values in a cube, provide alternative solutions for the same propagation. For example, for the AND–OR primitive of Fig. 6.13, two propagation D-cubes exist for propagating $\bar{D}$ on the c input to the primitive output. Since each of these $P_{\bar{D}}$ values has an X, there will be a total of four choices for propagating $\bar{D}$ from c to w. One of these choices is selected, and it is marked. If a failure happens later in the test generation process, a different input combination will be selected.

As a propagation example, consider propagating a D from the output of the AND–OR primitive of Fig. 6.16 through the c input of a similar gate to its output. The new diagram is shown in Fig. 6.17. According to Fig. 6.13, we can choose 0XDD or X0DD. In either case, D propagates to the output, and either a or b inputs must be set to 0. Satisfying $a = 0$ is again a justification problem for the gate that drives this line.

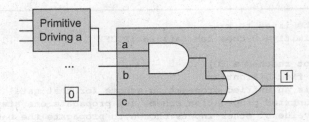

Fig. 6.16 Justification using primitive cubes

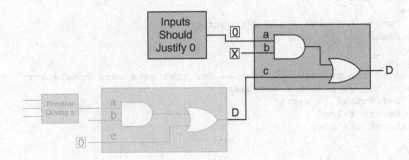

Fig. 6.17 Propagation using propagation D-cubes

6.1.3.5 D-Algorithm Procedure

The D-Algorithm consists of a sequence of propagations and justifications on primitives taken from D-frontier and J-frontier for generation of test for an injected fault. Figure 6.18 shows pseudo code for this algorithm.

The first **while** loop shown starts with a D at the site of the fault and propagates it to the output. Other propagations follow. When a propagation is successful, the D-frontier and J-frontier are updated. The D-frontier is updated with gates that have a D or $\overline{\text{D}}$ on their inputs, and the J-frontier is updated with gates whose output values are 0 or 1.

In the process of propagation, if a propagation fails, other propagation schemes (through other gate's fanouts) will be tried. If all schemes fail, the gate is removed from the D-frontier and backtracking to a previous step where a choice existed happens. If backtracking is not possible or there are no more gates in the D-frontier, the propagation phase reports a failure and exits.

The second **while** in Fig. 6.18 is the justification loop. This loop starts with several gates in the J-frontier. While the J-frontier is not empty, a gate is removed from it and justifying its output is tried with one of several available schemes (primitive cubes). If justification is successful, the J-frontier is updated. This continues until all gates from the J-frontier have been removed.

If a failure happens trying a gate justification scheme, a different scheme is tried (a different primitive cube). If none of a gate's justification schemes are successful, the algorithm backtracks one step to a previous gate. If backtracking reaches a node that has already been tried, the algorithm exits with a failure.

For exercising the procedure of Fig. 6.18, we use the same example circuit that used several times in Chap. 5. Figure 6.19 shows a circuit diagram and a fault for which a test is being generated. The diagram shows propagation paths with arrows going forward and justification paths with arrows going in the opposite direction.

Figure 6.20 shows a table that we use for showing steps taken in the D-Algorithm for generating test for l_8:SA0. We will show gates added and removed from the D- and J-frontiers, and test cubes as the algorithm progresses. On the left column, numbers show steps taken that we will refer to for

```
Initialize all the lines to x
Construct the primitive D-cube for failure (PDCF)

While (D or D' not reached a PO)
   Select a gate from D-frontier
   While (there is an untried propagation scheme for that gate)
      Select an untried propagation scheme and propagate one step
      Imply the value of other inputs chosen to propagate the D-cube
      If (propagated or implied values is not consistent with assigned values)
         Undo propagation
         Continue
      Else
         Update D-frontier and J-frontier
         Break
      End if
   End while
   If (none of the propagation schemes for that gate were consistent)
      Remove the gate from D-frontier
      If (D-frontier is empty)
         Return false
      Backtrack one step
   End if
End While
While (J-frontier is not empty)
   Select a gate from J-frontier
   While (there is an untried justification scheme for that gate)
      Select an untried justification scheme and justify one step
      If (justified values are not consistent with assigned values)
         Undo justification
         Continue
      Else
         Update J-frontier
         Break
   End while
   If (none of the justification paths for that element has worked)
      Backtrack one step
      If (revisiting a node)
         Return false
   End if
End while
Return true
```

Fig. 6.18 D-Algorithm

describing the procedure. These numbers are also shown on the arrows in Fig. 6.19 to indicate steps in the algorithm that cause the corresponding justifications and propagations.

Step 0:   Step 0 starts with all Xs in the test cube. We use lower case Xs for initial values in the table that there are by default, and bold uppercase Xs for those that are put in the test cube by exercising steps in the algorithm.

Step 1: l_8: SA0 is activated by a 1 on l_8, causing l_8 to become D. This value is justified by $l_4 = 1$ and $l_5 = X$. These values are justified by d input taking value 1, and e input becoming X. We mark this step as one that a choice exits, since l_8 of 1 can also be justified by $l_4 = X$ and $l_5 = 1$. The value D on l_8 causes FO1 for which a D input is known to be placed in the D-frontier.

Step 2:   FO1 is removed from the D-frontier and its D input is propagated to its outputs. This causes l_9 and l_{10} to become D. Note here that we are treating the fanouts structure as a primitive. Because of the simple relation between the input and outputs of this structure, we did not discuss it when

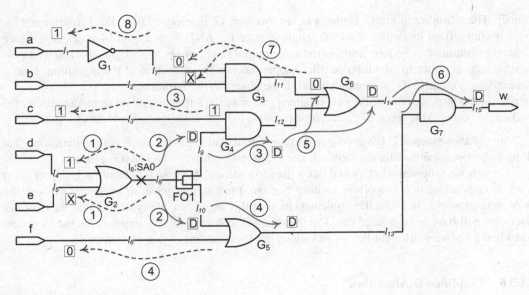

Fig. 6.19 D-Algorithm example

	J-frontier	D-frontier	a	b	c	d	e	f	l_1	l_2	l_3	l_4	l_5	l_6	l_7	l_8	l_9	l_{10}	l_{11}	l_{12}	l_{13}	l_{14}	l_{15}	w
0:			x	x	x	x	x	x	x	x	x	x	x	x	x	x	x	x	x	x	x	x	x	x
1:		FO1	x	x	x	1	X	x	x	x	x	1	X	x	x	D	x	x	x	x	x	x	x	x
2:		~~FO1~~ G4 G5	x	x	x	1	X	x	x	x	x	1	X	x	x	D	D	D	x	x	x	x	x	x
3:		~~G4~~ G5 G6	x	x	1	1	X	x	x	x	1	1	X	x	x	D	D	D	x	D	x	x	x	x
4:		~~G5~~ G6 G7	x	x	1	1	X	0	x	x	1	1	X	0	x	D	D	D	x	D	D	x	x	x
5:	G3	~~G6~~ G7	x	x	1	1	X	0	x	x	1	1	X	0	x	D	D	D	0	D	D	D	x	x
6:	G3	~~G7~~	x	x	1	1	X	0	x	x	1	1	X	0	x	D	D	D	0	D	D	D	D	D
7:	~~G3~~ G1		x	X	1	1	X	0	X	X	1	1	X	0	0	D	D	D	0	D	D	D	D	D
8:	~~G1~~		1	X	1	1	X	0	1	X	1	1	X	0	0	D	D	D	0	D	D	D	D	D

Fig. 6.20 Steps taken in D-Algorithm

discussing several other primitives. After l_9 and l_{10} receive their D values, G_4 and G_5 whose inputs have at least a D, and their outputs are X, will be placed in the D-frontier.

Step 3: Of the two gates in the D-frontier (G_4 and G_5), we select G_4 and use propagation D-cube for the AND gate to propagate D on its input to its output. For this purpose, we use a P_D for AND from Fig. 6.14. This propagation cube sets l_3 to 1 and l_{12} to D. The D on l_{12} puts G_6 that now has a D on one of its inputs in the D-frontier.

Step 4: Of the two gates in the D-frontier (G_5 and G_6), we select G_5 and find its appropriate propagation D-cube. As shown in Fig. 6.14, the P_D for this gate that we need is D0D, which sets l_6 to 0 and l_{13} to D. Since l_6 is directly driven by a primary input, setting f to 0 justifies l_6 value of 0. Since l_{13} received a D in this step, and is an input of G_7, this gate will be placed in the D-frontier.

Step 5: In this step, gate G_6 from the D-frontier is removed and based on P_D from Fig. 6.14 (OR gate, last row) l_{11} and l_{14} become 0 and D, respectively. With the latter assignment (l_{11} equal to 0), the output of G_3 is now 0 and is not justified by its inputs. This qualifies G_3 to go in our J-frontier.

Step 6: Since at this point, we still do not have a D or $\overline{D}$ on the circuit output, we continue with removing gates from the D-frontier and propagate D values. In step 6, G_7 is removed from the D-frontier, and using propagation D-cube, DDD, from Fig. 6.15, its input values are propagated to its output. This propagation causes l_{15} to receive value D. Line l_{15} drives w and this step puts a D on w.

Note that a D has propagated to the primary circuit output, propagation is complete, and justification phase begins.

Step 7:   The J-frontier in Step 7 contains G_3. In this step, G_3 is removed from the J-frontier and for justifying a 0 on its output, the 0X0 primitive cube for AND from Fig. 6.14 is selected. Since another primitive cube also justifies the same output requirement, this step is marked as a possible step to return to, should a conflict happen later in the execution of the algorithm. Line l_7 receiving a 0 causes G_1 whose output is l_7 to be placed in the J-frontier.

Step 8:   In this step, we remove the last remaining gate from the J-frontier and set its input to justify the 0 on the output. This sets l_1 to 1, and consequently a to 1, as shown in the last row of Fig. 6.20.

With the above step, a D has propagated to the output, and the J-frontier is empty, indicating that all propagation requirements are satisfied. For l_8:SA0 test, $abcdef$ = 1X11X0 is made.

Although we followed what looked like a pseudo code and discussed in detail each step of a test generation process, it is important to note that our presentation gave an overview of how the D-Algorithm works. In a specific implementation of this algorithm, many details that we have not discussed will have to be figured out. The use of specific data structures imposes the use of procedures that gives very different flavors to various implementations of this algorithm.

6.1.3.6   Simplified D-Algorithm

A simplification of D-Algorithm that requires a much simpler data structure and perhaps less programming effort is discussed in this section. The price paid for this simplification is the inability to consider multiple sensitized paths.

In the simplified D-Algorithm, alternative paths from site of fault to a primary output are identified, and one is selected in random for it to be sensitized. Gate input values along the path for sensitization of the selected path are determined, and the values are justified one at a time. If a justification fails, or input values justifying a required gate input value cause blocking of the selected path, then the path is rejected and another path is selected. This process continues until a sensitized path from site of fault to a primary output is justified by circuit primary inputs.

In Fig. 6.21, we are to generate a test for l_8:SA1. To activate this fault, l_8 must become 0, requiring d and e inputs to become 0. Starting from l_8, we have two paths to w. The lower path passes through G_5 and G_7, requiring l_6 and l_{14} to be 0 and 1, respectively. Justifying them is done similar to the D-Algorithm. The resulting test for l_8:SA1 is 01X000 for a, b, c, d, e, f inputs. Following the same procedure, test for l_8:SA0 becomes 01X1X0.

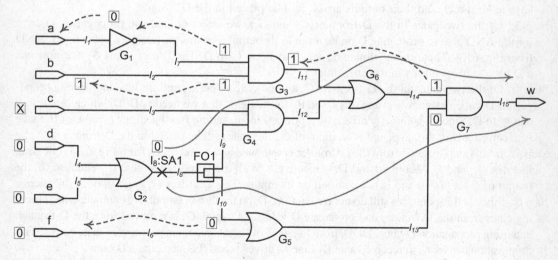

Fig. 6.21 Simplified D-Algorithm example

6.1.4   PODEM (Path-oriented Test Generation)

For finding a test vector, the D-Algorithm searches the internal lines of a circuit and sets proper values for fault propagation and justification. For test generation, however, we are looking for a primary input combination, and not a combination of values of the internal lines. In most logic circuits, there are far fewer primary inputs than internal lines, and limiting our search to the primary input combinations should find a test vector much faster than using circuit line values as our search space.

In addition to a larger search space, assigning values to circuit lines causes a problem with circuits that have XOR gates. A D or $\overline{D}$ on an XOR input propagates to the output (as D or $\overline{D}$), regardless of the value on its other input. The D-Algorithm does not keep this option open and sets a value on the XOR input that may have to be reversed. An example circuit illustrating the above-mentioned problems with D-Algorithm is shown in Fig. 6.22.

The discussion that follows refers to values in gray square boxes in Fig. 6.22. Test for l_8:SA0 in Fig. 6.22 begins with $l_1 = l_4 = 1$ for generating a D on l_8. This combination sets $a = b = 1$ and through implication sets l_5 to 1 also. With a D on l_8, two propagation D-cubes, D1$\overline{D}$ and D0D, are possible for the G_3 XOR. We choose the latter that sets $l_9 = 0$ and $l_{10} = D$. Then for G_4, D0D is selected that puts a D on w that completes the propagation phase. In the justification phase, a 0 for l_9 and a 0 for l_7 must be justified. For justifying a 0 on l_9, and given that l_5 is already set to 1, we have to use the primitive cube 110 that sets l_6 to 1. l_6 value of 1 is justified by l_3, and thus c being set to 1. Implication of this value sets l_7 to 1 which conflicts with the earlier requirement of this line being set to 0 for the OR gate propagation D-cube.

To resolve this issue, the D-Algorithm returns to the last place, a choice existed, which is selection of a propagation D-cube for G_3. This time, we select D1 that sets l_9 to 1. This value will be justified by $c = 0$, and $\overline{D}$ propagates to w.

In the above example, if we had limited ourselves to selecting input values for propagations and justifications related to l_8:SA0, we would start with $a = b = 1$ that activates the fault (see white squares). Then, we only had input c to set a value, for which we only had a choice of 1 or 0. Had we selected $c = 1$, we would have got a 1 on w that does not detect the fault. We would then reverse this decision and select $c = 0$, which would propagate D on l_8 as a $\overline{D}$ on w. Thus, test $abc = 110$ is found that detects l_8:SA0.

6.1.4.1   Basic PODEM

Consider a circuit with a fault to be detected that some of the circuit primary inputs have been set to activate the fault. PODEM [3] begins with setting the remaining inputs to X and propagating the Xs to all circuit lines and primary outputs.

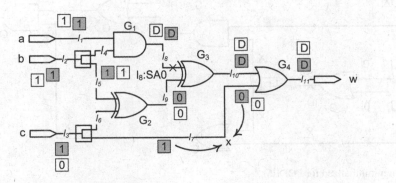

Fig. 6.22 Trying line combinations vs. input combinations

With the above initial setting, we should have:

(A) An activated fault (a D in the circuit) and
(B) At least a path of all Xs from the site of fault (the location of D) to a primary output.

PODEM selects an input randomly, sets a 0 or 1 to it, and propagates the value into the circuit, after which conditions A and B are checked. Three things can happen:

- *Both conditions still hold.* In this case, the input and its value are recorded, and another input is selected.
- *One or both conditions is (are) violated.* Then the input value is reversed and a different value is tried. In this case, if no more values remain to be tried (both 0 and 1 have been tried), a different input of the circuit will be selected.
- *A D or $\overline{D}$ propagates to a primary output.* In this case, the selected inputs and their assigned values form an input combination that detects the fault.

Figure 6.23 shows pseudo code for the basic PODEM as described above. Figure 6.24 shows an activated fault and a circuit initialized for application of PODEM. We will show steps necessary to reach a test for $l_8:SA1$ in this circuit.

The diagram of Fig. 6.25 shows input selection decision three and value assignments as PODEM steps are taken for generating a test vector for $l_8:SA1$ in Fig. 6.24. Bold numbers on the left of the diagram are the step numbers that are referenced below.

Fig. 6.23 Basic PODEM
pseudo code

```
BasicPODEM();
While PI Branch-and-bound value possible
   Get a new PI value;
   Imply new PI value;
   If error at a PO
      SUCCESS; Exit;
   Else
      IF error exists and X-path exists
         Keep PI value and effects;
      Else
         Reverse PI assignment;
      End if;
   End if;
End while;
```

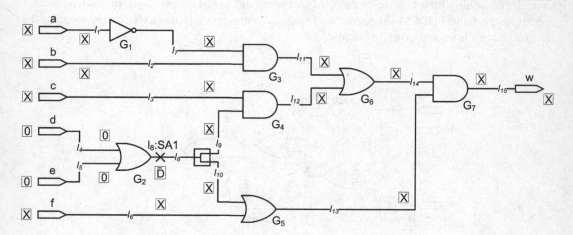

Fig. 6.24 Circuit initialized for PODEM

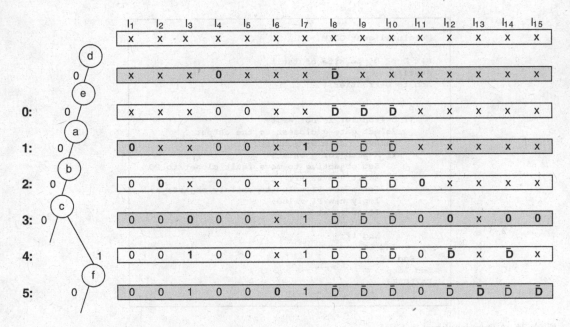

Fig. 6.25 PODEM decisions and steps for fault in Fig. 6.24

Step 0: Initially, inputs d and e are set to 0 to activate l_8:*SA1*. These values generate $\overline{D}$ on l_8, and several X-paths from l_8 to the circuit primary output exist. Propagation of $\overline{D}$ causes l_9 and l_{10} to become $\overline{D}$ as well.

Step 1: In the next step, after activation of fault, input a is selected and set to 0. This assignment causes lines l_1 and l_7 to become 0 and 1, respectively, and has no effect on the activated fault and the existing X-paths. Therefore, this assignment is accepted.

Step 2: In the next step, we select $b = 0$ and propagate this value in the circuit. This assignment causes lines l_2 and l_{11} to become 0 and again has no effect on the activated fault and the existing X-paths.

Step 3: Having both conditions still satisfied, we accept input value assignments we have made thus far, and go on to the next input, c. Setting c to 0, causes l_3, l_{12}, l_{14}, and l_{15} to assume 0 values. This causes elimination of X-paths from either l_9 or l_{10} to w, which means that we cannot accept $c = 0$.

Step 4: This step reverses assignment made in Step 3, and assigns $c = 1$. This changes l_3 to 1, and propagates $\overline{D}$ to l_{12} and l_{14} which is one step closer to w. We still have X-paths, and we still have our fault active. Therefore, this assignment is accepted.

Step 5: The next step shown in Fig. 6.25 sets input f to 0. This assignment changes l_{13} and l_{15} from X to $\overline{D}$. Since l_{15} is the output w and we have a $\overline{D}$ on w, the selected combination of inputs is a test for l_8:*SA1*.

6.1.4.2 A Smarter PODEM

The previous discussion showed a complete example of PODEM in which input sections were done on an arbitrary basis. A more efficient way of test generation would be to select inputs that are more likely to achieve propagation of D or $\overline{D}$ toward the output. In addition, the algorithm of Fig. 6.23 made no decision for which X-path to use, and allowed input values to propagate D or $\overline{D}$ to the primary outputs in any way possible. A smarter alternative would be to select an X-path that has a smaller chance of being blocked by the implication process.

```
MoreInteligentPODEM();

Set D or D' to site of fault;
Justify;
Add to D-Frontier;

Loop
    If D-Frontier is not empty
        Select gate g closest to the output;
        Remove g from D-frontier;
        If X-path for g exists
            Set objective to move fault closer to PO;
            Backtrace to PI;
            Get a new PI value to satisfy objective;
            Imply new PI value;
            If error at a PO
                SUCCESS; Exit;
            End if;
        End if;
    End if;
End Loop;
FAILURE;
```

Fig. 6.26 A more intelligent PODEM

```
Backtrace to PI; -- Find the best path to PI
Start with objective line value;
While line is not a PI
    If all gate inputs must be set (i.e., non-controlling values)
        Take the hardest to satisfy;
    Else if only one gate input is to be set (i.e., controlling)
        Take the easiest to satisfy;
    End if;
End while;

Easiest: Closer to input, better controllability, etc
Hardest: Farther from inputs, lower controllability, etc

Easiest: If it is satisfied, objective is met;
Hardest: If the objective is not to be satisfied we know sooner;
```

Fig. 6.27 Backtrace procedure

Figure 6.26 shows a pseudo code for a version of PODEM that makes more intelligent choices for selection of inputs and X-paths. Issues that distinguish this algorithm from that of Fig. 6.23 are:

- The algorithm selects an X-path based on distance to output, or observability (SCOAP).
- An *objective* is set to move fault closer to output [1]. An objective is setting a line value to 0 or 1.
- *Backtraces* from objective to primary inputs [1].

The algorithm shown in this figure uses the D-frontier that has the same definition as that in the D-Algorithm. For propagation of D or $\overline{D}$ values, a gate from the D-frontier is selected. By the definition of the D-frontier, such a gate has an X output and is likely to be on an X-path. An objective is set that enables propagation through the selected X-path. This objective is used to *backtrace* to a primary input to find the inputs that are likely to satisfy the objective.

The backtrace procedure is shown in Fig. 6.27. An objective is set by setting a line to 0 or 1. The line is then recursively backtraced toward the circuit's inputs. In many PODEM implementations, SCOAP parameters are used for estimating line controllability and observability.

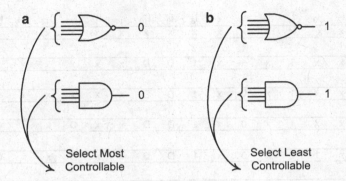

Fig. 6.28 Backtarce heuristics

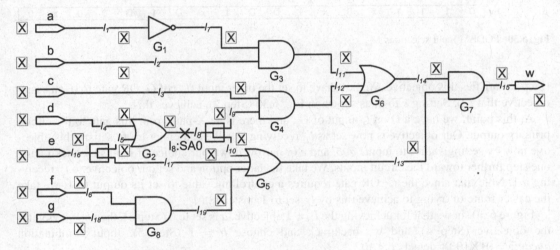

Fig. 6.29 PODEM example, making better choices

For a line value, the gate that drives it will be considered (see Fig. 6.28). If the required gate input value to make the output value happen is a control value, then the gate input that is easiest to control is selected.

If the required input value to set the output is a non-controlling value (1 for AND and 0 for OR), then the gate input that is least controllable is selected for further backtracing. The rational is that since in this case, all inputs must be set, we select the hardest to set, so that if a failure is going to happen, it happens sooner, so we select a different X-path.

We use circuit of Fig. 6.29 to show how procedures of Figs. 6.26 and 6.27 are used for a PODEM-based test generation. The fault we are trying to detect is l_8:SA0. In the following, we refer to Fig. 6.30 in which a decision diagram and corresponding line values of circuit of Fig. 6.29 are shown.

Fault l_8:SA0 requires a 1 on l_8 to be activated. This is achieved by setting $d = 1$, as shown in step 0 of Fig. 6.30. Going forward from l_8, from the site of fault, there are two X-paths: one through gates G_4, G_6, and G_7, and the other through gates G_5 and G_7. We use the latter X-path since it is shorter, and perhaps easier to reach circuit's primary output. Without having to go through elaborate computations, by observation we can tell that observability of l_{13} is better than that of l_{12}, which again supports our decision of taking the X-path through G_5.

The D value on l_8 reaches l_{10} which is the fanout branch. To make D reach l_{13}, the two inputs of G_5 must be set to 0. This is the case of Fig. 6.28b, which says that the hardest goal to achieve must be set as the objective. Thus, a 0 on l_{19} is set as our objective. As shown in Fig. 6.30, setting f to 0

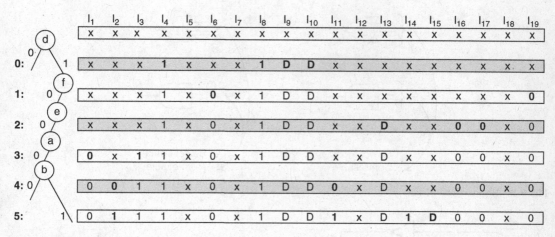

Fig. 6.30 PODEM input selections

(Step 1) satisfies this objective. We now have to set the other input (l_{17}) of G_5 OR gate to 0 (an easier objective that l_{19}). Setting e to 0, as shown in Fig. 6.30 (Step 2), achieves this.

At this point, we have a D on l_{13} input of G_7, and there is an X-path from this gate to the circuit primary output. Our objective is now set as l_{14} receiving a 1. We can either try to satisfy this objective now by setting values to inputs a, b, and c or narrow down our choice of inputs by backtracing one step further toward the circuit inputs. We take the latter option and set our objective to l_{11} receiving a 1. Note that since the G_6 OR gate requires a controlling value to set its output to 1, we take the easier route of trying to achieve this by l_{11} set to 1, instead of l_{12}.

Figure 6.30 shows that for achieving the $l_{11} = 1$ objective, a is set to 0 (Step 3), then $b = 0$ violates the objective (Step 4), and we backtrack and choose $b = 1$ (Step 5). Input combination $abcdefg = 01X100X$ detects l_8:SA0.

6.1.5 Other Deterministic Fault-oriented TG Methods

For about two decades, the D-Algorithm was regarded as the main algorithm for test generation, and most other methods were variations of this algorithm, or just different implementations. In the early 1980s, PODEM started a new line in test generation. Although PODEM uses many features and definitions of the D-Algorithm, it is conceptually different, because of assignment of values to the circuit primary inputs and the primary input search tree. Most of today's gate level test generation methods are based on D-Algorithm, PODEM, or a combination of both.

6.1.5.1 Fan

FAN (fanout-oriented test generation algorithm) [4–6] is based on PODEM. This algorithm optimizes the search for an input combination by adding several heuristics to PODEM.

In FAN, backtracing stops at the cone output of a fanout-free section of a CUT. This is justified, because inputs of such a cone do not affect other parts of the circuit and cannot, in any way, cause violation of an objective through other paths. This feature is similar to the way we handled our backtracing in the example of Fig. 6.29. After step 2, we decided to set our objective to $l_{11} = 1$. l_{11} is the output of a fanout-free section, whose inputs are a and b. Steps 3 and 4 in our example only set values to inputs of this section of the circuit.

Another feature of FAN is the use of multiple backtraces. For a gate that requires non-control values on its inputs, all such inputs will be backtraced simultaneously. Unique sensitized path is another optimizing feature in FAN. When the D-frontier contains only one gate, FAN finds a sensitized path from this gate to the circuit primary output. This eliminates unnecessary assignment of input values that would have to be reversed when they are found to block the only way that is to a circuit output.

6.1.5.2 Socrates

As FAN took off from PODEM and added optimizations to it, SOCRATES [7–9] started with FAN and added improvements to features of FAN that we discussed. Perhaps the most important feature of SOCRATES is its support for high-level RT level components like adders and multiplexers. Using RT level components as primitives eases backtracing and propagation of D values by taking bigger steps in the circuit.

6.1.6 Fault-independent Test Generation

Fault-oriented test generation algorithms, such as the D-Algorithm, PODEM, and all their variations, start at the site of fault and work forward and backward in the circuit to find a test. Fault-independent test generation programs, on the other hand, decide on a test vector and then decide what faults it detects.

A deterministic fault-independent test generation method is critical path tracing (CPT) [10–13]. This method is based on critical paths discussed in Sect. 4.3.5 (Chap. 4) for fault simulation. The CPT method starts at a primary output and finds critical paths that are justified by the inputs. The input vector forming the critical path is the test vector for all faults along the path. Figure 6.31 shows a simplified pseudo code for CPT test generation.

Starting with the output of a circuit, CPT sets a value on a line that is in the critical path, and based on this value finds critical paths leading to circuit primary inputs. Each critical line can take either a 0 or a 1, and selection of either can make different critical paths. When a choice of values for a line exists, a value is selected, and the line is marked. When a critical path reaches circuit primary inputs, faults along the path are marked as detected by vector at input. Then, the algorithm returns to the last place when a choice of 0 or 1 has not been tried.

Since by definition of critical paths, an output is on a critical path, CPT test generation starts with a 0 or 1 on an output. The recursive process mentioned above is executed until it reaches back to the output. At this time, the output is assigned the other logic value (1 or 0) and the process continues.

We use circuit of Fig. 6.32 to execute one pass of the above recursive procedure. In this pass, by selecting appropriate line values, we can find critical paths that cover more of the circuit lines; thus,

```
CPTG();

Select a PO, set it to 0;

    Recursively find critical paths, leading to inputs;
    Input values for CP constitute a test;
    Faults along the CP are detected;

Repeat for PO value of 1;
```

Fig. 6.31 Critical path tracing TG

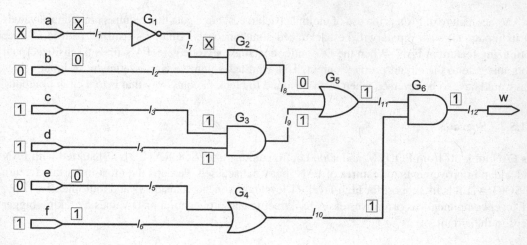

Fig. 6.32 CPT example

the generated test covers more number of circuit faults. Line values and critical paths that are generated are shown in this figure. The following paragraphs explain how values are assigned.

We start by assigning a 1 to circuit output. This makes l_{12} our first critical line. A 1 on l_{11} and a 1 on l_{10} justify $l_{12} = 1$, and make both lines l_{11} and l_{10} critical. Taking the upper path (l_{11} backward), l_8 and l_9 justify $l_{11} = 1$, and make l_9 a critical line. The 1 on l_9 is justified by $l_3 = l_4 = 1$ making these lines also critical, as marked in Fig. 6.32. This assignment makes inputs c and d equal to 1.

Backing up to G_5, for justifying a 1 on l_{11}, l_8 must be 0, that is justified by $b = 0$ and $a = $ X. This assignment does not create any new critical lines and just provides proper conditions for l_9 being critical.

Now we return to G_6, where we previously took the upper path through l_{11}. The lower input of G_6, line l_{10}, is also critical. This critical value is justified by $e = 0$ and $f = 1$, making l_6 also critical.

The above pass sets inputs a, b, c, d, e, and f to X, 0, 1, 1, 0, and 1 respectively. This input combination detects the following faults:

$$l_3\text{:}SA0,\ l_4\text{:}SA0,\ l_9\text{:}SA0,\ l_{11}\text{:}SA0,\ l_6\text{:}SA0,\ l_{10}\text{:}SA0,\ \text{and}\ l_{12}\text{:}SA0.$$

For detection of other faults, we return to places when choices existed for critical paths. The most recent place is the assignment of $l_5 = 1$ and $l_6 = 0$, instead of the earlier 0 and 1 values. This choice detects a new fault, l_5:SA0. When this is taken care of, we return to G_5, where a similar situation exists. When all choices have been tried, we go back to the w output and try $w = 0$.

CPT test generation is simple, fast, and requires a very simple data structure. The problem with this method is inability to handle fanout stems properly. As discussed in Sect. 4.3.5, determining if a fanout stem is critical requires simulation. This forward simulation pass requires a different data structure than that used for output to input path tracing and is a time-consuming process. CPT TG is most useful when dealing with fanout-free circuits.

6.2 Sequential Circuit Test Generation

Other than random techniques, all other algorithms we discussed in this chapter apply to combinational circuits. In spite of all the research done on this subject, deterministic fault-oriented test generation for sequential circuits still is in the research stages, and there are no good practical solutions. Practically, sequential circuits are turned into combinational circuits with design for test techniques, and combinational test generation methods are applied to them. These techniques will be discussed in Chap. 7.

This section discusses a sequential circuit test generation technique that expands the sequential circuit under test into several time frames and applies the D-Algorithm to this circuit. This technique is typical of most sequential circuit TG techniques. Such methods have a limited use in practice.

As shown in Fig. 6.33, the time frame expansion sequential circuit TG method [14] unrolls a sequential circuit by removing its flip-flops, and repeating its combination part several times. The flip-flop outputs become the state inputs of the combinational part and the flip-flop inputs are the combinational part state outputs. When several of the combinational parts are repeated, the state outputs of one become the state inputs of the next.

For test generation, a given fault is injected in one of the combinational slices, and propagation into the next slices and justification through the previous slices are done just like the D-Algorithm. Propagation continues until a D or a $\overline{D}$ appears at a primary output, and justification continues until state inputs of the flip-flops are all Xs. We illustrate this method by use of a simple example.

The example circuit is a Mealy machine that detects a sequence of 1011 on its a input. The circuit implementation of this example is shown in Fig. 6.34. The SA0 fault to be detected is shown in this figure.

The time frame expansion of this circuit is shown in Fig. 6.35. Initially, three time frames are constructed, that are numbered -1, 0, and +1; more frames will be added if required by propagation (+) or justification (−). All line values except the site of fault are initially set to X.

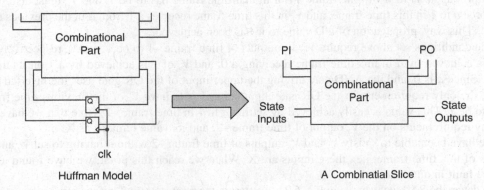

Fig. 6.33 Time frame expansion (each slice)

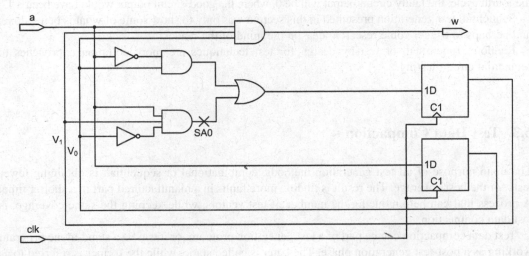

Fig. 6.34 Mealy 1011 detector

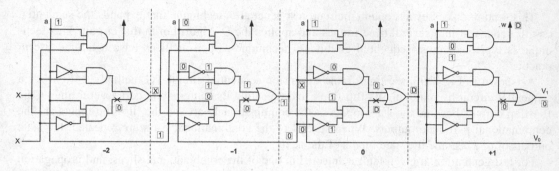

Fig. 6.35 Time frame expansion

Site of fault is set to its faulty value in all time frames. Fault is activated in circuit of time frame 0, and it is propagated forward to time frames ≥ 0 until it reaches to a primary output. Then, line values are justified in circuits of time frames ≤ 0, until we reach a time frame that justifies its state outputs without requiring to set its state inputs, i.e., leave them at X. At this time, test is found.

The SA0 in time frame 0 is activated by requiring a to be 1, and V_1 and V_0 to be 1 and 0, respectively. This activation puts a value D on the site of fault. Value D is propagated to V_1 of time frame 0 by setting the upper input of the OR gate to 0. This is easily achieved by the value of a being 1.

To propagate D to w in time frame +1, a in that time frame has to be 1, and V_0 must also be 1. We force a to 1 in this time frame, and V_0 in this time frame receives a 1 from a in the previous time frame. This way, propagation of a D value to w has been achieved.

The conditions set above require V_1, V_0 outputs of time frame -1 to be 1, and 0, respectively. V_0 of 0 is achieved by a in this time frame receiving a 0, and V_1 of 1 is achieved by a 1 on its upper input. Since a is 0, and the AND gate driving the upper input of the OR gate uses the inverted a as input, the only requirement for the OR gate input being 1 is to have V_0 of the previous time frame (-2) to be 1. This value is easily achieved by setting a to 1 in time frame -2. Note that we have not set any requirements on the V_1 output of time frame -2, and its value remains as X.

We have been able to satisfy V_1 and V_0 outputs of time frame -2 without having to set V_1 and V_0 inputs of this time frame, i.e., these inputs are X. When we reach this point, we have found a test for the fault in question.

To detect the SA0 fault shown in Fig. 6.34, we start in the reset state, and in four consecutive clocks, we apply 1011 to the a input (time frames -2, -1, 0, and +1 take values 1, 0, 1, and 1, respectively). In the fourth cycle, the faulty circuit output will be 0, where the good circuit output would have been a 1.

Sequential test generation presented in this section was only to show some of what is being done, and perhaps to trigger some research ideas in the mind of the reader.

Random, functional, or test by design for test techniques are more common approaches to sequential circuit testing.

6.3 Test Data Compaction

The main purpose of all test generation methods, combinational or sequential, is obtaining fewer tests for the best coverage. The result is finding more faults in a manufactured part in a shorter time. A process that can help reducing the number of test vectors, while keeping the same coverage, is test data compaction.

Test data compaction can be part of a test generation program, or it can be a stand-alone program working as a post-test generation phase. The latter is called static, while the former is referred to as dynamic test compaction [14, 15]. Test compaction can be applied to sequential or combinational circuit test vectors [14, 15, 16, 17, 18, 19]. Finally, test compaction can be for testing a single CUT or for concurrent testing of multiple CUTs. This section, discusses various applications and algorithms for test compaction.

6.3.1   Forms of Test Compaction

Dynamic test compaction becomes part of a test generation program [14, 15, 20]. In a deterministic test generation program, as tests are generated, a separate phase of the test generation compares test vectors and removes some test vectors or merges them. This requires changes made to the test generation program and often requires performing fault simulation for evaluation of merge or removal results [18]. Similarly, in random test generation, compaction becomes part of the test generation program.

In a static test compaction, mergence or removal of test vectors happens after all test vectors have been generated. As such, static compaction methods do not alter ATPG programs and are often less expensive to develop and apply.

Test compaction can be applied to combinational or sequential circuits. Combinational test compaction methods are easier to implement than those for sequential circuits. Furthermore, most sequential compaction methods require intensive fault simulations, which make them less desirable.

For a single CUT, test compaction looks for adjacent equal or compatible test vectors in the same test set. Merging or removing test vectors happens on such test vectors. Test compaction can also be applied between several test sets that are generated for different CUTs. Test compaction across two or more test sets looks for equal or compatible test vectors across different test sets. Merging and/or removing test vectors happen on different sets.

6.3.2   Test Compatibility

Compatibility is defined for two test vectors or two test sets. If two test vectors or test sets are compatible, one can be used in place of the other without loss of fault coverage.

6.3.2.1   Test Vector Compatibility

Test vectors consist of 1s, 0s, and Xs. In two test vectors, like values in the same bit positions are compatible. In addition, a 1 or a 0 in the same bit position as an X are also compatible. Compatible bits can be merged. Figure 6.36 defines the merge operator; dashes show incompatibility. Two test vectors, all bits of which are compatible, are called compatible test vectors and can be merged. Figure 6.37 shows several examples of compatible and incompatible test vectors.

A test set whose adjacent elements (test vectors) are compatible can be reduced in terms of its number of test vectors. This reduction compacts test sets generated for a CUT into a smaller test set. Figure. 6.38 shows a test set being compacted. In this figure, adjacent compatible test vectors are merged using the merge operator. In the first step, vectors 2 and 3 are merged into one vector. In Step 2, the merged 2 and 3 vector is further merged with vector 4.

$$
\begin{array}{c|ccc}
 & 0 & 1 & X \\
\hline
0 & 0 & - & 0 \\
1 & - & 1 & 1 \\
X & 0 & 1 & X
\end{array}
$$

Fig. 6.36 Merge operator

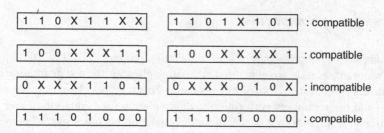

Fig. 6.37 Compatible test vectors

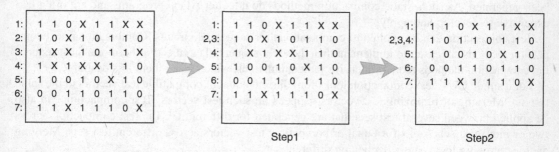

Fig. 6.38 Compaction by adjacent compatibility

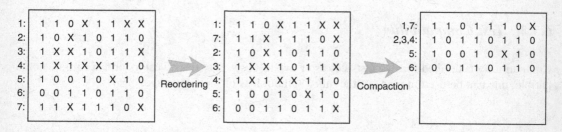

Fig. 6.39 Reordering before compaction

6.3.2.2   Test Vector Reordering

Combinational circuit tests can be reordered without any loss of coverage. Whereas in a sequential circuit, because of the internal states of the circuit, reordering test vectors changes the fault coverage. Where reordering does not affect fault coverage, i.e., tests for combinational circuits, further compaction can be achieved by placing compatible tests adjacent to each other. Figure 6.39 shows the effect of reordering before compaction.

Since test vectors 1 and 7 are compatible, they have been placed next to each other. The new adjacent vectors merge into one vector. Compaction after reordering results in one less test vector in the final test set (compare Fig. 6.38 and 6.39).

6.3.2.3 Test Set Compatibility

Test sets that are generated for different CUTs can also be merged [21]. This results in a compaction that reduces test time when the CUTs are concurrently tested. Figure 6.40 shows a scenario in which two CUTs are concurrently tested. If the test set for one CUT is a subset of another, test time for testing both is the same as the larger of the test times when testing the CUTs individually.

Test set compatibility can also be applied to different subsets of the same test set, for reducing the overall number of test vectors. In this case, merging of test vectors occurs in groups of multiple test vectors. Figure 6.41 shows a test set that has two subsets that can be considered for merging.

There are several ways in which two test sets (or two subsets of the same test set) can be merged. Below describes several possibilities of compatibility and thus merging.

Contiguous compatibility. Test sets *Test1* and *Test2* can be merged if starting in a certain position in one test set (*Test1*), all test vectors in the other test set (*Test2*) are compatible with the first test set. As shown in two examples in Fig. 6.42, after merging, the new test set includes all test vectors of both test sets with their original ordering. The two headed arrows in Fig. 6.42 show compatible vectors.

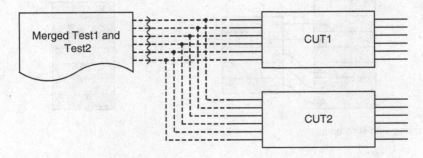

Fig. 6.40 Concurrent testing

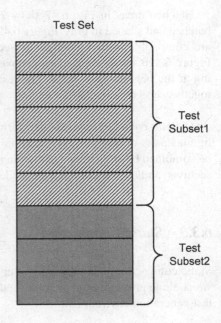

Fig. 6.41 Test subsets for
possible merging

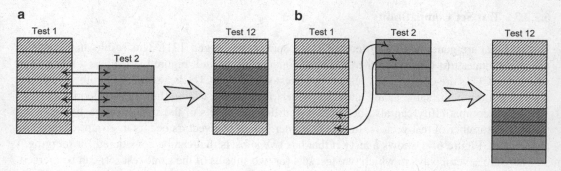

Fig. 6.42 Examples of test set contiguous compacting

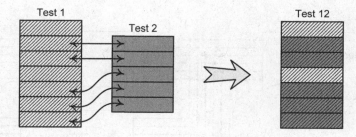

Fig. 6.43 Merging with stretching

Hatched areas in Fig. 6.42 show *Test1*, and gray areas show *Test2*. Merged vectors are both hatched and shaded in gray. Figure 6.42a shows contiguous compatibility where all vectors of *Test2* are compatible with a contiguous section of *Test1*. The resulting test set is as large as *Test1*. Figure 6.42b shows contiguous compatibility of the section at the end of *Test1* with a section starting at the beginning of *Test2*. The resulting merged test set is the size of the two test sets added together, less the compatible section.

Disjoint compatibility. Another form of compatibility is compatibility of vectors in disjoint contiguous parts of test sets. As shown in Fig. 6.43, test vectors with disjoint compatible sections can be combined by stretching and merging compatible sections. As shown, *Test2* is partitioned into two sections, and each section is independently merged with its compatible section in *Test1*.

6.3.3   Static Compaction

Static compaction methods apply after test generation. Because these methods are implemented as stand-alone programs and are independent of the ATPG programs, they are easily incorporated in a test generation environment.

6.3.3.1 Static Combinational Compaction

Several techniques are available for combinational circuit test compaction. An actual implementation of a test compaction program for combinational circuits may consist of a combination and/or modification of several techniques mentioned below.

Redundant vector elimination. The process of deterministic test generation can result in redundant test vectors detecting different faults. Reducing a test set by eliminating redundant test vectors does change the fault coverage. Implementation of this method does not require use of a fault simulation program [19].

Test vector merging. Compatible vectors in a test set can be merged without loss of fault coverage. Rules discussed in Sect. 6.3.2.1 and merge operation of Fig. 6.36 apply to this method of compaction. Furthermore, reordering test vectors discussed in Sect. 6.3.2.2 can create more adjacent compatible vectors that can be merged.

Test subset merging. A test set for a combinational circuit can be subdivided into several sections, and test set compatibility rules of Sect. 6.3.2.3 can be applied among these sections. This static method can be implemented as a stand-alone program in software programming languages. As in the previous method, fault coverage is not affected by test subset merging, and no fault simulation is required in this method.

Multi-CUT test set merging. Test subset merging described above can be applied among independently generated test sets for several CUTs. The procedure for merging is the same as merging subsets, but the way a merged test set is applied to multiple CUTs is different than having a single CUT. The exact details of applying a merged test set to multiple CUTs depend on many factors including the DFT method and input/output access mechanism for CUTs. If test set for a CUT is a subset of another, outputs corresponding to test vectors that are not related to the subset CUT must be ignored.

Test vector modification. The process of test generation (random or deterministic) creates tests that have a relatively small number of conflicts (according to merge operation of Fig. 6.36) in the same bit positions. Such vectors are referred to as close-compatible. Close-compatible test vectors may be merged by changing conflicting bits ⌈[19]. A method of test compaction is to find close-compatible test vectors, flip conflicting bits, and examine them for possible merging. The examination is done by performing fault simulation to check if the merged test vector still detects faults detected by all close-compatible vectors that have been merged. If so, merging is done; otherwise, test vectors are returned to their original form. This method of static combinational compaction requires the use of fault simulation, which can add a significant overhead to the test generation process.

Reverse order fault simulation. Whether we are doing random test generation, using deterministic methods, or a combination of both, it is generally true that easier faults are detected first, and the harder faults are left for test vectors that appear last in the list of test vectors [14]. Test vectors that detect harder faults are also likely to detect easier faults. However, since in our test generation processes, we perform fault dropping, tests vectors generated for harder-to-detect faults, which come at the end of a test set, do not get the opportunity to show how many faults they detect.

Reverse order fault simulation [14, 19] is a way of addressing this problem. By performing fault simulation on test vectors from the bottom of a test set, these vectors get a chance to show what faults of the original fault list they detect. By performing fault dropping, as we get near the beginning of a test set, we may be able to eliminate test vectors that are no longer efficiently detecting undetected faults.

A more elaborate method of selecting good test vectors is to examine all test vectors for faults that they detect, and select only those that detect faults not detected by other tests. Realizing what faults from an original fault list a test vector detects requires fault simulation for each test vector without faults dropping. This requires a significant processing time, and this elaborate test vector selection is rarely used.

6.3.3.2 Static Sequential Compaction

Test compaction in sequential circuits is more complex than in combinational circuits because of state dependency of a sequential CUT. Test reordering, merging, or elimination must always be accompanied with fault simulation, since the test vector independent of a CUT's state cannot determine the resulting coverage after compaction. In spite of this complexity, it is worth mentioning several methods that help reducing sequential circuit test vectors.

Sequence reordering. A simple method of test compaction for sequential circuits is to reorder test vectors [17]. This may help reducing the number of test vectors since test vectors at the end of a test set tend to detect harder faults. With first applying tests that detect harder faults, tests near the beginning of a test set may no longer be needed. Test reordering must always be accompanied with fault simulation.

Note that the reason test reordering can help compaction is similar to the way reverse fault simulation can help combinational test compaction. However, the former is always less effective due to dependence of test vector ordering on the internal state of a CUT.

Vector omission. Vector omission is a procedure that results in omission of test vectors, the presence of which in the test set does not improve fault coverage. The starting point for this procedure is a test set in a given order, each applied to the CUT in a clock cycle [17].

The procedure of vector omission examines every test vector for omission. It potentially eliminates a test vector from the set and finds the fault coverage for the new test set with the original test vector ordering. Measuring fault coverage is done by fault simulation for every fault of the circuit. If the new fault coverage is better than when omission was not done, the omission materializes. Otherwise, the next candidate for omission is considered. This procedure involves many fault simulation passes and is computationally intensive, but has relatively good results.

The procedure shown in Fig. 6.44 reduces the number of test vectors applied to a faulty circuit by only applying tests up to the test vector that detects the given fault. This reduced number of test vectors speeds up the fault coverage analysis that is needed after a test vector is marked for potential elimination. For this purpose, test vectors are identified by sequential numbers representing their locations in the original test set. Then, a list of faults and test vectors that detect them is generated. Simulating a faulty circuit for checking if its fault is detected is only done with test vectors that appear in the test set before the test that detects the faults.

The first nested loops in Fig. 6.44 goes through all circuit faults and forms the *detectedBy* array. This array, which is indexed by circuit faults, contains the test vector sequence number that detects the fault. The second set of nested loops in this figure contains a loop that marks every test for elimination, one that injects every circuit fault, and one that examines the fault for detection.

As discussed, the two outer loops are straight forward, i.e., every test vector is examined for elimination from the test set, and every fault, is checked for detection by this reduced test set. The inner-most loop examines test vectors of the test set that has a potentially eliminated test for detection of the injected fault. This loop takes advantage of the *detectedBy* array and only examines test vectors that come before the vector that originally detected the injected fault.

If by examining only this subset of test vectors, the injected fault is still detected, it means that elimination of test vector in the outer loop has no effect on detecting this fault. If a fault was undetected by the original test set, examining whether it is covered by the test set with eliminated vectors applies to the entire test vectors, and not just a subset. Therefore, it is possible that the reduced test set with some vectors eliminated gets a better coverage than the original one.

As an example implementation for this procedure, we have developed a Verilog testbench for compacting test data obtained by random test generation for the Residue-5 circuit of Chap. 5. This testbench illustrates some details, like those associated with clock timing and loop conditions, that were not evident in the abstract pseudo code of Fig. 6.44.

Figure 6.45 shows the outline of the testbench for Residue-5. As before, a good circuit and a faultable model are instantiated in the *Tester()* module. The outputs *woG* and *woF* will be compared

```
For every fault i
|   Inject fault;
|   For every test j
|                    |
|                    |   Apply t_j, check for detection
|                    |   If t_j detects f_i, then
|                    |       detectedBy[i] = j;
|                    |   Else clock CUT, try next test
|                    |
|   End for
|   Remove fault
End for


For every test m
|   Mark m for elimination
|   For every fault i
|   |   Reset circuit;
|   |   Inject fault i;
|   |   For test j from 1 to detectedBy[i]
|   |   |   Exclude elimination candidate m;
|   |   |   Apply tj, check for detection
|   |   |   Calculate coverage;
|   |   End for
|   End for
|   Compare with old coverage;
|   Decide to keep or eliminate test m;
End for
```

Fig. 6.44 Vector elimination procedure

during fault simulations that follow. As shown near the end of code in Fig. 6.45, *clk* signal is applied to good and faulty circuits. Applying data to the circuits is synchronized with this clock.

Creating faulty models, evaluating test vectors, and the elimination of test vectors happen in the **initial** block that is shown in this code. In this block, circuit is reset, fault and test files are read into local arrays (*wireName* and *testArray*), and after execution of the procedure of Fig. 6.44, reduced test vectors are written, and reports are generated.

The dual nested loop in Fig. 6.45 implements the upper part of the procedure of Fig. 6.44, and the triple nested loop implements test vector elimination that appears in the lower part of Fig. 6.44. The details of these sections are shown in Figs. 6.46 and 6.47, respectively.

The main task of the nested loops in Fig. 6.46 is generating the *detectedBy* array. This array is indexed by fault numbers and contains the test number that detects the fault. For generating this, the outer loop iterates for every circuit fault, and after resetting the circuit, it injects fault *i* in the circuit. Prior to fault injection, *reset* is asserted and held high until **posedge** of *clk* is detected.

After fault injection, the inner loop iterates for test vectors in the test set until the injected fault is detected. In the loop, a test is applied, and outputs of the good and faulty circuits are compared. If they are different, fault is detected and test vector loop exits. If the outputs are the same, the injected fault is not detected by the current test. In which case, the circuit is clocked @(**posedge** *clk*), and the next test vector is tried.

When fault *i* is detected, *detectedBy[i]* is set to the test vector id, *j*. If fault is not detected and the end of test file is reached, the *detectedBy* entry for that fault receives a -1 value. At the end of this section of code, fault coverage for the original test set is also calculated.

Figure 6.47 shows the implementation of the lower part of the procedure in Fig. 6.44 for the Residue-5 circuit. The outer loop puts a 1 in the *omitted* vector in the location of the test vector that is a candidate for omission. The second loop injects every circuit fault from the *wireName* array

```verilog
module Tester();
   . . .
   residue5_net GUT(global_reset, clk, reset,in, woG);
   residue5_net FUT(global_reset, clk, reset,in, woF);
   . . .
   initial begin

       global_reset = 0;   //initialization

       faultFile = $fopen("res5.flt","r");
       testFile = $fopen("res5.tst","r");
       // read data into: wirename[] and testArray[];

       numDetectedFaults = 0;
       for (i = 0; i < numFaults; i = i + 1) begin
           . . .
           for (j = 0; ((j < numTests)&&(!faultDetected));j = j + 1) begin
               . . .
           end
       end

       coverage = $itor(numDetectedFaults)*100/$itor(numFaults);
       . . .
       for (m = 0; m < numTests; m = m + 1) begin
           . . .
           for (i = 0; i < numFaults; i = i + 1) begin
               . . .
               while ((j < numTests) && (n <= detectedBy[i]) . . .) begin
                   . . .
               end
           end
           . . .
       end

       resultFile = $fopen("selectedTests.tst","w");
       // write from testArray[];

       //display results
       $display("Old Coverage: %f", oldCoverage);
       $display("New Coverage: %f", coverage);

       $fclose(faultFile); $fclose(testFile); $fclose(resultFile);
   end

   always #100 clk = ~clk;

endmodule
```

Fig. 6.45 Residue-5 test compaction testbench

using the **$InjectFault** PLI function. As before, complete resetting is performed prior to fault injection. The **while** loop in Fig. 6.47 that we refer to as the *Test* loop checks to see if the injected fault is still detected without considering tests marked by "1" in the *omitted* array (note the **if** statement in this loop).

The condition of the **while** loop is the key point here. Note here that while all tests from 0 to *numTests* are considered, the condition "$n \leq detectedBy[i]$" stops the continuation of this **while** loop beyond the test vector that detected the injected fault in the original test set.

While faults are checked for detection, the *Test* loop in Fig. 6.47 calculates the fault coverage, while test vector *m* is eliminated. After exiting from the *Fault* loop, this new fault coverage is

```
module Tester();
    . . .
    initial begin
        . . .
        numDetectedFaults = 0;

        for (i = 0; i < numFaults; i = i + 1) begin
            //initialization
            faultDetected = 1'b0;

            reset = 1; @(posedge clk); reset = 0; //circuit reseting

            $InjectFault (wireName[i], stuckAtVal[i]);

            for (j = 0;((j < numTests)&&(!faultDetected));j = j + 1) begin

                indata = testArray[j];
                #10;

                if(woG != woF) begin
                    detectedBy[i] = j; // fault i is detected by test j
                    faultDetected = 1;
                    numDetectedFaults = numDetectedFaults + 1;
                end else @(posedge clk); // prepare for next test
            end

            if(!faultDetected) detectedBy[i] = -1; //undetected fault

            $RemoveFault(wireName[i]);
        end

    end
    . . .
endmodule
```

Fig. 6.46 Creating list of faults and tests that detect them

compared with the previously calculated fault coverage, and if it is not less than that, the test vector that had become a candidate for omission is omitted.

We started this example with 108 faults, 81 two-bit test vectors and 89.81% fault coverage. In a Verilog simulation environment, the testbench described in Fig. 6.45 was able to reduce the test set to only 36 test vectors, with the same coverage. Figure 6.48 shows the report generated by our testbench.

6.3.4 Dynamic Compaction

Dynamic compaction methods engage in the test generation program [22]. Because of the overhead on the ATPG programs, they are less often used than stand-alone static compaction methods.

A simple dynamic test compaction method for combinational circuits is redundant vector elimination [20]. In this method, as tests are generated, they are checked against existing ones and eliminated if they have already been added to the test set.

```
module Tester();
    . . .
    initial begin
        . . .
        for (m = 0; m < numTests; m = m + 1) begin

            omitted[m] = 1'b1; // candidate for omission

            numDetectedFaults = 0;
            for (i = 0; i < numFaults; i = i + 1) begin
                //initialization
                faultDetected = 1'b0;
                reset = 1 @(posedge clk); reset = 0; //circuit reseting

                $InjectFault (wireName[i], stuckAtVal[i]);

                j=0; n=0;
                while((j<numTests)&&(!faultDetected)&&((n <= detectedBy[i])
                   || (detectedBy[i] == -1))) begin
                    if(omitted[j] != 1'b1) begin
                        indata = testArray[j];
                        #10;
                        if(woG != woF) begin
                            faultDetected = 1'b1;
                            numDetectedFaults = numDetectedFaults + 1;
                        end else @(posedge clk);
                        n = n + 1;
                    end
                    j = j + 1;
                end
                $RemoveFault(wireName[i]);
            end
            //calculating coverage
            newCoverage = $itor(numDetectedFaults)*100/$itor(numFaults);
            if(newCoverage >= coverage) begin
                coverage = newCoverage;
                $display("New Coverage: %f", coverage);
            end else omitted[m] = 1'b0;
        end
    end
    . . .
endmodule
```

Fig. 6.47 Omission and reevaluating

For sequential circuit test generation, the dynamic compaction method [19] removes test vectors with low efficiency between similar states of the circuit. This method uses random test generation for generating test vectors for a sequential circuit. As a random test is generated, its coverage in the sequence is calculated and the circuit state that this test puts the circuit in is recorded. At a later time, when another random test puts the state of the machine in the same state, faults covered between the two times the same state was observed are counted. If the number of detected faults is lower than a threshold, all test vectors between the repeated states are ignored and the faults are returned to the non-detected fault list.

This method requires extensive use of fault simulation and a large database for retaining circuit states. Nevertheless, it is a technique that serves well as an example for dynamic sequential test compaction.

```
# The system time (Hour:Minute:Second:MilliSecond) is: 14:57:57:465
# number_of_faults:         108
# number_of_tests:          81
# Original Coverage: 89.814815
# New Coverage: 89.814815
# New Coverage: 89.814815
# number_of_faults:         108
# number_of_tests:          36
# Best Coverage: 89.814815
# Old Coverage: 89.814815
# end...
# The system time (Hour:Minute:Second:MilliSecond) is: 14:58:07:385
```

Fig. 6.48 Compaction results of Residue-5 circuit

6.4 Summary

This chapter presented deterministic test generation algorithms and test compaction. Chapter 5 focused on random test generation and developed complete programs for random test generation. Although the use of deterministic test generation programs were mentioned, no specific details were given. The first part of this chapter, on the other hand, mentioned the use of two-phase test generation programs and placed the deterministic test generation as the second phase of such a program.

Section 6.2 then focused on algorithms for deterministic test generation. We presented base algorithms, based on which many of the existing deterministic TPG programs are developed.

Section 6.3 presented test compaction. Perhaps this can be regarded as the third phase of an ATPG program. We discussed combinational and sequential methods. The most practical and most efficient test compaction programs are the static ones made for combinational circuits.

References

1. Roth JP (1966) Diagnosis of automata failures: a calculus and a method. IBM J Res Dev 10(4):278–291
2. Roth JP, Bouricius WG, Schneider PR (1967) Programmed algorithms to compute tests to detect and distinguish between failures in logic circuits. IEEE Trans Electron Comput EC-16(5):567–580
3. Goel P (1981) An implicit enumeration algorithm to generate tests for combinational logic circuits. IEEE Trans Comput C-30(3):215–222
4. Fujiwara H (1985) FAN: A Fanout-oriented test pattern generation algorithm. In: Proceedings of the international symposium on circuits and systems, pp 671–674, July 1985
5. Fujiwara H, Shimono T (1983) On the acceleration of test generation algorithms. In: Proceedings of the international fault-tolerant computing symposium, pp 98–105, June 1983
6. Fujiwara H, Shimono T (1983) On the acceleration of test generation algorithms. IEEE Trans Comput C-32(12):1137–1144
7. Schulz MH, Auth E (1988) Advanced automatic test pattern generation and redundancy identification techniques. In: Proceedings of the international fault-tolerant computing symposium, pp 30–35, June 1988
8. Schulz MH, Auth E (1989) Improved deterministic test pattern generation with applications to redundancy identification. IEEE Trans Comput-Aided Des 8(7):811–816
9. Schulz MH, Trischler E, Serfert TM (1988) SOCRATES: A highly efficient automatic test pattern generation system. IEEE Trans Comput-Aid Des CAD-7(1):126–137
10. Abramovici M, Breuer MA, Friedman AD (1994) Digital systems testing and testable design. IEEE Press, Piscataway, NJ. Revised printing
11. Abramovici M, Menon PR, Miller DT (1984) Critical path tracing: an alternative to fault simulation. IEEE Des Test Comput 1(1):83–93

12. Menon PR,Levendel YH, Abramovici M (1988) Critical path tracing in sequential circuits. In: Proceedings of the international conference on computer-aided design, pp 162–165, Nov. 1988
13. Menon PR,Levendel YH, Abramovici M (1991) SCRIPT: A critical path tracing algorithm for synchronous sequential circuits. IEEE Trans Comput-Aided Des 10(6):738–747
14. Bushnell ML, Agrawal VD (2000) Essentials of electronic testing for digital, memory, and mixed-signal VLSI circuits. Kluwer, Dordecht
15. Miczo A (2003) Digital logic testing and simulation, 2nd Ed. Wiley, New York
16. Jha N, Gupta S (2003) Testing of digital systems. Cambridge University Press, Cambridge
17. Pomeranz I, Reddy SM (1996) On static compaction of test sequences for synchronous sequential circuits. Proceedings of third design automation conference, pp 215–220
18. Nsiao MS, Rudnick EM, Patel JH (1997) Fast algorithms for static compaction of sequential circuit test vector. Proceedings of 15th IEEE VLSI test symposium, pp. 188–195
19. Hamzaoglu I, Patel JH (2000) Test set compaction algorithm for combinational circuits. IEEE Trans Comput Aided Des Integr Circuits Syst 19(8):957–963
20. Rudnick EM, Patl JH (1996) Simulation based techniques for dynamic test sequence compaction. IEEE/ACM international conference on computer aided design, pp 67–73
21. Niermann TM,Roy RK,Patel JH, Abraham JA (1992) Test compaction for sequential circuits. IEEE Trans Comput Aided Des Integr Circuits Syst 11(2):260–267
22. Goel P, Rosales BC (1979) Test generation and dynamic compaction of tests. Proceedings of test conference, pp 189–192

Chapter 7
Design for Test by Means of Scan

Most test generation schemes look at a CUT as a black box, the only available nodes of which for testers to control are its primary inputs, and to observe one are its primary outputs. This limited controllability and observability of circuits under test (CUT) means complex test generation algorithms for combinational circuits, and near-impossible test generation for the sequential circuits.

To overcome this difficulty in testing, digital circuits must become more testable by incorporation of design for test (DFT) techniques. For this purpose, designers must get involved in the test process by incorporating testability hardware in their designs and evaluating their designs for testability. DFT techniques offer ways of making internal structure of a design more controllable and easier to observe. Because such tasks are handled by designers, hardware description languages play an important role in facilitating insertion and evaluation of hardware structures that are put in a circuit for making it more testable.

This chapter discusses DFT techniques. We begin by covering some of the basics, followed by a section on options we have for hardware structures that make a circuit more testable. This section leads to some of the established scan insertion techniques. The section that follows this highlights the use of HDLs in modeling DFT techniques, evaluating them, and verifying their operation.

7.1 Making Circuits Testable

A circuit is testable if tests can be generated for it efficiently, it can be tested with a high fault coverage, and the time it takes to test the manufactured part is reasonable. Testability is a combination of controllability and observability. A circuit becomes more testable by making it more controllable and more observable [1, 2].

7.1.1 Tradeoffs

Improvements in controllability and observability are almost always done by inserting additional hardware in a design. This means that DFT techniques that make circuits more testable always introduce additional hardware that results in more pins, more delays, more power consumption, and obviously a hardware overhead.

What we are getting instead is a better coverage and reduced test time, and in some cases, DFT techniques gain access to the internal structures of a circuit that would otherwise be impossible to test.

Reducing the time it takes to test a manufactured part is the most important concern of DFT techniques. This goal is achieved by introducing extra test hardware in the CUT. Reducing this hardware, its power consumption during test, the delay it introduces in the design, and the extra pins that are needed to take test to the manufactured part are the parameters a design engineer must optimize.

Z. Navabi, *Digital System Test and Testable Design: Using HDL Models and Architectures*, 213
DOI 10.1007/978-1-4419-7548-5_7, © Springer Science+Business Media, LLC 2011

7.1.2 Testing Sequential Circuits

Perhaps one of the most important contributions of DFT is making sequential circuits testable. The previous chapter showed how complex it is to generate a complete test for a sequential circuit. It also showed that even if tests are generated, application of test requires many clock cycles to move a circuit into states that activate faults.

DFT techniques alter a sequential circuit model in such a way that combinational test techniques can be used for it. With this alteration, test generation schemes, random or deterministic, can be used to generate test for the altered circuit. DFT techniques define ways in which tests generated for a combinational model of a sequential circuit can be applied to the actual sequential circuit [3].

7.1.2.1 Sequential Circuit Huffman Model

Huffman model, as discussed in Chap. 2, is a useful model for sequential circuit testing. As shown in Fig. 7.1, a sequential circuit is modeled by a combinational circuit having feedback through delay elements. In synchronous sequential circuits, the delay elements are clocked flip-flops, and each feedback is a state variable of the circuit.

Circuit primary inputs and primary outputs only apply to the combinational part. Synchronous or asynchronous set, reset, and other control inputs only apply to the feedback flip-flops. This model completely separates a CUT's registers from its combinational part. The combinational part consists of all circuit buses, logic units, arithmetic functions, and discreet logic parts. The feedback registers consist of control flip-flops, data registers, and small register files, including register signals such as set, reset, enable, and parallel load.

Although memory parts of a system can also fit in this model, for test purposes, this is not recommended. Memories must be separated from logic parts of a system, and memory buses (data and address) must be treated as inputs and outputs of the above sequential circuit model.

7.1.2.2 Unfolding Sequential Model

As discussed above, the usefulness of the above model is separation of registers (sequential part) and combinational part of a digital system. For test purposes, this separation enables application of test methodologies to the combinational part of the circuit, and treating registers separately.

Since the majority of a circuit's logic gates are contained in the combinational part, testing this part, while considering register part inputs and outputs, covers the majority of a circuit's faults.

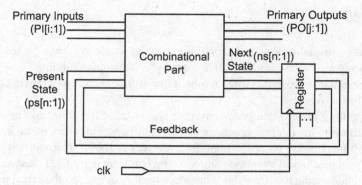

Fig. 7.1 Huffman model for test

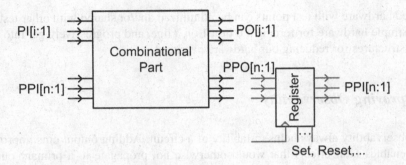

Fig. 7.2 Unfolded circuit model

Also, since flip-flops are treated as primitive building blocks in most technologies, testing a circuit for internal flip-flop faults is a secondary issue in most logic test systems.

A circuit model that separates combinational and register parts of a digital system and puts the focus of testing on the combinational part is obtained by unfolding the circuit model of Fig. 7.1. As shown in Fig. 7.2, by unfolding a sequential circuit, its registers are separated and ignored. Register outputs become pseudo inputs for the unfolded circuit, and register inputs become pseudo outputs of this circuit.

For test purposes, only the upper block in Fig. 7.2 is considered. Fault collapsing, test generation, and evaluation of test vectors are all done on this circuit model. Note that faults on pseudo inputs and pseudo outputs (PPI, PPO) of the unfolded model are the same as those of the feedback register outputs and inputs, and testing the unfolded model already includes testing the register ports.

Testing register clock, set, reset, etc. are not dealt with when considering the unfolded model and must be treated separately. Test results obtained from the unfolded model are used by test equipment for testing the actual circuit. A Verilog testbench that models this test equipment or test environment of the actual circuit is called a virtual tester. Section 7.4 elaborates on this topic.

7.1.3 *Testability of Combinational Circuits*

Combinational circuits can also benefit from DFT techniques. With an additional hardware inserted into a combinational circuit, it can be made to give a better coverage, reduce the number of test vectors required to test it, or achieve both at the same time. DFT techniques alter a combinational circuit for better controllability and observability. The model of the altered circuit is used for test generation.

7.2 Testability Insertion

By simply adding a jumper, or an extra input or output pin, we have taken steps in making our circuit more testable. This obviously adds additional hardware to our circuit, and the use of such testability provisions must carefully be planed. Limit on the number of extra input/output pins for testing, gained test time, ease of testing, and issues such as timing and power consumption must be considered when choosing DFT techniques.

This section discusses various choices that we have for hardware structures for making a circuit more testable [4]. In general, this section shows how a "test point" can be added to a circuit and how the associated hardware with test points can be minimized and/or shared with other test points. We start with a simple hardware for forcing a 0 or a 1 on a line, and progressively get into multiplexer and register structures for reducing our hardware overhead.

7.2.1 Improving Observability

Improving observability always helps testability of a circuit. Adding output pins improves observability, and enables fault effects that would otherwise not propagate to a primary output have a chance to show themselves through newly added output pins. The end result is fewer test vectors to test for faults of a CUT, thus less test application time for the final manufactured chip.

Line observability values are measured by SCOAP or are probability based. Lines with observability values below a certain threshold (higher SCOAP parameter values) can be considered as candidates for becoming extra output pins.

Improving testability by means of observability can also be achieved by making state flip-flop outputs observable by pulling them out as primary outputs. Figure 7.3 shows the Mealy machine implementation of the previous chapter in which flip-flop outputs are pulled as primary outputs. Starting in state 10, while $a = 1$, the SA0 fault shown can be detected in one clock cycle on the new $Out\ V_1$ output. Provisions for forcing this circuit into a certain state (e.g., state 10) are made in Fig. 7.3. Such techniques include 0 or 1 insertion and will be discussed in the next sub-section.

In an RT-level design, control signals going out from circuit's controller and controlling flow of data through buses and logic units, as well as logic and ALU control inputs are appropriate places where extra output pins can be added.

Other candidates for observing via new circuit outputs are bus and multiplexer select inputs, multiplexer and decoder enable inputs, tri-state controls, register load input, count up and count down signals, shift-register mode control inputs, and feedback lines.

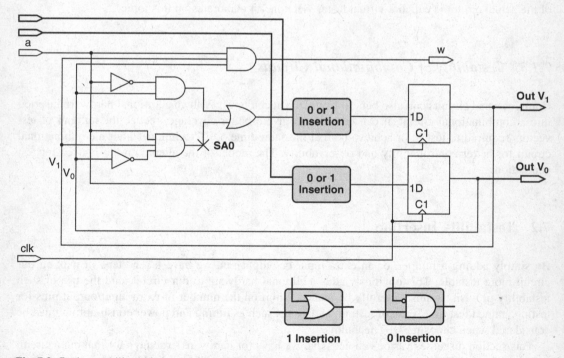

Fig. 7.3 Basic testability techniques (observability, controllability)

7.2.2 *Improving Controllability*

Controllability parameters identify hard to reach places in a combinational circuit. As a first step in adding controllability to a combinational circuit, low controllability lines should be considered. In addition, by directly controlling fanout stems and branches from a circuit's inputs, problems in test generation contributed by reconvergent fanouts can be resolved.

In sequential circuits, flip-flop inputs are good candidates for being directly controlled by circuit inputs. In addition, controlling flip-flop reset and other control inputs help driving a sequential circuit into a given state.

In RT-level designs, direct control of control signals coming from a circuit's controller to its data-path enables testing of individual data operations independently. For example, being able to externally control the *load accumulator* control signal of a datapath enables testing the loading of the accumulator register and the bus structures that drive it.

Other places where external control can contribute to testability of a circuit include multiplexer and bus control inputs, tri-state control inputs, arithmetic and logic unit select inputs, and counter and shift-register mode control signals.

Unlike observability that can easily be achieved by adding the line to be observed to our primary outputs, there are several ways controllability can be achieved. Figure 7.4 shows several such methods. Shown here are several hardware structures that intercept a line that would normally go between two logic parts. The hardware inserted in the line between *Logic A* and *Logic B* uses primary inputs to control value going to *Logic B*.

Figure 7.4a shows 0-insertion logic. The line labeled *Insert 0* is the controlling line that can be directly driven by a primary input. The circuit is in its normal mode of operation when *Insert 0* is 0. To drive a 0 into the input of *Logic B*, *Insert 0* must be set to 1. Figure 7.4b shows a 1-insertion logic

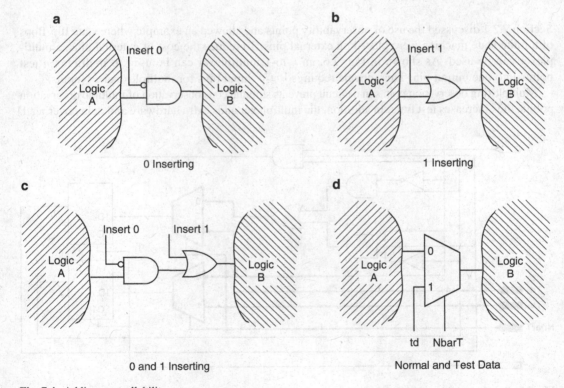

Fig. 7.4 Adding controllability

that works like the 0-insertion, except that asserting *Insert 1* puts a 1 at the input of *Logic B*. For being able to control input of *Logic B* with either a 0 or a 1, logic shown in Fig. 7.4c should be used. In combinational circuit testing, 0-insertion and 1-insertion structures are useful in places with low 0- and 1-controllability, respectively.

Figure 7.4d shows a more general case of driving a line in a circuit with a given test data (*td*). With use of a multiplexer, signal *NbarT* puts the circuit in normal (when 0) or test (when 1) mode. In the test mode, instead of data from *Logic A*, *td* that can be driven by a primary input acts as an input to *Logic B*.

If hardware structures shown here are independent and do not share pins with other test structures, 0- and 1-insertions require one primary input pin, while the other two structures shown in Fig. 7.4 require two pins. In all cases shown, in addition to the gates used and the required input pins, timing delays are also added between the two logic parts. The multiplexer (Fig. 7.4d) has the largest delay and is the most general case.

Figure 7.3 that we used for demonstrating improvement in observability also shows controllability hardware added to the flip-flop inputs. Recall from the discussion in Sect. 7.2.1 that making flip-flop outputs observable makes testing for the fault shown possible by starting in state $V_1 V_0 = 10$, and clocking the circuit once while *a* is 1. To force this state into the flip-flops, 1-insertion and 0-insertion structures should be used at the inputs of flip-flop 1 and flip-flop 0, respectively.

Testing circuit of Fig. 7.3 becomes easier by use of multiplexers at its flip-flop inputs. Figure 7.5 shows multiplexers at flip-flop inputs sharing the *NbarT* test control input. To force a given state into the circuit, *NbarT* should be driven with a 1 and the desired state value should be placed on Td_1 and Td_0 test data inputs.

7.2.3 Sharing Observability Pins

Section 7.2.1 discussed the use of observability points and showed an example where state flip-flops of a finite state machine were placed on external pins. To reduce the cost of external pins, a multiplexer can be used. As shown in Fig. 7.6, an *n*-to-1 multiplexer can be used to multiplex *n* test points into one output pin. This scheme requires $\log_2 n$ input pins for multiplex select inputs.

Multiplexing observation points in a circuit prevents simultaneous observation of multiple observation points, thus increases test time. In addition, the multiplexer adds extra hardware and delay overhead.

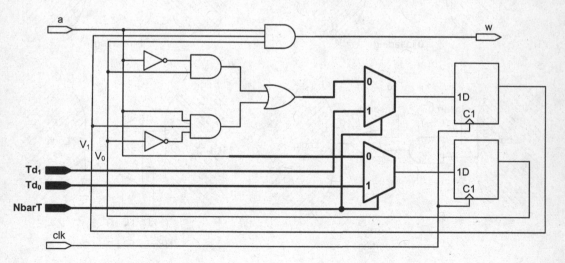

Fig. 7.5 Put circuit into any desired state

Fig. 7.6 Multiplexing observability points

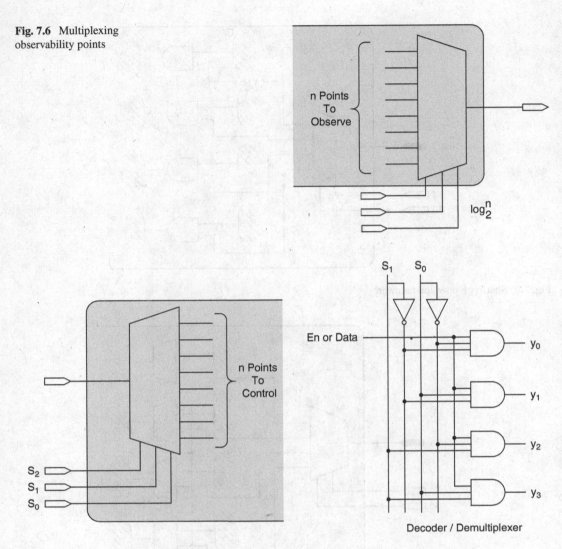

Fig. 7.7 Demultiplexer/decoder, reduce pins for control

The hardware shown can be repeated if multiple activation of several observation points are required. For observing parallel buses, same bit positions of several buses to simultaneous can be grouped, and the hardware shown in Fig. 7.6 can be used for each group. In this case, the select inputs can still be shared, but the actual output data pin must be repeated for each group.

7.2.4 Sharing Control Pins

To reduce pins required for controlling n internal lines of a circuit, a 1-to-n demultiplexer is used. The demultiplexer has several select inputs that select the output that the data input drives.

A demultiplexer is a decoder with an enable input. Figure 7.7 shows a demultiplexer symbol and logic structure for a 1-to-4 demultiplexer, which is no different than a 2-to-4 decoder.

Figure 7.8 shows a 2-to-4 decoder used for controlling four 1-insertion lines. The decoder enable input is used as *NbarT* input. When this input is 0, none of the decoder outputs is 1, and

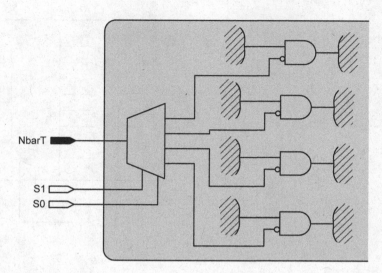

Fig. 7.8 Sharing 1-insertion hardware

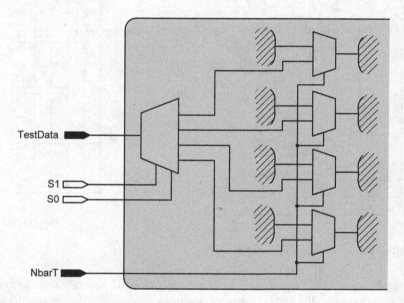

Fig. 7.9 Sharing test data insertion hardware

circuit works in normal mode. For test mode, *NbarT* becomes 1, and according to value of *S1*, *S0*, one of the 1-insertion points becomes active. With this hardware, for controlling n lines of a CUT, $\log_2 n + 1$ pins are needed.

For multiple test data insertions (Fig. 7.4d), a demultiplexer helps reducing input pin requirements. Figure 7.9 shows a 1-to-4 demultiplexer providing test data for four multiplexers for test data insertions. With this hardware, for controlling n lines of a CUT, $\log_2 n + 2$ pins are required.

As shown in Fig. 7.9, when *NbarT* is 0, the multiplexers between logic parts (shaded areas) connect the left logic to the right logic, and the test hardware is transparent. In the test mode (*NbarT* = 1), *S1* and *S0* select *TestData* for driving the input of one of the left-hand side logic blocks.

Demultiplexing control points prevents simultaneous assignment of values to the control points that share pins. Therefore, simultaneous assignment of test data to bits of parallel buses is not possible unless the above hardware is repeated for each bus bit that needs to be controlled. In this case, select inputs (e.g., *S1*, *S0*) and *NbarT* can still be shared by controllability hardware for different bus bits. Figure 7.10 shows extension of Fig. 7.9 for parallel buses. This figure shows *TestData0* through *TestDatan* providing test data for four *n*-bit buses, *A*, *B*, *C*, and *D*. Select inputs *S1* and *S0* select which bus the test data will go to.

7.2.5 Reducing Select Inputs

Although multiplexing observation points of Fig. 7.6 and demultiplexing control points of Fig. 7.7 significantly reduce test pin counts, for a large number of test points, we still face the problem of having many pins for multiplexer or demultiplexer select inputs. For the price of still longer test time, we can resolve this problem by adding a counter to the CUT. The counter clock and count inputs will be driven by external pins and the counter output selects point to be controlled or observed. For controlling or observing a line, the counter counts up, and when it reaches the number of the test point, other test signals will be issued. Figure 7.11 shows a counter used for selection inputs of a control point demultiplexer.

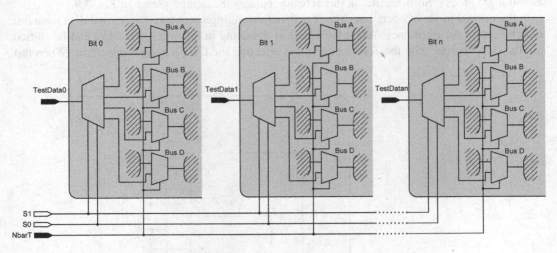

Fig. 7.10 Simultaneous control of bits of buses

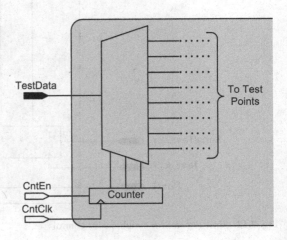

Fig. 7.11 Selection input counter

To select a test point, the counter counts up to the location of the test point on the multiplexer or demultiplexer. At the test application time, *TestData* drives the test point selected by the counter. The price we are paying for the reduction in the number of test pins is longer test time.

7.2.6 Simultaneous Control and Observation

DFT techniques discussed above were effective in reducing the number of test pins. On the other hand, none of these techniques allowed simultaneous activation of control or observe points, unless separate test pins were used and hardware structures were repeated (e.g., Fig. 7.10).

7.2.6.1 Simultaneous Control of Test Points

To allow simultaneous control for several test points and still keeping the number of pins low, a shift-register can be used to take test data serially, and apply all the bits in parallel in the test mode. Figure 7.12 shows a shift-register used for taking serial test data to be applied simultaneously to several logic blocks. Shift-register in this scheme replaces the demultiplexer of Fig. 7.9.

In the normal mode of operation, *NbarT* is 0 and the multiplexers take the normal data from left logic blocks to the right ones. While the circuit is operating in this mode, test data can be shifted into the shift-register with the *TestClock* without affecting the CUT's normal operation. When test

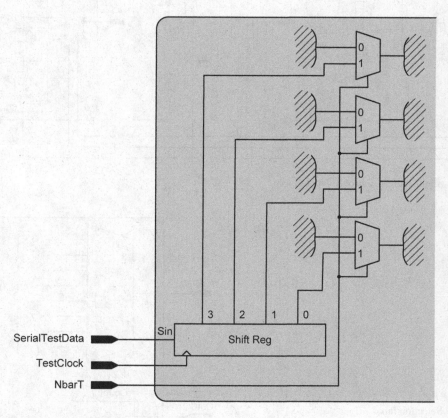

Fig. 7.12 Shift-register for simultaneous control

data are completely shifted in, the CUT is put in the test mode (i.e., *NbarT* = 1), which allows the test data from the shift-register to be used for input of logic blocks that are being tested (right-hatched blocks in Fig. 7.12).

Because test data shifting takes several clock cycles, we overlap the normal operation of CUT with shifting test data into the shift-register. In this case, when data are being shifted, *NbarT* must be 0, so that changing test data at the shift-register output does not interfere with the normal data. In general, serial shifting of test data is a time-consuming process, but, as we will see later in this chapter, it is the only way testing can be done. One way to improve this is by use of a faster clock for the test clock. Since shift-register is the only recipient of the test clock, circuit delays do not apply, and this clock can be much faster than the normal CUT's clock.

7.2.6.2 Simultaneous Observation of Test Points

The concept of shift-registers mentioned above can also be used for simultaneously observing several lines in a CUT. Figure 7.13 shows a shift-register used for simultaneous collection of values from several test points and shifting them out serially. The device or ATE that is reading the output must perform serial to parallel conversion for having data from all test points at the same time.

In the test mode, when the response of CUT is ready and available at the shift-register input, the *load* input of the shift-register is enabled and the shift-register is clocked. This operation loads CUT's line values, or lines to be observed, into the shift-register. Following this operation, the shift-register *load* input is deasserted, which puts the shift-register in the serial mode. After application of *n* clocks in this mode, the line values being observed have been shifted out, and the shift-register is ready for the next parallel load of the test output data.

7.2.6.3 Isolated Serial Scan

The two concepts of using a shift-register for controlling test points and one for observing them can be combined and the same shift-register can be used for both purposes. This testability method is referred to as Isolated Serial Scan and is illustrated in Fig. 7.14.

For the normal mode of operation, the *NbarT* input of CUT in Fig. 7.14 is 0. This causes no interference from the test logic except perhaps the delays from the multiplexers. While in this

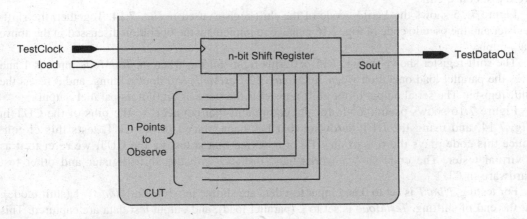

Fig. 7.13 Using a shift-register for simultaneous collection of line values

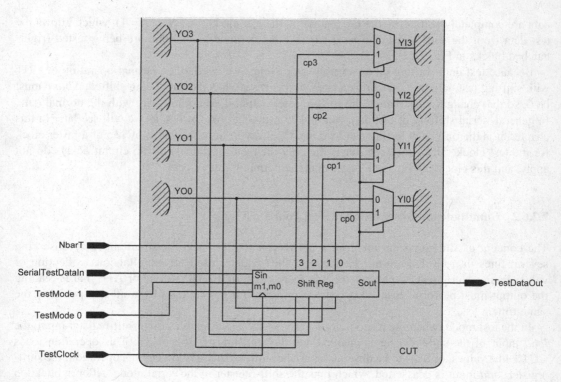

Fig. 7.14 Isolated Serial Scan

mode, the shift-register is put in the serial mode and new test data are shifted in, while collected data from last testing to be observed (from *YO3*, *YO2*, *YO1*, and *YO0*) are being shifted out. The input test data bits that are synchronized with the test clock are placed on the shift-register inputs. At the end of shifting, the CUT is put in the test mode by asserting *NbarT*. This applies the test data in the shift-registers (through the multiplexers) to the inputs of the part of CUT being tested. While this is happening, and the part of CUT is propagating this test data, the shift-register mode is changed to parallel loading. Enough time is given for test data to propagate and shown its affect on *YO* internal output lines. At this time, the shift-register is clocked just once to collect the output. This is followed by the serial mode that shifts in a new test data while shifting out the data that were just collected.

Figure 7.15 shows the Verilog code of the shift-register used in Fig. 7.14. Together, the shift-register and the pseudo code of Fig. 7.16 manage to implement the operation discussed in the above paragraph.

The shift-register shown in Fig. 7.15 performs right shifting when its *TestMode* input is 1 and does the parallel load operation when *TestMode* is 2. *TestMode* = 0 does nothing, and 3 resets the shift-register. The serial output of the shift-register is the same as bit 0 of its parallel output.

Figure 7.16 shows pseudo code for the equipment that connects to the pins of the CUT in Fig. 7.14, and using the DFT hardware that has been placed in the CUT, tests this circuit. Since this code plays the role of an ATE or a device that is testing the CUT, we refer to it as a virtual tester. The code shown works hand-in-hand with the shift-register and other test hardware in CUT.

For testing, *NbarT* is set to 0 and input test data are shifted in while *TestMode* = 1 (shift mode). At the end of shifting, *TestMode* is set to 2 (parallel load), and output test data are captured. This **while** loop repeats for as long as CUT is being tested. Each time a new test data are shifted in.

```
module shiftreg # (parameter size = 4) (input [size-1:0] parin, input
                                        TestClock, SerialTestDataIn,
                                        TestClock, input [1:0] TestMode,
                                        output TestDataOut,
                                        output reg [size-1:0] parout);
      always @ (posedge TestClock) begin
      case (TestMode)
          0: parout <= parout;
          1: parout <= {SerialTestDataIn, parout[size-1:1]};
          2: parout <= parin;
          3: parout <= 0;
      endcase
    end
    assign SerialTestDataOut = parout[0];
endmodule
```

Fig. 7.15 Isolated Serial Scan shift-register

```
Virtual Tester:Isolated Serial Scan
While (testing) begin
   N̄/T = 0;
   For (i=1; i<= numbTestPoints; i++) begin
      N̄/T = 0; TestMode = 1
      Place serial test data on SerialTestDataIn;
      Clock shift-register (TestClock)
      Collect TestDataOut;
   End for
N̄/T = 1; TestMode =2;
Clock shift-register (TestClock)
End
End Isolated Serial Scan
```

Fig. 7.16 Isolated Serial Scan virtual tester

Of all the hardware structures we discussed above, Isolated Serial Scan was the most complete and most complex to operate. This structure provides a background for standard scan architectures that are discussed next.

7.3 Full Scan DFT Technique

With the background provided in Sect. 7.2, we are now ready to discuss standard practices in DFT. The most common DFT technique is full scan that we discuss here [5]. This technique can be explained as one that is very similar in concept to the isolated scan of the previous section.

The isolated scan has an overhead of a shift-register that has to be inserted in the circuit under test. Inclusion of this register still does not solve the problem of test generation for sequential circuits, and we still have to treat the CUT as a sequential circuit with a few extra test points for adding controllability and observability. Full scan takes care of the hardware overhead and sequentiality of CUT by incorporating the required shift-register in the CUT's state flip-flops. As we will see, this reduces the hardware overhead, and at the same time CUT becomes virtually a combinational circuit.

7.3.1 Full Scan Insertion

For scan insertion, we start with the Huffman model of a sequential circuit as shown in Fig. 7.1. As discussed in Sect. 7.1, unfolding a sequential circuit (Fig. 7.2) and applying our testing to the combinational part of a circuit covers all logic and register interconnection faults. Full scan testing takes the Huffman model of Fig. 7.1, and by inserting a shift-register in its register structures, makes a virtual model of the unfolded circuit of Fig. 7.2.

7.3.1.1 Scan Register

Figure 7.17 shows scan insertion in the sequential circuit model of Fig. 7.1. The register part of the Huffman model now includes a shift mode that serially loads *STDI* (Serial Test Data In) into the shift-register. In the normal mode, the register in Fig. 7.17 performs parallel loading.

Figure 7.18 shows Verilog code for the register of Fig. 7.17. The general operation of this register is similar to the one used for isolated scan, the Verilog code of which was shown in Fig. 7.15. Only the signals for control of serial and parallel modes are different.

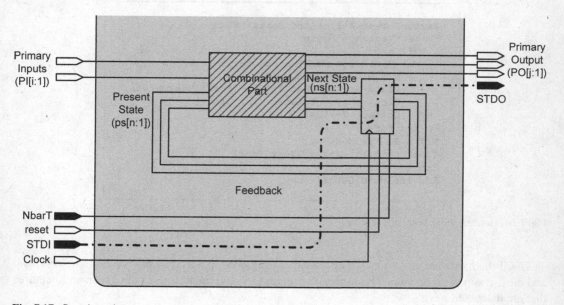

Fig. 7.17 Scan insertion

```
module FBRegister #(parameter size = 4)(input [size-1:0] ns, input
                                         STDI, Clock, NbarT, Reset, output STDO,
                                         output reg [size-1:0] ps);
    always @ (posedge Clock) begin
        if (Reset == 1'b0) begin
            if (NbarT == 1'b0) ps <= ns;
            else ps <= {STDI, ps[size-1:1] }; //shift mode
        end
        else
            ps <= 0;
    end
    assign STDO = ps[0];
endmodule
```

Fig. 7.18 Testable Huffman model feedback register

The feedback shift-register in Fig. 7.18 works in normal mode by loading *ns* (next state) into *ps* (present state) when *NbarT* is 0. In test mode (*NbarT* = 1), the register becomes a shift-register with *STDI* serial input and *STDO* serial output.

Here, we have used a simple signaling for the control of the shift-register, in which *NbarT* is used for both normal mode (when 0) and shift mode (when 1). The exact signaling depends on the type of the flip-flops used and their data selection mechanisms. These issues will be discussed in Sect. 7.3.2.

7.3.1.2   Test Procedure

The test procedure for the testable Huffman model of Fig. 7.17 only involves testing the combinational part, and the modified register is used as a mechanism for providing controllability of *ps* and observabilty of *ns*. The test architecture shown makes *ps* inputs the pseudo primary inputs and *ns* the pseudo primary outputs of the combinational part.

In the normal mode (*NbarT* = 0), the register loads *ns* into *ps*. In the test mode, *NbarT* becomes 1 and test data that is to be applied to *ps* input of the combinational logic is shifted into the register. When all data bits are shifted, the first part of test data becomes ready at the circuit's *ps* input, which is its PPI in test mode. At this time, the second part of the test data will be applied to the circuit's primary inputs through external pins. The combinational circuit in Huffman model takes the two-part test data (PI and PPI) and propagates it to its outputs. The primary outputs of the circuit will be available on circuit pins immediately. This is collected and stored.

We then put the circuit in the normal mode of operation by setting *NbarT* to 0, and clocking the circuit only once. This causes the *ns* part of the output of the combinational circuit, which is its PPO, to get clocked into the register (in parallel). This PPO will now be shifted out, so that together with the circuit's PO forms the complete output of the combinational part. While this shifting is happening, we will also shift in a new test data into the shift-register for the next round of testing.

Timing and implementation of this testing process depends on the type of flip-flops and the corresponding selection logic structures that we use. The next section discusses this.

7.3.2   Flip-flop Structures

The Verilog code of Fig. 7.18 assumes a simple flip-flop structure and a multiplexer for selection of normal and shift modes. In addition to this structure, there are other structures that are more efficient in terms of timing and gate structures, which we will discuss here.

7.3.2.1   Latches and Flip-flop

A static clocked latch is formed by a cross-coupled gate structure and gates for implementing a clocking mechanism. Figure 7.19 shows an SR-latch, its equivalent NAND structure, and a corresponding symbol.

The cross-coupled structures provide the memory, and the other two gates handle the clocking. The symbol shown indicates dependency of *S* and *R* on the clock. Latches are transparent, meaning that when clock is active, *S* and *R* inputs directly affect the *Q* output. Unless complemented by other logic structures or other latches, transparent latches cannot be used in feedback paths in sequential circuits (e.g., Fig. 7.1).

A D-type latch and its corresponding symbol are shown in Fig. 7.20. This is also a transparent latch, and the *D* input drives the *Q* output while *C* is 1.

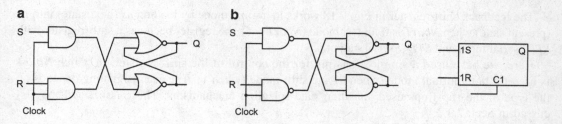

Fig. 7.19 Basic latch

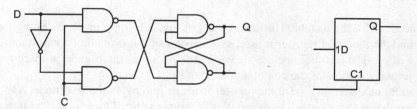

Fig. 7.20 D-latch

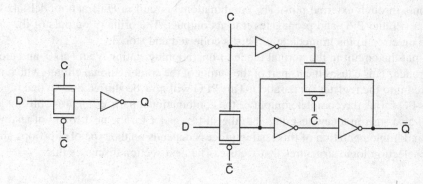

Fig. 7.21 CMOS latches

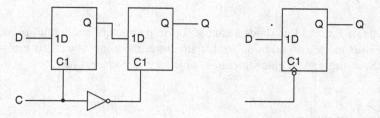

Fig. 7.22 D-type flip-flop

In CMOS, dynamic or pseudo static latches can be built with fewer gates and transistors. Figure 7.21 shows dynamic and pseudo static latches. Test flip-flop architectures that we present here and in the following discussions, that are based on the static latches, also apply to dynamic and pseudo static latches with minor changes.

A flip-flop is built by using two latches with complementary clocks. Flip-flops are not transparent and can be used in sequential circuit feedback paths. Instead of using the complement of the flip-flop clocks (as shown in Fig. 7.22), CMOS uses two nonoverlapping clock phases. Figure 7.22 shows a D-type flip-flop and its symbol. The output of the flip-flop receives the input D after the falling edge of the clock. Although the symbol shown is for triggering on the clock edge, rather than after the clock edge, this symbol is often used for both flip-flop timings. The flip-flop structure shown here is often referred to as a master-slave flip-flop, where the left latch is referred to as master and the right one as slave.

7.3.2.2 Multiplexed Test Data

The flip-flop of Fig. 7.23 provides a close correspondence with the Verilog description of Fig. 7.18. This flip-flop can be used as a scan element for the feedback register in Fig. 7.17. Figure 7.23 also shows gate-level details of the multiplexer with enable input that we use for flip-flop reset.

The flip-flop shown here has a multiplexer that uses *NbarT* to select between *DataIn* (when 0) and *SerialIn* (when 1). The multiplexer active low enable input provides a synchronous active high reset input. Figure 7.24 shows how three such flip-flops are used as a scan register for the testable model of Fig. 7.17. As before, the hatched areas in this diagram represent the combinational part of the circuit, here that of Huffman model.

In normal mode of operation, when *NbarT* is 0, *ns[2:0]* loads into *ps[2:0]* on the falling edge of *Clock*. In the test mode when *NbarT* is 1, serial data bits on *STDI* clock into the feedback register in three consecutive clocks, after which time they become available on *ps[2:0]*. The right-most bit is the serial output (*STDO*).

The Verilog code for the flip-flop structure of Fig. 7.23, including the multiplexer, is shown in Fig. 7.25. For modeling multi-bit feedback registers, we can use a Verilog code similar to the scan code in Fig. 7.18, or we can individually cascade flip-flops of Fig. 7.25 to form the right size register. For a behavioral description of a scan-inserted circuit, Verilog code of Fig. 7.18 may be more appropriate, while in a netlist where low-level detailed simulations may be needed, forming the feedback register by cascading individual flip-flops may be more useful.

The flip-flop discussed here is simple, but has the problem of multiplexer delay that adds to the logic delay. In fact, this structure increases the worst-case delay of the circuit for which scan is inserted. This reduces the speed of the normal system clock, and thus, a slower overall operation.

7.3.2.3 Dual Clocking

As shown in Fig. 7.17, with each normal mode clock, the *ps* output of the feedback register, that is the input of the combinational circuit, must travel through the entire combinational part to affect this part's *ns* output [6]. This involves a delay, only after which the register can be clocked again.

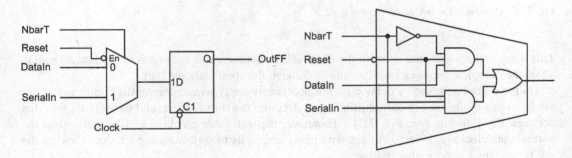

Fig. 7.23 Multiplexed scan element

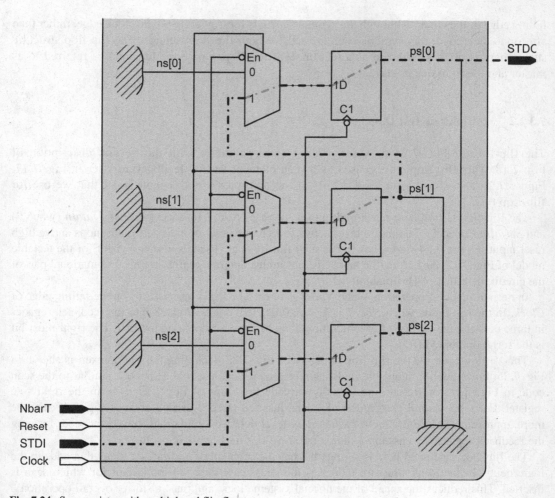

Fig. 7.24 Scan register with multiplexed flip-flops

```
module MuxedFF (input NbarT, Reset, DataIn, SerialIn, Clock, output reg OutFF);
   always @ (negedge Clock) begin
      if (Reset) OutFF <= 1'b0;
      else OutFF <= ~NbarT ? DataIn : SerialIn;
   end
endmodule
```

Fig. 7.25 Multiplexed scan element Verilog code

This delay is the worst-case delay of the CUT, and its normal clock speed has to be slow enough to allow the complete propagation of *ps* into *ns* through the combinational part.

On the other hand, such a delay does not necessarily apply when running the circuit in the test mode. In this mode, we are only shifting serial data into the shift-register, and the only logic is that between flip-flop bits (see Fig. 7.24). Therefore, the test mode clocking can be faster than the normal data clocking. In large circuits with many serial bits to shift-in, using a faster clock for the test time gains a good saving in time.

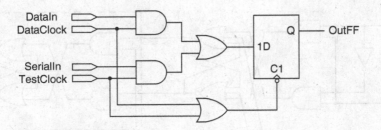

Fig. 7.26 Scan flip-flop with dual clocking

```
module DualClockFF (input DataIn, DataClock, SerialIn, TestClock, output reg OutFF);
   wire Clock;
   assign Clock = DataClock | TestClock;
   always @ (negedge Clock) begin
      if (DataClock) OutFF <= DataIn;
      else if (TestClock) OutFF <= SerialIn;
   end
endmodule
```

Fig. 7.27 Dual clock scan flip-flop Verilog code

The flip-flop of Fig. 7.26 uses dual clocks: one for normal and one for the test mode. The OR gate at the flip-flop clock input causes it to be clocked with either *DataClock* or *TestClock*. The AND–OR logic at the flip-flop input selects *DataIn* when *DataClock* is 1 and selects *SerialIn* when *TestClock* is 1. While either clock is 1, the proper data appear at the flip-flop D-input, and after the clock becomes 0, the data at D are clocked into the flip-flop.

Figure 7.27 shows the Verilog code of the scan flip-flop with the dual clocking system. In this code, an internal clock signal (*Clock*) is made by ORing the two flip-flop clocks. In the **always** statement, that is sensitive to this clock, the data and test clocks are used for conditioning what gets clocked into the flip-flop. Because we have used this procedure and not the exact gate-level equivalent of circuit of Fig. 7.27, our Verilog model does not represent the timing details of the actual circuit.

The problem with this clocking scheme is the hazard that may occur in the logic at the flip-flop D-input. In addition, the problem of introducing the logic gates at the inputs of the circuit flip-flops and increasing the worst-case delay of the circuit still remains.

7.3.2.4   Two-port Flip-flops

To reduce the flip-flop D-input logic delay, the clocking scheme of Fig. 7.28 can be used. The figure shows gate-level details of the flip-flop complementary clock latches. The logic of the right-hand side latch (slave) remains the same as that of Fig. 7.19, and the other latch (master), combines its clocking logic with its required selection logic. To make explanation of the logical operation of the flip-flop in Fig. 7.28 easier, we use the AND–NOR logic of Fig. 7.19 instead of its NAND equivalent shown in the same figure.

In a timing diagram, Fig. 7.29 shows the operation of the two-port flip-flop of Fig. 7.28. For loading *DataIn* into the flip-flop, *ClockA* and *ClockB* are asserted alternatively, while *ClockC* remains at 0. For loading *SerialIn*, *ClockC* and *ClockB* are applied in alternative orders, and in this case *ClockA* is inactive.

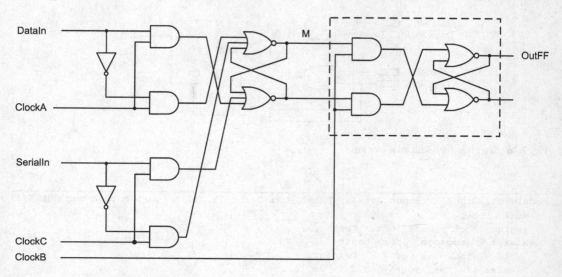

Fig. 7.28 Two-port three-clock flip-flop

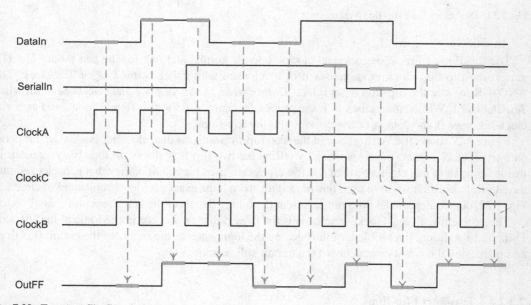

Fig. 7.29 Two-port flip-flop timing

When *ClockA* is asserted, while *ClockB* is 0, data on *DataIn* is latched into the master latch and appears on *M*. When *ClockB* is asserted, data on *M* is latched into the flip-flop output. The situation is similar, when *ClockA* is inactive and *ClockB* toggles.

Figure 7.30 shows a symbol for the structure of Fig. 7.28 that is based on two latches. The master latch is a two-port latch and the slave is a standard D-latch such as that of Fig. 7.20.

Figure 7.31 shows the register part of Fig. 7.17 implemented with the flip-flop of Fig. 7.30. Although shift and normal mode signals are different than the multiplexed flip-flop design of Fig. 7.24, the overall operation remains the same. For normal operation, *ClockA* and *ClockB* are

Fig. 7.30 Two-port flip-flop symbol

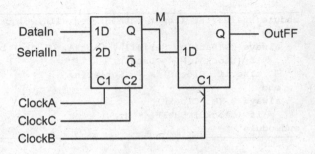

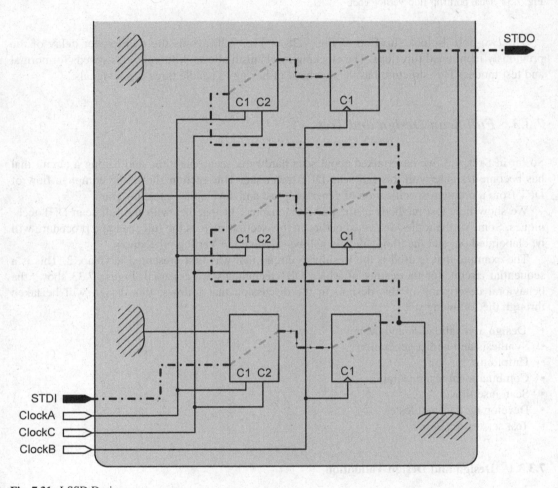

Fig. 7.31 LSSD Design

used, and in test mode, *ClockC* and *ClockB* become complementary clocks. The design shown here is called LSSD (Level Sensitive Scan Design) and was first used by IBM in 1977. Scan path for shifting serial test data into feedback registers is highlighted in this figure.

Figure 7.32 shows Verilog code of the flip-flop discussed above. Signal names apply to those shown in Fig. 7.30. This code is based on description of two latches. The master latch uses two clocks, and slave uses *ClockB*. The output is changed when *ClockB* becomes 1.

```
module DualPortFF (input DataIn, SerialIn, ClockA, ClockB, ClockC, output reg OutFF);
    reg M;
    always @ (DataIn, SerialIn, ClockA, ClockC) begin
        if (ClockA) M <= DataIn;
        else if (ClockC) M <= SerialIn;
    end
    always @ (M, ClockB)
        if (ClockB) OutFF <= M;
endmodule
```

Fig. 7.32 Dual port flip-flop Verilog code

The two-port 3-clock flip-flop of Fig. 7.28 or Fig. 7.30 avoids the multiplexer delay of the previously mentioned flip-flops. The clocking mechanism allows different clock speeds for normal and test modes. This structure has the overhead of having to handle three clock signals.

7.3.3   Full Scan Design and Test

So far in Sect. 7.3, we have talked about scan hardware, scan operation, and testing a circuit that has become testable with this full scan DFT technique. This section shows the complete flow of DFT from a problem specification to generating test and developing a test program.

We show how test methods of the previous chapters fit together with the full scan DFT techniques. Some of the topics discussed earlier in this section such as the full scan test procedure will be elaborated on, and the timing details will be explained in Verilog procedures.

The example that is used is the Residue-5 circuit that was first presented in Chap. 2. This is a sequential circuit, for the testing of which DFT techniques are essential. Figure 7.33 shows the behavioral description of this design. In the discussion that follows, this design will be taken through the following steps:

- Design and validation and design
- Synthesis and netlist generation
- Unfolding
- Combinational test generation
- Scan insertion
- Developing a virtual tester
- Test set verification

7.3.3.1   Design and Design Validation

As mentioned, our example design is the Residue-5 circuit discussed in this and several other chapters in this book. This hardware is described in Verilog (Fig. 7.33). The coding style is according to the Huffman model of Fig. 7.1. The register part specifies an asynchronous resetting mechanism, which means that the reset signal does not participate in the combinational part of the circuit. Keeping reset and other flip-flop control signals away from the combinational part is good for postmanufacturing testing.

The design described in an HDL must be validated. For this purpose, a testbench for functional testing of the design must be developed. This topic has been covered in Chap. 2 and will not be repeated here.

Although we have not taken advantage of this opportunity, the combined design and test environment that we are presenting here allows a testbench for design validation to be used as a template

```verilog
module residue5(input clk, reset, input[1:0] in, output[2:0] out);
reg[2:0] nState, pState;
parameter zero = 3'b000, one = 3'b001, two = 3'b010, three = 3'b011,
          four = 3'b100;

   always@(posedge clk, posedge reset)
      if(reset) pState = zero;
      else pState = nState;

   always@(pState, in) begin

      case(pState)
      zero:
         case(in)
            2'b00: nState = zero;
            2'b01: nState = one;
            2'b10: nState = two;
            2'b11: nState = three;
         endcase

      one:
         case(in)
            2'b00: nState = one;
            2'b01: nState = two;
            2'b10: nState = three;
            2'b11: nState = four;
         endcase
      . . .

      four:
         case(in)
            2'b00: nState = four;
            2'b01: nState = zero;
            2'b10: nState = one;
            2'b11: nState = two;
         endcase
      endcase

   end//always
   assign out = pState;
endmodule
```

Fig. 7.33 *residue5* partial Verilog code

for a testbench or a test program for postmanufacturing testing. Starting with a testbench for design validation, and gradually improving it for postsynthesis and eventually for manufacturing test is referred to as "testbench migration". In testbench migration, a testbench begins with just functional test data for verifying the presynthesis design of a circuit and gradually changes to include test data generated by ATPG programs.

7.3.3.2 Synthesis and Netlist Generation

The next step after design validation is synthesis. Using an FPGA-based synthesis program and a netlist converter, we have successfully synthesized the Verilog code of Fig. 7.33. The result is shown

```
module residue5_net(clk, reset, in, out);
input clk;
input reset;
input [1:0]in;
output [2:0]out;
wire
wire_1,
wire_2,
. . .
in_0_0,
in_0_1,
. . .
out_0_0,
out_0_1,
. . .

pin #(2) pin_0 ({in[0], in[1]}, {in_0, in_1});
pout #(3) pout_0 ({out_0_7, out_1_7, out_2_5}, {out[0], out[1],
                  out[2]});
fanout_n #(8, 0, 0) FANOUT_3 (in_0, {in_0_0, in_0_1, in_0_2, in_0_3,
                              in_0_4, in_0_5, in_0_6, in_0_7});
fanout_n #(7, 0, 0) FANOUT_4 (in_1, {in_1_0, in_1_1, in_1_2, in_1_3,
                              in_1_4, in_1_5, in_1_6});
. . .
notg #(0, 0) NOT_1 (WIRE_3, in_1_0);
notg #(0, 0) NOT_2 (WIRE_4, out_2_0);
. . .
and_n #(3, 0, 0) AND_14 (wire_25, {wire_6_5, wire_3_4, out_2_4});
or_n #(4, 0, 0) OR_2 (wire_21, {wire_25, wire_24, wire_23, wire_22});
dff INS_1 (out_0, wire_1, clk, reset, 1'b0, 1'b1, NbarT, Si, 1'b0);
dff INS_2 (out_1,wire_13, clk, reset, 1'b0, 1'b1, NbarT, Si, 1'b0);
dff INS_3 (out_2,wire_21, clk, reset, 1'b0, 1'b1, NbarT, Si, 1'b0);

endmodule
```

Fig. 7.34 Postsynthesis *residue5* netlist

in Fig. 7.34. This netlist uses primitives that are compatible with our utility PLI functions for fault collapsing, fault simulation, and test generation. This netlist, the components of which are shown in Appendix B, is automatically generated from the circuit's behavioral description discussed in Appendix F.

As expected, the netlist includes a logical block consisting of basic gates with a feedback through three flip-flops.

This is compatible with the Huffman model of Fig. 7.1. An example for logic feedback through the flip-flops can be seen by tracing wire *wire_13* that goes to the D input of *INS_2* flip-flop.

Before going to the next step of design, it is necessary to perform postsynthesis simulation of this netlist and make sure it is a correct translation of the behavioral model of Fig. 7.33.

7.3.3.3 Unfolding

Once we have verified that our synthesis has been performed correctly, we start the test generation process. The circuit in Fig. 7.34 is a sequential circuit, and test generation methods and programs for sequential circuits are not very efficient in terms of fault coverage. For this reason, we convert our CUT to a combinational circuit by unfolding it, as presented in Fig. 7.2.

Figure 7.35 shows the *residue5* netlist after being unfolded. As discussed, unfolding means removing flip-flops and making their outputs and inputs pseudo primary inputs and pseudo primary

```
module res5_Unfold (/*PI,PPI*/{in,out_2, out_1, out_0},
                    /*PO,PPO*/{out,HM_1, HM_13, HM_21});
input [1:0]in;
input out_0, out_1, out_2; //ppIn
output [2:0]out;
output HM_1, HM_13, HM_21; //ppOut
wire
wire_2,
. . .
in_0_0,
in_0_1,
. . .
out_0_0,
out_0_1,
. . .
wire_3_2,
wire_3_3,
. . .
PPI0,
PPI1,
PPI2,
PPO1,
PPO13,
PPO21;

pin #(2) pin_0 ({in[0], in[1]}, {in_0, in_1});
pin #(3) pin_1 ({out_0, out_1, out_2}, {PPI0, PPI1, PPI2});//Pseudo PI
pout #(3) pout_0 ({out_0_7, out_1_7, out_2_5}, {out[0], out[1],
                   out[2]});
pout #(3) pout_1 ({PPO1, PPO13, PPO21}, {wire_1, wire_13, wire_21});
fanout_n #(8, 0, 0) FANOUT_3 (in_0, {in_0_0, in_0_1, in_0_2, in_0_3,
                              in_0_4, in_0_5, in_0_6, in_0_7});
. . .
notg #(0, 0) NOT_1 (wire_3, in_1_0);
notg #(0, 0) NOT_2 (wire_4, out_2_0);
. . .
and_n #(3, 0, 0) AND_14 (wire_25, {wire_6_5, wire_3_4, out_2_4});
or_n #(4, 0, 0) OR_2 (PPO21, {wire_25, wire_24, wire_23, wire_22});

//dff INS_1 (out_0, wire_1, clk, reset, 1'b0, 1'b1, NbarT, Si,1'b0);
//dff INS_2 (out_1, wire_13, clk, reset, 1'b0, 1'b1, NbarT, Si,1'b0);
//dff INS_3 (out_2, wire_21, clk, reset, 1'b0, 1'b1, NbarT, Si,1'b0);

endmodule
```

Fig. 7.35 Unfolded *residu5* netlist

outputs. Figure 7.35 shows that *out_0*, *out_1*, and *out_2* that used to be flip-flop outputs are now mapped to *PPI0*, *PPI1*, and *PPI2* that are circuit's pseudo primary input. Similarly, former flip-flop inputs *wire_1*, *wire_13*, and *wire_21* are now mapped to *PPO1*, *PPO13*, and *PPO21* pseudo primary outputs. The new signal mentioned above also appears on circuit port list as inputs and outputs.

7.3.3.4 Combinational TPG

The netlist in Fig. 7.35 represents a combinational circuit and uses our standard primitives. Chapter 5 showed several random TG methods and Verilog testbenches for implementing them. We have used the AECpt algorithm of Sect. 5.3.2.2 for test generation for this module. The Verilog testbench for this purpose is similar to the code shown in Fig. 5.27 and is not repeated here. The testbench we used generates the collapsed fault list before test generation.

The results from the test generation Verilog program are as follows.

- Number of faults: 104
- Number of test vectors: 26
- Bits per test vector: 5 (2 PI, 3 PPI)
- Fault coverage: 100%

Recall from Chap. 5 that a testbench was developed for sequential test generation of the Residue-5 circuit (Fig. 5.30). That testbench generated 7 test vectors applied over 45 clock cycles (effectively, 45 tests), and resulted in only 82.4% fault coverage. The unfolding of this circuit enabled us to achieve a 100% coverage.

7.3.3.5   Scan Insertion

To facilitate application of tests generated by the procedure discussed above to the actual CUT, a scan for accessing state flip-flop inputs and outputs is inserted in the CUT. With the insertion of this scan, the block diagram of the netlist of *residue5* that was originally modeled as in Fig. 7.1 becomes as that shown in Fig. 7.36. The netlist corresponding to this figure will be shown next.

For this purpose, the postsynthesis netlist of Fig. 7.34 is modified to include the necessary scan flip-flops and signals. The synthesis tool that generated the original netlist used flip-flop types that already included serial shift facilities that were not used in this netlist. Figure 7.37 shows this flip-flop with *NbarT*, serial input (*Si*), and standard control signals. This description corresponds to the flip-flop notation used in Fig. 7.36 for the feedback flip-flop. Note: The block diagram notation for the flip-flops in Fig. 7.36 shows two flip-flop D inputs marked by 1,2D and 1,3D. The notation specifies that both D inputs are controlled by the clock signal number 1. The upper input requires mode 2 to be active and the lower input needs mode 3. Modes 2 and 3 are determined by a 1 or a 0 on the lower-left input of the flip-flop. We will use this notation in flip-flops of this chapter.

The netlist shown in Fig. 7.38 takes advantage of shift features of the flip-flop of Fig. 7.37. This netlist has additional scan control inputs *NbarT* and *Si*.

The *NbarT* input connects to *NbarT* inputs (shift control) of the state flip-flops (*INS_1*, *INS_2*, and *INS_3*, shown in the last part of code of Fig. 7.38). The *Si* input connects to flip-flop 0 (*INS_1*), the output of which goes to the input of the next, eventually forming a chain of three scan flip-flops. Signal *out_2* that is the output of the last flip-flop drives the *So* serial output signal.

7.3.3.6   Developing a Virtual Tester

Figure 7.38 represents a testable circuit. This netlist implements the original desired functionality of the design, as well as our inserted test hardware. Once manufactured, it has to be tested with a test plan that depends on the test architecture that we have developed, i.e., full scan. The test program running on an ATE implements this test plan.

In this section, we use a Verilog testbench to imitate an ATE. The test program running on the ATE that is written in C/C++ or other high-level software languages will be written here in Verilog procedural code. As previously discussed, we refer to this Verilog testbench as a *virtual tester*.

The block diagram of the virtual tester testing the full scan version of the Residue-5 circuit is shown in Fig. 7.39. The main task of the virtual tester is to read predetermined test data from an external file, apply it to the CUT, get the output of the CUT, and compare the response with the

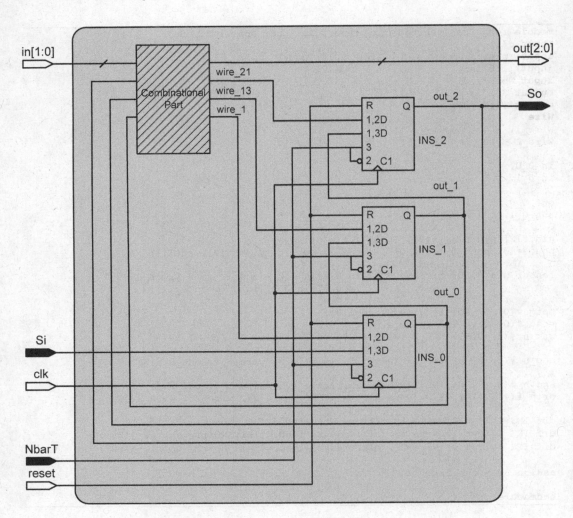

Fig. 7.36 *residue5* with scan chain

```
module dff #(parameter tphl = 0, tplh = 0) (Q, D, C, CLR, PRE, CE,
                                            NbarT, Si, global_reset);
    input D, C, CLR, PRE, CE, NbarT, Si, global_reset;
    output reg Q;
    reg val;

    always @(posedge C or posedge PRE or posedge CLR) begin
        if(CLR || global_reset)
            val = 1'b0;
        else  if(PRE)
            val = 1'b1;
        else  if(NbarT)
            val = Si;
        else  if(CE)
            val = D;
    end

    always@(posedge val) #tplh Q =  val;
    always@(negedge val) #tphl Q =  val;
endmodule
```

Fig. 7.37 Flip-flop with scan facilities

```
module res5_ScanInserted(clk, reset, in, out, NbarT, Si, So);
input clk;
input reset;
input [1:0]in;
input NbarT, Si;
output So;
output [2:0]out;
wire
wire_1,
wire_2,
. . .
in_0_0,
in_0_1,
. . .
out_0_0,
out_0_1,
. . .
pin #(2) pin_0 ({in[0], in[1]}, {in_0, in_1});
pout #(3) pout_0 ({out_0_7, out_1_7, out_2_5}, {out[0], out[1],
                  out[2]});
fanout_n #(8, 0, 0) FANOUT_3 (in_0, {in_0_0, in_0_1, in_0_2, in_0_3,
                             in_0_4, in_0_5, in_0_6, in_0_7});
. . .
notg #(0, 0) NOT_1 (wire_3, in_1_0);
notg #(0, 0) NOT_2 (wire_4, out_2_0);
xor_n #(2, 0, 0) XOR_2 (wire_13, {out_1_3, wire_14});
. . .
and_n #(3, 0, 0) AND_14 (wire_25, {wire_6_5, wire_3_4, out_2_4});
. . .
and_n #(3, 0, 0) AND_14 (wire_25, {wire_6_5, wire_3_4, out_2_4});
or_n #(4, 0, 0) OR_2 (wire_21, {wire_25, wire_24, wire_23, wire_22});

dff INS_1 (out_0, wire_1, clk, reset, 1'b0, 1'b1, NbarT, Si, 1'b0);
dff INS_2 (out_1, wire_13, clk, reset, 1'b0, 1'b1, NbarT, out_0, 1'b0);
dff INS_3 (out_2, wire_21, clk, reset, 1'b0, 1'b1, NbarT, out_1, 1'b0);

assign So = out_2;

endmodule
```

Fig. 7.38 Scan-inserted circuit under test

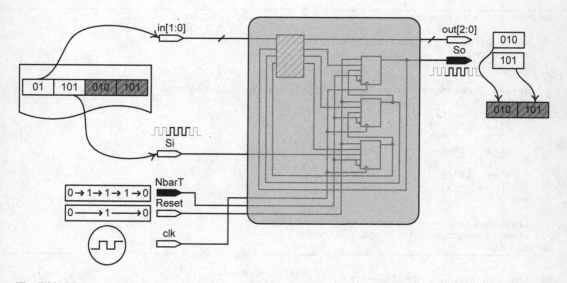

Fig. 7.39 Virtual tester for *residue5*

expected response from the external file. Since we are developing a virtual tester, and our CUT is really not a faulty circuit, our virtual tester also has the responsibility of injecting fictitious faults in the circuit to see if the test set that is provided detects them. We take this process one step further, and inject all circuit faults obtained by fault collapsing in the postsynthesis netlist of Fig. 7.34.

As shown in Fig. 7.39, the parallel data read from the external test file have two parts: one is directly applied to *in[1:0]* input of CUT and the other is serialized and applied through *Si*. The timing of these data is such that when all serial bits have been shifted in the scan chain, the parallel data must be applied to *in[1:0]*.

As shown, the output also has two parts. The first part becomes available on *out[2:0]* immediately after all inputs (i.e., parallel and serial) have been applied. Then, the state outputs (pseudo outputs) become available after the flip-flops have been clocked and then shifted out through *So*.

The Verilog testbench in Fig. 7.40 implements the test environment shown in the block diagram of Fig. 7.39. This testbench reads external data files and through control of clocking of the flip-flops and timing of inputs and outputs applies this test data to the inputs of our CUT, and collects the corresponding outputs. It then compares output data with expected outputs available in the test data file. The paragraphs that follow discuss the manner in which clock data and test inputs of the scan-inserted Residue-5 are controlled that leads to testing this circuit with input/output data that are obtained from the unfolded combinational model of Residue-5.

```verilog
module   Tester;

   parameter nff = 3;
   parameter in_size = 2;
   parameter out_size = 3;
   parameter st_size = 3;
   parameter stIndex = 8; /*in_size + out_size + in_size*/
   parameter line_size = st_size * 2 + in_size + out_size;
   reg [numOfPIs+ numOfPOs + 2 * numOfDffs - 1 : 0] line;
   . . .

   res5_ScanInserted FUT(clk, reset, PI, PO, NbarT, si, so);

   always #10 clk = ~clk;

   initial begin

     $FaultCollapsing(Tester.FUT, "Res5.flt");


   while( !$feof(faultFile)) begin

       while((!$feof(testFile))&&(!detected)) begin
         . . .
       end//test
     . . .
   end//fault

     $fclose(faultFile);
     $display("number of faults = %f",numOfFaults );
     $display("number of detected faults = %f", numOfDetected );
     $display("Coverage = %f", numOfDetected * 100.0  /  numOfFaults);

     $stop;
   end//initial

endmodule
```

Fig. 7.40 Virtual tester for full scan Residue-5

The testbench in Fig. 7.40 is a generic tester module, at the top of which parameters specifying the number of circuit inputs, outputs, and flip-flops are specified. Other necessary declarations follow the parameter specifications. Variable *line* is declared such that a line from test data file can be read into this variable. Instantiation of the scan-inserted CUT and generation of a periodic clock are also shown here.

The procedural code of the tester starts with the **initial** statement and, immediately following this statement, performs fault collapsing and opens the fault file for subsequent reading. We then have two **while** loops for fault injection and test data application. Display tasks reporting test results appear at the end of the **initial** block.

Figure 7.41 shows the fault injection (the outer **while** loop). In an actual test program for testing a manufactured part, this loop does not exist. In postmanufacturing test, we are testing for existing faults and not intentionally injecting faults. After every fault injection, the loop in Fig. 7.41 resets the CUT and prepares it for testing. Testing is done in the inner **while** loop shown here.

Figure 7.42 shows the details of the inner loop of the procedural code of Fig. 7.40. This **while** loop is responsible for reading a line of test data from *testFile*, applying the input part of it to the circuit, and checking circuit's response with the output part of the line of test data.

A line read from *testFile* has an input part and an output part. Arrangement of inputs and outputs in *line* is shown in Fig. 7.43. The input part has a part that applies directly to the circuit's primary

```
module  Tester;
   . . .
   initial begin
      . . .

      while( !$feof(faultFile)) begin

         testFile   = $feof("Res5.tst","r");
         status = $fscanf(faultFile,"%s s@%b\n",wireName, stuckAtVal );

       $InjectFault ( wireName, stuckAtVal );

         global_reset = 1'b1; reset = 1'b1; #1;
         global_reset = 1'b0; reset = 1'b0;

         PI = 0;
         cur_expected_st = 0;
         detected = 1'b0;

         while((!$feof(testFile))&&(!detected)) begin
            . . .
         end//test

         if(detected == 0) $display("NOT DETECTED = %s s@%b", wireName,
         stuckAtVal );

         $RemoveFault(wireName);
         numOfFaults = numOfFaults + 1;
         $fclose(testFile);

      end//fault
      . . .
   end//initial

endmodule
```

Fig. 7.41 Fault injection in Residue-5

```
module Tester;

    initial begin

        while( !$feof(faultFile)) begin
            . . .
            while((!$feof(testFile))&&(!detected))begin
                status = $fscanf(testFile,"%b\n",line);

                pre_expected_st = cur_expected_st;

                expected_PO = line[out_size - 1:0];
                cur_expected_st = line[st_size + out_size -1 :out_size];
                PI = line[st_size+out_size+in_size-1:st_size+out_size];

                @(posedge clk);

                NbarT = 1'b1;
                index = stIndex;

                repeat(nff) begin
                    si = line[index];
                    saved_st[index - stIndex] = so;
                    @(posedge clk);
                    index = index + 1;
                end

                SampledPO = PO;
                NbarT = 1'b0;
                #5;

                if({pre_expected_st, expected_PO} != {saved_st, SampledPO})
                begin
                    numOfDetected = numOfDetected + 1;
                    detected = 1;
                end

            end

        end

    end//initial

endmodule
```

(left brace labels: Fault, Test)

Fig. 7.42 Test application and response collection

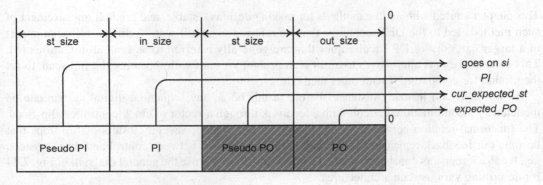

Fig. 7.43 Arrangement of stimulus and response in *line*

inputs and another part that must be shifted into scan flip-flops (*Pseudo PI*). Similarly, the output part itself has two parts: one part becomes available immediately when proper inputs (*PI*) and scan data are shifted into the flip-flops (*PO*) and a second part that will be shifted out when serial data belonging to the next test vector are being shifted in (*Pseudo PO*). Understanding this timing is crucial in collecting the right outputs and comparing relevant responses.

The *Test* loop in Fig. 7.42 reads stimuli and response data, applies it to the CUT, compares the outputs, and reports if a test detects the injected fault. Data that contain stimuli and response have the format shown in Fig. 7.43. Variable names storing various parts of *line* data are indicated in this figure.

As shown in Fig. 7.42, before *cur_expected_st* (current expected state) is overwritten with new data from *line*, it is saved in *pre_expected_st* (previous expected state). The *pre_expected_st* is saved because the actual state flip-flop contents of the previous test will not become available until current serial data are shifted into the state flip-flops, and that shifted-out data is what needs to be compared with *pre_expected_st*.

The *PI* part of the current data read from *line* is immediately applied to the scan-inserted Residue-5. (See *PI* in port connections of *res5_ScanInserted* in Fig. 7.40.)

The **repeat** loop in Fig. 7.42 takes the pseudo PI part of line and shifts it into the circuit's state flip-flops. Meanwhile, flip-flop outputs are serially shifted out and saved in *saved_st*. During the shift operation, the circuit is in test mode (*NbarT* = 1), and each shift is accompanied with the positive edge of clock. Shifting repeats *nff* (number of flip-flops) times.

After shifting completes, with PI at the circuit's primary inputs, and pseudo PI part of *line* in the state flip-flops, the circuit produces the output that corresponds to the input data of *line* that was just read. This *PO* is saved in *SampledPO*.

In the last part of the loop in Fig. 7.42, the collected pseudo output and the primary output of the circuit are concatenated and compared with the expected state and the expected primary output.

7.3.3.7 Test Set Verification

The test set for Residue-5 circuit was developed using its unfolded model with no registers, and using combinational test generation methods. We obtained a 100% coverage from this test set. The above procedure, in which this same test set is applied to the actual sequential model of Residue-5, verifies the test set obtained from the combinational model.

7.4 Scan Architectures

This chapter started with ad hoc methods for making designs testable, and gradual improvement of such methods led to the DFT technique that we referred to as full scan. Actually, full scan is part of a larger category of DFT techniques that are generally referred to as scan architectures [5]. This section presents alternatives to a full scan design. We start with repeating the full scan, to set the ground for comparing it with other methods.

As we have said in many instances before in this book, any sequential digital system can be modeled by a combinational circuit with a feedback through a vector of clock-controlled flip-flops. The functional relation between the flip-flops may form one or several clusters of flip-flops that become our feedback registers. In this model, no line is drawn between control and data registers, i.e., feedback registers consist of data and control flip-flops. We use the general diagram of Fig. 7.44 for describing various scan architectures.

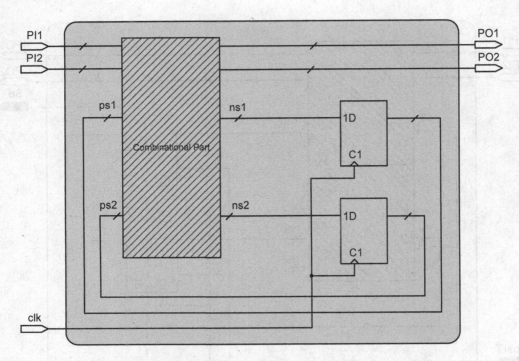

Fig. 7.44 Huffman model with multiple vector inputs, outputs, and states

7.4.1 Full Scan Design

In the full scan DFT techniques, feedback registers are given the additional capability of acting as shift-registers in the test mode. Full scan chains all the registers together and provides a serial-in and a serial-out ports. This method enables serial access for controlling all flip-flop outputs (circuit's present state) and observing all flip-flop inputs (circuit's next state). Full scan refers to the fact that all circuit flip-flops are included in the scan chain. Figure 7.45 shows addition of full scan to the model of Fig. 7.44. The problem with full scan is the long chain of flip-flops that test data have to be shifted into that reflects on the test time [7].

7.4.2 Shadow Register DFT

An alternative design to full scan design is the use of shadow registers. This technique duplicates the feedback registers. It uses one set for normal operation of the circuit and another set for test purposes. This method reduces the test time by overlapping the time of test data preparation and response collection with the normal operation of the circuit [6].

7.4.2.1 Shadow Architecture

Figure 7.46 shows the addition of a shadow register to the model of Fig. 7.44. The new set of registers receives next-state outputs from the combinational part of the circuit and produces outputs that are multiplexed on the present state inputs of the combinational part.

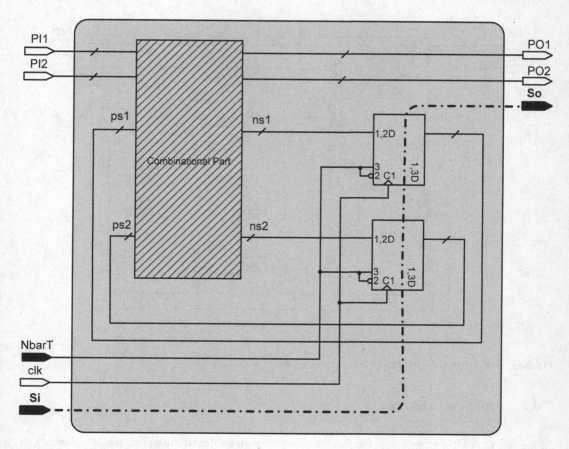

Fig. 7.45 Full scan DFT technique

7.4.2.2 Shadow Test Procedure

In the normal mode ($NbarT = 0$), the circuit has its normal inputs, and normal feedback registers provide data for the present state of the circuit. While in this mode, the shadow registers are put in their serial shift mode. Therefore, simultaneous with the normal operation of the circuit, the test data on Si are shifted into the shadow registers with $Tclk$ test clock. When all test data have been shifted in, $NbarT$ is asserted, which puts the circuit in the test mode and disables shifting serial data into the shadow register. In this mode, the normal circuit clock (clk) is disabled and (see the AND gate on clock) test states ($TS1$ and $TS2$) will drive the circuit's state inputs ($ps1$ and $ps2$).

In the test mode, the $ps1$ and $ps2$ inputs of the combinational part are driven by test data. We will also drive $PI1$ and $PI2$ with their corresponding test data. This will set all inputs of the combinational part of the CUT to previously prepared test data. With test inputs provided to all combinational block inputs, the response becomes available on $PO1$, $PO2$, $ns1$, and $ns2$. The primary output part of the response ($PO1$ and $PO2$) can be read at this time, but $ns1$ and $ns2$ must be clocked into the shadow register and shifted out in order to be observed. Clocking the circuit while $NbarT$ is 1 loads $ns1$ and $ns2$ into the shadow registers (see 2, 3 control inputs of the shadow registers, mode 2 allows serial shift , and mode 3 takes the normal flip-flop input).

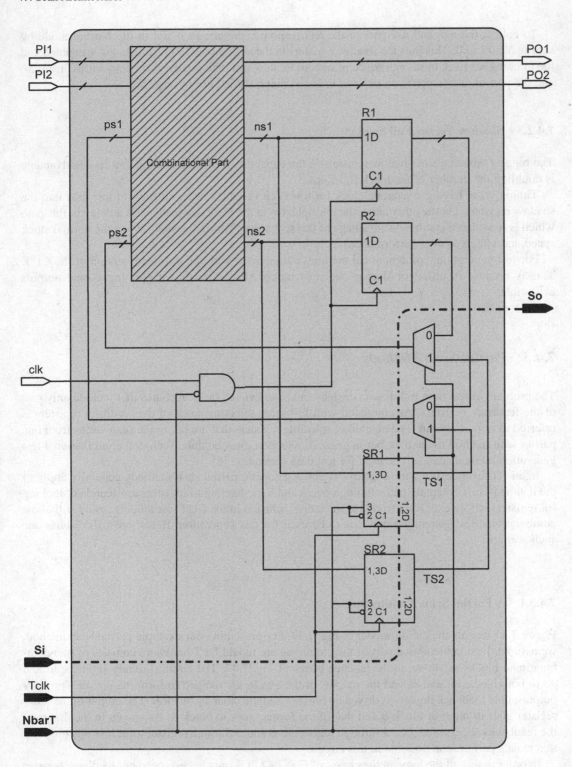

Fig. 7.46 Shadow register

To collect the *ns1* and *ns2* part of the test response, the circuit is put in the normal mode by setting *NbarT* to 0. This puts the shadow registers in the shift mode. In this mode, *clk* is enabled and the circuit goes back to normal mode of operation, new test data inputs start being shifted from *Si*, and the part of test response on *ns1* and *ns2* start being shifted out through *So*.

7.4.2.3 Shadow Versus Full Scan

The biggest advantage of shadow registers is the capability of online testing, and its disadvantage is doubling the number of feedback flip-flops.

Timing wise, having a separate clock for test (*Tclk*) enables faster shifting of test data into the shadow registers. On the other hand, the multiplexers in the feedback path cause a delay in this path which is considered another disadvantage of this technique. This delay slows down the normal clock speed and affects system performance.

For test generation, combinational methods can be used with the unfolded version of the CUT. This is because insertion of shadow registers makes all combinational part inputs and outputs accessible.

7.4.3 Partial Scan Methods

The problem of test time in full scan designs can be alleviated by scan chains that include only part of the feedback registers. As compared with full scan, selecting some of the feedback registers is referred to as partial scan. The method of selecting registers that are put in the scan varies from one partial scan method to another, but in general, selection must be done such that combinational test generation methods can still be used for test data generation [8].

In an RT-level design with a controller and a datapath, partial scan methods generally apply to the datapath. In a datapath, paths through buses and logic leading to registers are searched, looking for registers whose exclusion from a scan chain (inclusion in the CUT test model) would still allow combinational test generation methods to be used for test generation. Below we will discuss one such scenario.

7.4.3.1 A Partial Scan Architecture

Figure 7.47 repeats the circuit model of Fig. 7.44. For presenting our example partial scan method, we assume the combinational part of the circuit we are to add DFT hardware consists of two combinational blocks as shown in the hatched boxes of Fig. 7.47. The circuit primary inputs split and go to both blocks, *A* and *B*, and the outputs of the blocks are merged to form the circuit's primary outputs. One feedback register is driven by block *A* and the other by block *B*. The output of the latter register goes to input of block *B* and that of the former goes to block *A*. As shown in the diagram, the feedbacks are crossed, i.e., *A* through register to *B* and *B* through register to *A*. It is this property that makes partial scan possible in this circuit.

In order to test all the logic in the circuit of Fig. 7.47, it is only necessary to put feedback register *R2* in a scan path. This scan is shown by a heavy dotted line in Fig. 7.47. Applicability of this method to datapath of an RT-level design becomes more clear by turning the diagram of Fig. 7.47 into that of Fig. 7.48.

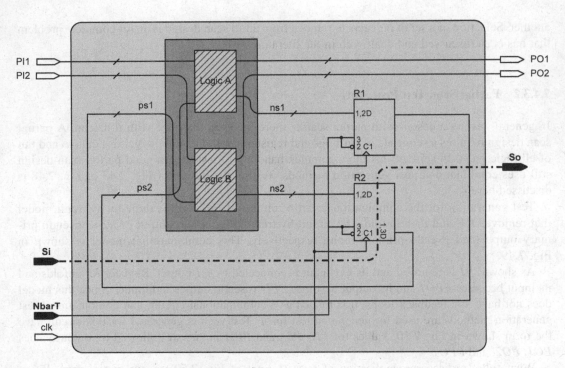

Fig. 7.47 Partial scan, starting with Huffman model

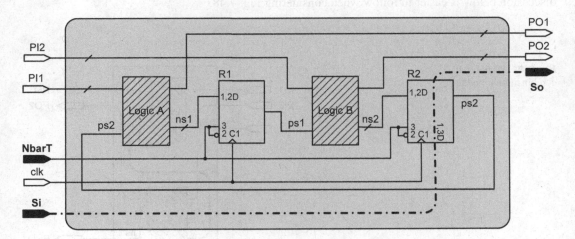

Fig. 7.48 Partial scan datapath

Combinational blocks *A* and *B* can be a mixture of logic units and buses in a datapath. Although it may be difficult to map an entire datapath into an architecture like that of Fig. 7.48, it is not hard to find sections of a large datapath that fit this model. In addition, with minor modifications to a datapath, this and other sub-architectures that qualify for various partial scan designs can be formed.

In the above discussion, we selected a register to be removed from full scan and found that this removal would still qualify the resulting circuit model for combinational test generation. In a large RT-level design, there may be many such choices, some of which may conflict with one another. Selection of a set of registers to remove from a full scan design is an *np*-complete problem that has been discussed and dealt with in the literature [10].

7.4.3.2 Partial Scan Test Procedure

In general, testing a design with partial scan is more involved than one with full scan. A partial scan design requires a sequence of shiftings and register parallel loading to get test data to and out of all logic parts. In spite of this more complex handling of test data, a good partial scan design still uses combinational test generation methods. Testing the design of Fig. 7.47 or Fig. 7.48 is discussed here.

Test generation for the combinational part of circuit of Fig. 7.47 is done by a circuit model that removes *R1*, and then unfolds the circuit by treating input and output of *R2* as pseudo primary output and pseudo primary input, respectively. This combinational model is shown in Fig. 7.49.

As shown, *R1* is removed and its *ns1* input is connected to *ps1* output. Register *R2* unfolds and its input becomes *PPO1* and its output becomes *PPI1* (pseudo output and input). Since this model does not have any feedback loops, it is treated as a combinational circuit, and combinational test generation methods are used for generating test for it. Test vectors generated for this circuit have the form shown in Fig. 7.50. Values for *PI1*, *PI2*, and *PPI1* inputs are followed by output values *PO1*, *PO2*, and *PPO1*.

What follows discusses application of tests (format of Fig. 7.50) to our partial scan design (Fig. 7.47 or Fig. 7.48), and collection of outputs. Although Figs. 7.47 and 7.48 are equivalent, the discussion below is easier to follow when considering Fig. 7.48.

Fig. 7.49 Partial scan combinational model

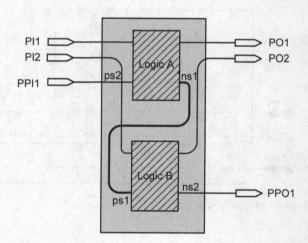

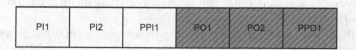

Fig. 7.50 Test vector arrangement

In the normal mode of operation, *NbarT* is 0 and circuit is performing its normal operation. With each clock, *ps1* receives *ns1* (register *R1*), and *ps2* receives *ns2* (register *R2*). In the test mode, *R1* is disabled and *R2* is put in the shift mode. When all data bits belonging to *PPI1* segment of a test vector have been shifted into *R2* (which means that *ps2* in Fig. 7.48 has proper test data), the *PI1* segment is applied to the *PI1* input of the circuit. At this time, *PO1* test response is collected and *ns2* is generated as well.

We then disable *R2* and enable *R1* parallel loading. This is happening while combinational block *A* (Fig. 7.48) has proper *PI1* and *ps2* test inputs. Clocking the circuit at this time will load *ns1* in *R1*, and thus applies it to block *B* via *ps1*.

The other part of input of block *B* is *PI2* that is read from the test vector and applied to this input. Now, block *B* has all its test data applied to its inputs. The *PO2* output from block *B* is read and collected.

Of the three parts of the test vector response, we have two parts ready (*PO1* and *PO2*). We now put the circuit back in the test mode, and while new test data shift in *R2*, test response of *B* in *R2*'s *ns2* input will serially be collected. After the shifting, the three parts of the output (*PO1*, *PO2*, and *PPO1*) are ready to be compared with the expected response.

The procedure discussed above is summarized as shown below. This procedure is for the test mode and begins by setting *NbarT* to 1.

- While *NbarT* is 1, clock *R2*, shift *PPI1* test data into *R2*, disable *R1*; also of previous test serially collect *PPO1*.
- Apply *PI1* test data to *PI1* input, collect response from *PO1*.
- Set *NbarT* to 0, enable *R1*, clock once.
- Apply *PI2* test data to *PI2* input, collect response from *PO2*.
- Keeping *NbarT* at 0, clock once.
- Return to step 1.

7.4.3.3 Partial Scan Versus Full Scan

Obviously, partial scan reduces test time by having fewer bits to shift in. On the other hand, partial scan has a more complex test procedure as discussed above. The main problem with partial scan is that there is no unique partial scan method, and not all circuits can necessarily take advantage of a partial scan method. For finding proper registers to scan, a topological processing of the circuit is necessary. Configuring a circuit for partial scan must be paralleled with extracting a test procedure that works with the scan design.

A partial scan method such as the one described here fits well with pipeline architectures. In this case, the test procedure becomes dependent on depth of the pipeline.

7.4.4 Multiple Scan Design

The problem of long scan chain can be moderated by using multiple independent or parallel scan chains. In multiple independent scan chains, each scan register has its shift, load, and clock control, whereas in multiple parallel scan chains, all scan registers are controlled by the same set of signals. If registers to be scanned can be put into groups of equal number of cells, then they can be regarded as parallel scan registers with the same set of control signals. However, if the number of flip-flops in the scan chains in a design are not the same, then they need independent shift and clock enable control signals [9].

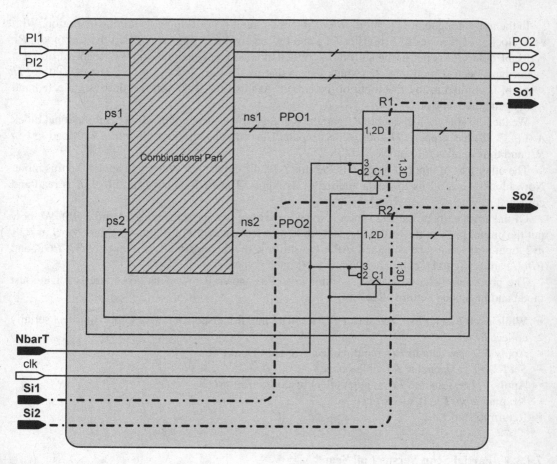

Fig. 7.51 Multiple parallel scan chains

7.4.4.1 Multiple Scan Architecture

Figure 7.51 shows the familiar model that we have been using for demonstrating our scan design. In this figure, *R1* and *R2* are put into two separate scan chains with *Si1* and *Si2* inputs and *So1* and *So2* outputs. We assume here that registers *R1* and *R2* have the same length. Therefore, the scan chains have become two parallel registers with the same clock and *NbarT* control inputs.

7.4.4.2 Multiple Scan Test Procedure

Test generation for a multiple scan design is based on the unfolded model and is no different than test generation for a full scan design. The difference is in application of test vectors through serial test inputs.

Input test data for the circuit of Fig. 7.51 consist of that for *PI1*, *PI2*, *PPI1*, and *PPI2*, and test responses are for *PO1*, *PO2*, *PPO1*, and *PPO2*. For testing this circuit that has two scan chains for *PPI1* and *PPI2*, individual test data bits from the corresponding test data segments are read and are applied simultaneously to *Si1* and *Si2*. When *NbarT* is 1, *R1* and *R2* are in the shift mode and shift test data into the registers. While shifting-in is taking place, previous results are collected from *So1* and *So2*. When shifting is complete, *NbarT* is set to 0, parallel test data for *PI1* and *PI2* are applied to the CUT, and *PO1* and *PO2* are read.

7.4.4.3 Compared with Full Scan

Multiple scan significantly reduces test time, with no overhead on test generation procedure. The overhead is on extra test pins, and in case of independent scans, on test clock and normal mode controls.

7.4.5 Other Scan Designs

Depending on the architecture of circuit under test, test data length, test procedure, and other such factors, many variations of scan are possible.

A variation of serial scan is random access scan. In this case, the feedback flip-flops in Fig. 7.44 are put in a memory array. In normal mode of operation, the flip-flops work in parallel mode, and in test mode, individual flip-flops are addressed, read out through scan output, and set to their test values through the scan input. Addressing scan flip-flops in test mode is done by row and column addresses that can be shifted-in, counted to, or a mixture of both.

This method offers fast access to the scan flip-flops, but has a large hardware overhead for implementation of the memory array. Addressing scan flip-flops and extra pins for this purpose are cause of extra hardware and possibly test time reduction.

Often datapath and control parts of a circuit require different DFT methods. For separating control and data parts of a circuit, the isolated scan method can be used. We can then use full scan for the control part, and perhaps partial scan for the data part. Such arrangements are architecture dependent, and too many to enumerate. Hardware designers must be aware of the possibilities and decide on the DFT method to use based on the circumstances. Often, there is no single best solution.

7.5 RT Level Scan Design

The previous section discussed various scan designs for the general sequential circuit model of Fig. 7.44 (Huffman model). In this section, we show how these methods apply to an actual RT-level design with a datapath and a controller.

A complete RT level design can either be partitioned into its datapath and controller, and each part be treated for scan separately, or all the registers and control flip-flops can be lumped together and treated according to the Huffman model of Fig. 7.44. In the example that we present in this section, we take the latter approach and treat all flip-flops similarly. The example that we use is the Adding Machine first presented in Chap. 2.

7.5.1 RTL Design Full Scan

As shown in Fig. 2.30, the Adding Machine has two 8-bit registers for *AC* and *IR*, and it has a 6-bit register for the program counter. With the addition of two control flip-flops, this design has a total of 24 flip-flops. Scan insertion in this design is done as shown in Fig. 7.52. The serial input is the left-most bit of *AC*, and serial output is taken from the least significant control bit.

Although we are showing an RT-level view of the circuit, the actual scan is inserted in the netlist of this circuit. The netlist is obtained by synthesizing the behavioral description of the Adding Machine. The scan is inserted in this netlist manually.

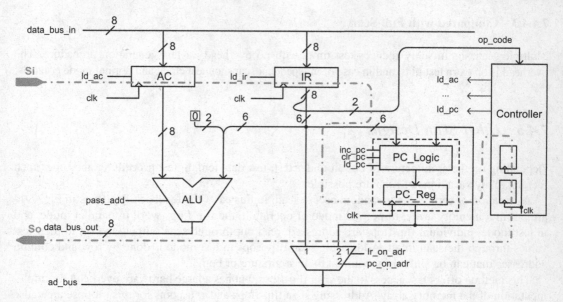

Fig. 7.52 Full scan Adding Machine

With the scan inserted as shown, *data_bus_in* becomes the circuit's primary input, and *data_bus_out* and *ad_bus* together form the primary output. The testable circuit has 24 pseudo primary input and 24 pseudo primary output that are register outputs and inputs, respectively.

For testing this circuit, an unfolded model of the circuit was created, as was done for the Residue-5 example of Sect. 7.3. This combinational netlist was used for test generation by HOPE for 976 faults. HOPE generated 64 tests and obtained coverage of 80.53%. Each test vector has 32 bits on the input side and 38 bits on the output side. From the generated 32 bits, 8 bits are used for *data_bus_in* and the remaining 24 bits are used for the scan path. The outputs are 8, 6, and 24 for *data_bus_out*, *ad_bus*, and the scan path, respectively.

The testable model of the Adding Machine was obtained by putting together the circuit flip-flops in one chain. Figure 7.53 shows the virtual tester for this full scan testing of the testable netlist of this circuit. Other than the circuit under test being different and the number of shifts in the scan flip-flops, this tester is no different from that of the Residue-5 circuit discussed in Sect. 7.3. Figure 7.53 only shows the loop that is responsible for applying test vectors to the scan-inserted netlist.

The testbench shown reads 64 test vectors from *testFile* and applies them to the full scan circuit for every 976 circuit faults. Eight bits of every test vector are applied in parallel and the other 24 are shifted-in serially by the **repeat** loop of Fig. 7.53. This loop also shifts out 24 output bits. The **if** statement shown in Fig. 7.53 compares concatenation of the serially collected output and 14 parallel outputs with the 38 bits of the expected outputs from *testFile*.

7.5.2 RTL Design Multiple Scan

This section shows implementation of a multiple scan design in the Adding Machine of Chap. 2. As in the previous discussion on full scan, although the circuit diagram looks different than the Huffman model of Fig. 7.44, the circuit is still structured as such, and the multiple scan method of Sect. 7.4 can be applied to this circuit.

```
module    Tester;
    . . .

    CPU_ScanInserted FUT(clk, PI, PO, NbarT, Si, So);
    always #200 clk = ~clk;

    initial begin
        . . .
        while( !$feof(faultFile)) begin
            . . .
          while((!$feof(testFile))&&(!detected))begin
              status = $fscanf(testFile,"%b\n",line);

              pre_expected_st = cur_expected_st;

              expected_PO     = line[out_size - 1:0];
              cur_expected_st = line[st_size + out_size -1 :out_size];
              PI = line[st_size+out_size+in_size-1:st_size+out_size];

              NbarT = 1'b1;
              #delay;

              index = stIndex;
              repeat(24) begin
                  si            = line[index];
                  @(posedge clk);
                  saved_st[index - stIndex] = so;
                  index         = index + 1;
              end

              NbarT = 1'b0;
              @(posedge clk);
              SampledPO = PO;

              if({pre_expected_st, expected_PO} != {saved_st, SampledPO})
              begin
                  numOfDetected = numOfDetected + 1;
                  detected = 1;
              end
              #5;

          end//test
          . . .
        end//fault
        . . .
    end//initial
endmodule
```

Fig. 7.53 Full scan Adding Machine virtual tester

We use three equal scan chains: one covers *AC*, second covers *IR*, and the third covers all of *PC* plus two control flip-flops. Since the size of the scan chains is the same, we can use multiple parallel scan chains that simplifies scan controls. Figure 7.54 shows the three scan chains in the RTL diagram of the Adding Machine.

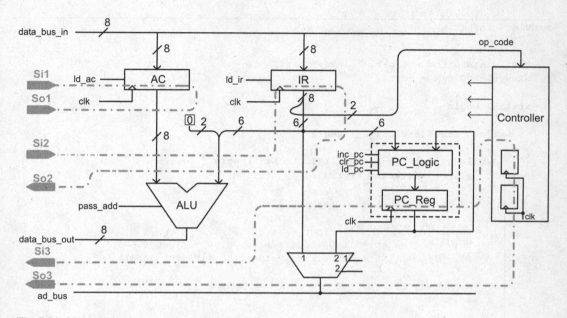

Fig. 7.54 Multiple parallel scans for Adding Machine

```
module CPU_M_ScanInserted(clk, {reset, data_bus_in}, {adr_bus,rd_mem,
                          wr_mem,data_bus_out}, NbarT, ir_Si, ac_Si,
                          pc_Si, ir_So, ac_So, cntrl_So);
  . . .
  dff INS_1 (wire_6,wire_96,clk,1'b0,1'b0,ir_en,NbarT,ir_Si, 1'b0);
  dff INS_2 (wire_2,wire_98,clk,1'b0,1'b0,ir_en,NbarT,wire_6, 1'b0);
  . . .
  dff INS_8 (wire_26,wire_173,clk,1'b0,1'b0,ir_en,NbarT,wire_54, 1'b0);
  //
  dff INS_9 (wire_130,wire_180,clk,1'b0,1'b0,ac_en,NbarT,ac_Si, 1'b0);
  dff INS_10(wire_135,wire_186,clk,1'b0,1'b0,ac_en,NbarT,wire_130, 1'b0);
  . . .
  dff INS_16(wire_52,wire_208,clk,1'b0,1'b0,ac_en,NbarT,wire_50, 1'b0);
  //
  dff INS_17 (wire_65,wire_211,clk,1'b0,1'b0,pc_en,NbarT,pc_Si, 1'b0);
  dff INS_18(wire_71,wire_216,clk,1'b0,1'b0,pc_en,NbarT,wire_65, 1'b0);
  . . .
  dff INS_24 (wire_4,wire_249,clk,1'b0,1'b0,pc_en,NbarT,wire_18, 1'b0);
  assign ir_So = wire_26;
  assign ac_So = wire_52;
  assign cntrl_So = wire_4;
endmodule
```

Fig. 7.55 Inserting three scan chains

The scan chains are manually inserted in the netlist that is the result of synthesizing the Verilog code of the Adding Machine. Insertion of scan makes changes in the ports of the netlist and arrangement of its flip-flops. Figure 7.55 shows partial code of the modified netlist.

As shown, *NbarT* and three serial inputs (*ir_Si*, *ac_Si*, and *pc_Si*) are added to the inputs, and three serial scan outputs (*ir_So*, *ac_So*, and *cntrl_So*) are added to the output ports of the scan-inserted netlist. The three bracketed sections in Fig. 7.55 show the three scan chains. All

```
module    Tester;
    . . .
    CPU_M_ScanInserted FUT(clk, PI, PO, NbarT, ir_Si, ac_Si, pc_Si, ir_So,
                            ac_So, cntrl_So);


    always #200 clk = ~clk;

    initial begin
        . . .
        while( !$feof(faultFile)) begin

            while((!$feof(testFile))&&(!detected)) begin
                status = $fscanf(testFile,"%b\n",line);
                pre_expected_st = cur_expected_st;
                expected_PO = line[out_size - 1:0];
                cur_expected_st = line[st_size + out_size -1 :out_size];
                PI = line[st_size+out_size+in_size-1:st_size+out_size];
                NbarT = 1'b1;
                #delay;

                index = stIndex;
                repeat(8) begin
                    ir_Si = line[index];
                    ac_Si = line[index+8];
                    pc_Si = line[index+16];
                    @(posedge clk);
                    saved_st[index - stIndex]      = ir_So;
                    saved_st[index+8 - stIndex]    = ac_So;
                    saved_st[index+16 - stIndex]   = cntrl_So;
                    index = index + 1;
                end

                NbarT = 1'b0;
                @(posedge clk);
                SampledPO = PO;

                if({pre_expected_st, expected_PO} != {saved_st, SampledPO})
                begin
                    numOfDetected = numOfDetected + 1;
                    detected = 1;
                end
                #5;

            end//test
            . . .
        end//fault
        . . .
    end//initial
endmodule
```

Fig. 7.56 Test program for multiple parallel scans

flip-flops use *NbarT* as input. The first chain begins with *ir_Si* and ends with *wire_26 (INS_1 to INS_8)* that becomes the *ir_So* output (see *assign* statements near the end of the code).

The serial input of the second scan is *ac_Si*, and its serial output is *wire_52 (INS_9 to INS_16)* that becomes *ac_So*. Finally, the third scan chain begins with *pc_Si* and ends with *cntrl_So (INS_17 to INS_24)*.

For developing a test program for this scan design, we have to consider that our design has three equal-length scan chains using the same clock, and the same normal/test mode (*NbarT*) control input. A virtual tester imitating the behavior of an ATE and its test program for testing our design is shown in Fig. 7.56. This figure only shows instantiation of the three-scan-chain circuit and the procedure for application of test data. Other parts of the testbench that are basically the same as code discussed in Sect. 7.3 are not shown here. As shown, instantiation of the netlist, the partial code of which is shown in Fig. 7.55, includes *NbarT* as well as serial inputs and outputs of the three scan chains.

The test procedure is enclosed in the **initial** block shown in Fig. 7.56. All but the **repeat** loop of this procedure is the same as that of Fig. 7.53. The **repeat** loop here takes every 8 bits of *line* that has the input vector and assigns them to *ir_Si*, *ac_Si*, and *pc_Si*. This happens while *NbarT* for all three chains is 1 (shift mode). After the three assignments, the registers are clocked (simultaneously), after which outputs from *ir_So*, *ac_So*, and *cntrl_So* are collected in *saved_st*. These output values are positioned in the *saved_st* vector, 8 bits apart to allow output bits of the same chain to be adjacent to each other.

7.5.3   Scan Designs for RTL

It is easy to plan various scan designs for an RT level design by considering its RTL view. Once planning is done, an estimate for the test time can be obtained. The implementation of the planned scan design is done on the synthesized output of the RT-level HDL code. Developing a test program is made easier when thinking of the circuit in terms of its Huffman model after scan insertion.

Other possibilities for scan design for our RTL example of this section are partial scan, and multiple independent scans. Alternatively, if we partition the circuit into its datapath and a controller, separate scan design can be implemented for each part. Splitting a circuit into several parts can be implemented by isolated scan.

When planning different scan strategies, the number of test vectors, length of each test vector, and fault coverage are the main parameters to consider.

7.6   Summary

This chapter discussed DFT techniques. We started with some ad hoc methods that evolved into isolated scan. All such methods apply to combinational as well as sequential circuits. The scan design, that we started discussing in Sect. 7.3, was primarily focusing on sequential circuits, and its main purpose was to turn a sequential circuit into a combinational circuit, such that combinational TG methods could be used for generating tests.

Other scan designs that we discussed basically followed the same rule. In partial scan methods, the introduction of pseudo inputs and pseudo outputs does not completely turn a sequential model into a combinational one, and registers still exist in the obtained model. In spite of this, the obtained model can still be used for combinational test generation. The deviation from a pure combinational model affects the test procedure where extra clock pulses compensate for registers that are left in the model.

To show application of scan methods to RT-level designs and demonstrate the applicability of such schemes, the last section used a small RT-level design and a Verilog testbench to test its inserted scan. Larger designs can be partitioned into several subcomponents, where each subcomponent may require a different form of scan. Separating a design into its subcomponents can be done by a shift-register for scanning in and scanning out signal values in the interface of the subcomponents.

References

1. Abramovici M, Breuer MA, Friedman AD (1994) Digital systems testing and testable design. IEEE Press, Piscataway, NJ, revised printing.
2. Wilkins BR (1986) Testing digital circuits, an introduction. Van Nostrand Reinhold, Berkshire, UK.
3. Eichelberger EB, Lindbloom E, Waicukauski JA, Williams TW (1991) Structured logic testing. Prentice-Hall, Englewood Cliffs, NJ.
4. Agrawal VD, Mercer MR (1982) Testability measures – What do they tell us? In: Proceedings of the International Test Conference, Nov 1982, pp 391–396.
5. Willaims MJY, Angell JB (1973) Enhancing testability of large-scale integrated circuits via test points and additional logic. IEEE Trans Comput C-22(1):46–60.
6. Miczo A (2003) Digital logic testing and simulation, 2nd edn. Wiley, Hoboken, NJ.
7. Cheng K-T, Lin C-J (1995) Timing-driven test point insertion for full-scan and partial-scan bist. In: Proceedings of the International Test Conference, Oct 1995, pp 506–514.
8. Cheng K-T, Agrawal VD (1990) A partial scan method for sequential circuits with feedback. IEEE Trans Comput 39(4):544–548.
9. Narayanan S, Gupta R, Breuer MA (1993) Optimal configuring of multiple scan chains. IEEE Trans Comput 42(9):1121–1131.
10. Jha NK, Gupta S (2003) Testing of digital systems, Cambridge University Press, Cambridge, UK.

Chapter 8
Standard IEEE Test Access Methods

A different DFT method than the scan of the previous chapter is boundary scan that primarily targets the boundary of a CUT, instead of the scan whose focus is on the inside of chip or a core. Boundary scan that has become an IEEE standard (IEEE std.1149.1) does not interfere in the design of a core, and its main purpose is to isolate the core being tested from other devices on a board or chip. This chapter discusses architecture, application, and operation of this IEEE standard. We use BS-1149.1 to refer to this standard.

Section 8.2 discusses the architecture and hardware of BS-1149.1. We use Verilog code to cover the details of the major parts of the architecture. BS-1149.1 uses instructions for testing a chip's internal and external connections and isolating it from its surroundings. Section 8.3 covers the related instructions and roles they play in testing a core or a chip. We wrap up our presentation of the 1149.1 standard by a complete example in Sect. 8.5. Another standard related to 1149.1 is the boundary scan description language (BSDL) that will be discussed in Sect. 8.6. This chapter also presents several arrangements for using multiple boundary scan chains on a chip or a board.

8.1 Boundary Scan Basics

Due to complexity of digital components, and multilayer printed circuit boards, in-circuit testing by bed-of-nails probing technique for isolating components is no longer an easy solution. Furthermore, testing complex components and multicore chips is not possible by off-chip test methods. It is thus required to be able to access various components and/or cores and be able to isolate them from each other, without a significant increase in the cost of test [1].

In 1985, joint test action group (JTAG) that consisted of designers, manufactures, and test engineers was formed. This group established a set of specifications for shifting serial test data into a board, for testing it. Later in 1990, these specifications become the IEEE std.1149.1. This standard set a unique set of rules to follow for test engineers, ATE developers, and test program developers [1].

With the guidelines and constraints that this standard provided, a flexible, but yet standard set of design rules for designing test access mechanism at the board level was created. The standard was usable in all digital designs, and eased the development of ATE hardware and software by setting a standard.

The IEEE std.1149.1 is also known as boundary scan, since it mainly consists of a scan register on the ports of a component for testing its interconnects and core logic. Boundary scan standard eliminates the need for probing a component's pins with a physical probe. The method improves controllability and observability within a PCB or a chip. Figure 8.1 shows how this test standard is incorporated into a chip along with the chip's core logic.

Z. Navabi, *Digital System Test and Testable Design: Using HDL Models and Architectures,*
DOI 10.1007/978-1-4419-7548-5_8, © Springer Science+Business Media, LLC 2011

Fig. 8.1 General structure of
BS-1149.1

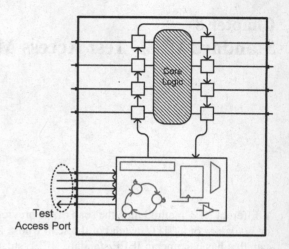

Test
Access Port

Boundary scan uses a chain of scan flip-flops to shift test data into the inputs of a core logic being tested, and uses the same mechanism to move test response out. The scan flip-flops isolate a core from its neighboring cores, and testing of each core is done independently.

The boundary scan standard works in two modes of noninvasive and pin-permission. In the non-invasive mode, independent from board or chip core logic, the test hardware (BS-1149.1) communicates with the outside world for bringing in test data, or transmitting response out of the system. This is done while the rest of the system performs its normal functions [1].

In the pin-permission mode, the BS-1149.1 hardware takes over input and output pins of a core logic for testing its interconnects. In this mode, the core logic is disconnected from its environments and is only operated by test logic. After completion of a pin-permission mode operation, it is important for the test hardware to be put back in the noninvasive mode to avoid bus conflicts while the system performs its normal functions.

8.2 Boundary Scan Architecture

Figure 8.1 shows an overall view of how boundary scan is incorporated on a chip. As shown, test hardware that consists of controllers, registers, and decoders sits on a chip along with the chip's core logic. The boundary scan test hardware also has a scan register that wraps around the core logic to control its communication with the outside world. The core logic and BS-1149.1 hardware form a single testable package like a chip on a board or a core on a chip.

Figure 8.2 shows the main details of the boundary scan hardware. Decoder, register cells, a state machine, ports, and other hardware details are shown in this figure. In this section, hardware structures shown in this figure along with their Verilog codes will be discussed.

8.2.1 Test Access Port

As shown in Fig. 8.2, BS-1149.1 adds several pins to the normal inputs and outputs of a core for test data and test control. There are a total of four or five (one is optional) such signals. These pins are for test purposes only, and cannot be used by the core logic for its normal functionalities.

TMS, *TCLK*, and *TRST* are control pins. *TMS* (Test Mode Select) is used for putting the test protocol in a given state for data or for instruction. *TCLK* (Test Clock) is the main test clock input that runs all the corresponding test hardware. *TRST* (Test Reset) is an optional pin and, if used, resets the test hardware into its noninvasive mode [2].

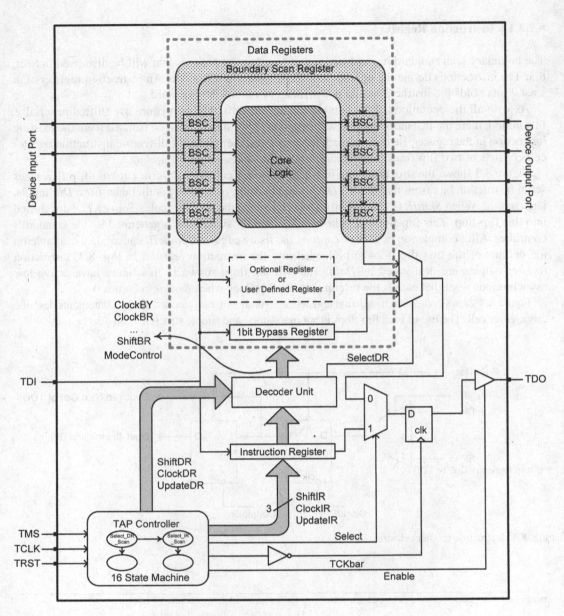

Fig. 8.2 Details of boundary scan standard

There are also two data pins, *TDI* and *TDO*. *TDI* (Test Data In) is for shifting serial test data and instruction into the chip, and thus, into the BS-1149.1 registers. *TDO* (Test Data Out) is the serial output of this standard test protocol [2].

It is important to note that pins that are not used should be left floating high. This prevents interference of such pins in the normal functionality of core and test logic.

8.2.2 BS-1149.1 Registers

Several registers form the main hardware of the BS-1149.1 standard. These are basically shift registers with special hardware to comply with the standard's various modes of operation. The registers are categorized into instruction and data registers [2].

8.2.2.1 Instruction Register

The boundary scan standard has instructions with certain bit patterns that will be discussed in Sect. 8.3. The instructions define the operation of the standard in test mode. An instruction register of at least 2 bits holds the instructions and is a mandatory part of this standard.

As with all the operations of the boundary scan standard, the instructions are shifted in serially. In addition, there are instances that a new instruction being shifted must be isolated from the existing instruction in the register. Because of such requirements, the standard full-feature instruction register cell consists of two flip-flops, one for *shift* or *capture* and another for *update*.

Figure 8.3 shows the structure of an instruction register cell. The *shift* (or *capture*) flip-flop takes serial instruction bits from its *Sin* (or *TDI*) input, i.e., *shift*, or it takes parallel data from *Din* inputs, i.e., *capture*. When *ShiftIR* is 1, serial instruction bits from the previous cell's *Sout* or *TDI* are shifted into this flip-flop. This flip-flop is clocked by the *ClockIR* signal that is generated by the standard's controller. After completion of *shift* or *capture*, the rising edge of *UpdateIR* causes data available on the outputs of the first flip-flops to be loaded into the instruction register. In Fig. 8.3, instruction register outputs are designated by *Dout*. The two flip-flops shown in this figure have active low asynchronous reset that causes the resetting of the flip-flops when *RstBar* becomes 0.

Figure 8.4 shows code of an instruction register cell. As shown, two clocked **always** statements describe this register cell. The use of both flip-flops is not mandatory, and simple structures can also be used.

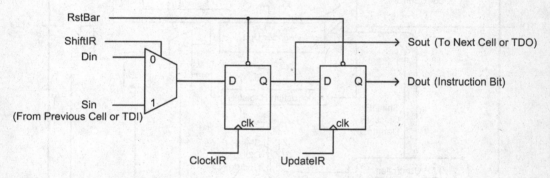

Fig. 8.3 Instruction register cell structure

```verilog
module InstructionRegister1bit (Din, Sin, ShiftIR, UpdateIR,
                                ClockIR, RstBar, Sout, Dout);
   input Din, Sin, ShiftIR, UpdateIR;
   input ClockIR, RstBar;
   output Sout;
   output reg Dout;
   wire D_DF1;
   reg Q_DF1;
   assign D_DF1 = ShiftIR ? Sin: Din;
   always @(posedge ClockIR, negedge RstBar)
      if(!RstBar) Q_DF1 <= 0; else Q_DF1 <= D_DF1;
   always @(posedge UpdateIR, negedge RstBar)
      if(!RstBar) Dout <= 0; else Dout <= Q_DF1;
   assign Sout = Q_DF1;
endmodule
```

Fig. 8.4 Instruction register Verilog code

8.2.2.2 Data Registers

The instruction that is loaded in the instruction register causes one of the data registers to go between *TDI* and *TDO* serial input and serial output. Any register that can logically be placed in the *TDI*, *TDO* serial path is referred to as a data register. A data register may have only a single cell, or as many as the core logic's input–output pins. Data register cell structure is similar to that of the instruction register. However, based on the specific data register and applications, other cell type may also be used. Description of various BS-1149.1 data registers follows.

Bypass register. The bypass register is a mandatory boundary scan data register, and it is used to bypass a core from scan chain so that serially shifted data can reach the target core quicker. The bypass register is a single-bit register and its cell structure only uses the *shift* or *capture* flip-flop as shown in Fig. 8.5. The bypass register can be loaded with a 0 through its *Din* input. BS-1149.1 uses this feature for chain integrity. In the bypass mode (*ShiftBY* = 1), *TDI* is clocked into the bypass register on the rising edge of *ClockBY*. Figure 8.6 shows the Verilog code of this register. This is a subset of the Verilog code of Fig. 8.4.

Device identification register. Device identification register (DIR) is an optional register in the 1149.1 standard. If used, this is a 32-bit register that contains an identification code for the core logic that it is a part of. The structure of this register is similar to that of the instruction register shown in Fig. 8.3. Core logic id that is stored in this register can be shifted out serially that becomes available for chain integrity testing.

Boundary scan register. The boundary scan register that is considered as a data register in BS-1149.1 is the most important of all the registers of this standard. The boundary scan register is placed on the boundary of the core logic that is being tested. The register cells go between external pins (interconnects) and ports of the core logic. This mechanism improves controllability and observability of core logic's inputs and outputs.

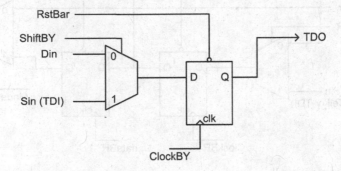

Fig. 8.5 Bypass register cell structure

```
module ByPassRegister (Din, Sin, ShiftBY, ClockBY, RstBar, TDO);
    input Din, Sin;
    input ShiftBY, ClockBY, RstBar;
    output reg TDO;
    wire D_DF;
    assign D_DF = ShiftBY ? Sin: Din;
    always @ (posedge ClockBY, negedge RstBar)
        if (!RstBar) TDO <= 0; else TDO <= D_DF;
endmodule
```

Fig. 8.6 Bypass register Verilog description

Figure 8.7 shows a typical boundary scan register cell. Signals *Din* and *Dout* go between the interconnect and the core logic. If this cell is used on a core logic's input, *Din* connects to the interconnect and *Dout* to the port of the core logic. For an output port, this arrangement is reversed. Signals *Sin* and *Sout* are used for shifting serial data that enter the register on the *TDI* port and exit the register on the *TDO* port of the standard.

Like the instruction register cell, the boundary scan register cell has a *shift* or *capture* flip-flop and an *update* one. The *update* flip-flop holds the contents of the boundary scan register. Two multiplexers in this structure are for proper routing of data in various modes of operation of this cell.

In the normal mode of operation, the *ModeControl* of the output multiplexer is 0, and *Din* connects to the *Dout*. In this mode, *ShiftBR*, *ClockBR*, and *UpdateBR* have no effect on *Dout*, i.e., *Dout* receives data directly from *Din*. This mode allows the noninvasive operation, where serial test data can be shifted in, and clocked into the *update* flip-flops, without interfering in the normal operation of the core logic. Another noninvasive operation is capturing data on *Din* in the *capture* flop-flops and then shifting it out. In the pin-permission test mode, the select input of the multiplexer on the output side of the register cell (*ModeControl*) becomes 1, which causes the *update* flip-flops to be connected to *Dout* cell outputs.

As mentioned, if used as an input cell, *Dout* connects to the input of the core logic, and if used as an output cell, *Dout* will be connected to the interconnect. If the core logic has a bidirectional pin, then we use three boundary scan cells. One for the input side, with its *Dout* connected to the core's input, one on the output side with *Dout* connected to the interconnect, and the third cell for driving the bidirectional tristate buffer. This arrangement is shown in Fig. 8.8. Verilog code corresponding to a boundary scan cell is shown in Fig. 8.9.

User-defined registers. Users are able to add their custom user-defined registers. The only restriction is that, when used for test purposes, such registers must logically be placed between *TDI* and *TDO* ports for shift path consistency.

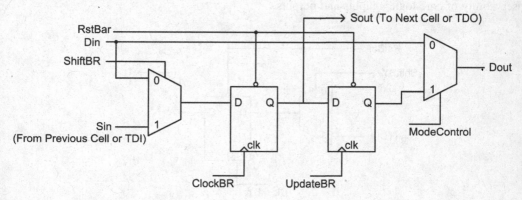

Fig. 8.7 Boundary scan register cell

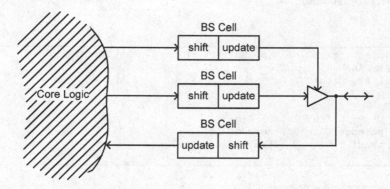

Fig. 8.8 Using a BS cell in bidirectional ports

```
module BSRegister1bit(Din, Sin, ShiftBR, UpdateBR, ClockBR,
                      RstBar, ModeControl,Sout, Dout);
   input Din, Sin, ShiftBR, UpdateBR;
   input ClockBR, RstBar, ModeControl;
   output Sout, Dout;
   wire D_DF1;
   reg Q_DF1, Q_DF2;
   assign D_DF1 = ShiftBR ? Sin: Din;
   always @(posedge ClockBR, negedge RstBar)
      if (!RstBar) Q_DF1 <= 0; else Q_DF1 <= D_DF1;
   always @(posedge UpdateBR, negedge RstBar)
      if(!RstBar) Q_DF2 <= 0; else Q_DF2 <= Q_DF1;
   assign Dout = ModeControl ? Q_DF2: Din;
   assign Sout = Q_DF1;
endmodule
```

Fig. 8.9 BS cell Verilog code

8.2.3 TAP Controller

All boundary scan operations are controlled by a simple controller that has sixteen states. The controller uses *TCLK* for the clock, *TMS* for its input, and, if used, *TRST* for resetting it. The controller is called the test access port (TAP) controller, and by issuing its control signals it controls the operation of the instruction and various data registers. Figure 8.10 shows the BS-1149.1 TAP controller. Going from one state to another is controlled by *TCLK* synchronized 1 and 0 values on *TMS*. The controller states are arranged in two columns, where the left column states are for controlling the data registers, and the right column states for the instruction register.

The TAP controller has two starting states, seven data register control states, and seven instruction register control states. Holding *TMS* high for five or more clocks always returns control to the reset state. This feature prevents performing unwanted operations for unexpected glitches on *TMS*. Figure 8.11 shows the synthesizable Verilog code of the TAP controller. State names and signals are according to the diagram of Fig. 8.10. States of the controller will be discussed below [2].

Test_Logic_Reset state. As shown in Fig. 8.10, the first state of the controller is the reset state. This state is entered either by issuing *TRST* (not shown) or by *TMS* being one for five consecutive clocks. While in this state, a *RstBar* reset signal is issued to all the boundary scan components to reset them to their initial states. This signal loads a null pattern in the instruction register to prevent the test logic from interfering in the normal operation of core logic.

Run_Test_Idle state. The next state of the TAP controller that is entered if *TMS* is 0 is *Run_Test_Idle*. As for as the operation of BS-1149.1 is concerned, this state is an idle state. However, contents of instruction and data registers remain the same as what they were in the previous state. In this state the core logic can perform its own self-test operations. When *TMS* becomes 1, the boundary scan exits this idle mode and enters the next state.

Select_DR_Scan state. The *Select_DR_Scan* state of the TAP controller is a temporary state in which the controller either continues with data register operations or gets ready for performing operations related to the instruction registers.

Select_IR_Scan state. As with the *Select_DR_Scan*, the *Select_IR_Scan* is also a temporary state in which either the reset state or the instruction register states are decided.

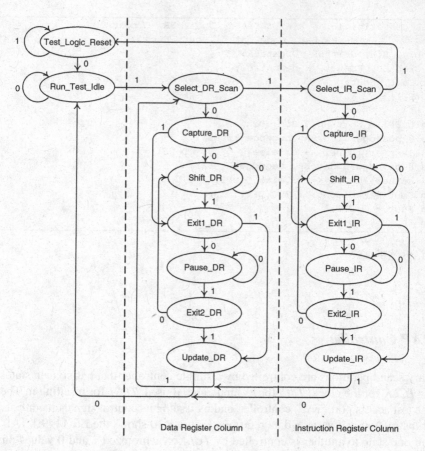

Fig. 8.10 TAP controller

Capture_IR (*Capture_DR*) states. In the *Capture_IR* state, the instruction register *ClockIR* is issued that causes it to perform a parallel load (see Fig. 8.3). This will load "01" in the least significant bits of the instruction register, and if the instruction register is longer than 2 bits, the rest of the bits will receive a predefined value that is treated as a null instruction.

In the *Capture_DR* state and other data register-related control states, a specific data register that is selected by the instruction register is targeted. *Capture_DR* state issues *ClockDR* of the selected data register, i.e., boundary scan, identification register, bypass, etc.

For the boundary scan data register, *Capture_DR* state causes it to capture data on *Din*. For an input cell, this means data from the interconnects are loaded into the *capture* flip-flops, and for an output cell, capture state captures the output from the core logic in this flip-flop.

The next state after *Capture_IR* (*Capture_DR*) is *Shift_IR* (*Shift_DR*) that is entered if *TMS* is 0. If shifting is not required, the TAP controller moves toward exiting the data register branch of the control states.

Shift_IR (*Shift_DR*) states. In the *Shift_IR* the instruction register is placed between TDI and TDO. While in this state, rising edges of *ClockIR* cause captured data to be shifted out on *Sout* and new serial data moved from *Sin*. In this state a new instruction bit pattern that appears on *TDI* will be shifted in the *capture* flip-flops. After completion of shifting the proper bit pattern, the instruction is loaded when TAP controller goes into the *Update_IR* state.

```
module TAPController (TMS, TCLK, RstBar, sel, Enable, ShiftIR,
                      ClockIR, UpdateIR, ShiftDR, ClockDR,
                      UpdateDR);

 . . .

    always @(posedge TCLK) begin
      case (TAP_STATE)
        Test_Logic_Reset :
            if (TMS == 1'b0) TAP_STATE = Run_Test_Idle;
            else if(TMS == 1'b1) TAP_STATE = Test_Logic_Reset;
        Run_Test_Idle:
            if (TMS == 1'b1) TAP_STATE = Select_DR_Scan;
            else if (TMS == 1'b0) TAP_STATE = Run_Test_Idle;
        Select_DR_Scan :
            if (TMS == 1'b0) TAP_STATE = Capture_DR;
            else if (TMS == 1'b1) TAP_STATE = Select_IR_Scan;
        Capture_DR:
            if (TMS == 1'b0) TAP_STATE = Shift_DR;
            else if (TMS == 1'b1)TAP_STATE = Exit1_DR;
        Shift_DR:
            if (TMS == 1'b1) TAP_STATE = Exit1_DR;
            else if (TMS == 1'b0) TAP_STATE = Shift_DR;
        Exit1_DR:
            if (TMS == 1'b0) TAP_STATE = Pause_DR;
            else if (TMS == 1'b1) TAP_STATE = Update_DR;
        Pause_DR:
            if (TMS == 1'b1) TAP_STATE = Exit2_DR;
            else if (TMS == 1'b0) TAP_STATE = Pause_DR;
        Exit2_DR:
            if (TMS == 1'b1) TAP_STATE = Update_DR;
            else if (TMS == 1'b0) TAP_STATE = Shift_DR;
        Update_DR:
            if (TMS == 1'b0) TAP_STATE = Run_Test_Idle;
            else if (TMS == 1'b1) TAP_STATE = Select_DR_Scan;
        Select_IR_Scan:
            if (TMS == 1'b0) TAP_STATE = Capture_IR;
            else if (TMS == 1'b1)TAP_STATE = Test_Logic_Reset;
        Capture_IR:
            if (TMS == 1'b0) TAP_STATE = Shift_IR;
            else if (TMS == 1'b1) TAP_STATE = Exit1_IR;
        Shift_IR:
            if (TMS == 1'b1) TAP_STATE = Exit1_IR;
            else if (TMS == 1'b0) TAP_STATE = Shift_IR;
        Exit1_IR:
            if (TMS == 1'b0) TAP_STATE = Pause_IR;
            else if (TMS == 1'b1) TAP_STATE = Update_IR;
        Pause_IR:
            if (TMS == 1'b1) TAP_STATE = Exit2_IR;
            else if (TMS == 1'b0) TAP_STATE = Pause_IR;
        Exit2_IR:
            if (TMS == 1'b1) TAP_STATE = Update_IR;
            else if (TMS == 1'b0) TAP_STATE = Shift_IR;
        Update_IR:
            if (TMS == 1'b0) TAP_STATE = Run_Test_Idle;
            else if (TMS == 1'b1) TAP_STATE = Select_DR_Scan;
      endcase
    end // end always
```

Fig. 8.11 TAP controller Verilog code

```
    always @(negedge TCLK) begin
        RstBar = 1'b1;
        Enable = 1'b0;
        ShiftIR = 1'b0;
        ShiftDR = 1'b0;
        ClockIR = 1'b1;
        UpdateIR = 1'b0;
        ClockDR = 1'b1;
        UpdateDR = 1'b0;

        case (TAP_STATE)
            Test_Logic_Reset:
                RstBar = 1'b0;
            Shift_IR: begin
                Enable = 1'b1;
                ShiftIR = 1'b1;
                ClockIR = 1'b0;
             end
            Shift_DR: begin
                Enable = 1'b1;
                ShiftDR = 1'b1;
                ClockDR = 1'b0;
            end
            Capture_IR:
                ClockIR = 1'b0;
            Update_IR:
                UpdateIR = 1'b1;
            Capture_DR:
                ClockDR = 1'b0;
            Update_DR:
                UpdateDR = 1'b1;
        endcase
    end // end always
 . . .
endmodule
```

Fig. 8.11 (continued)

The *Shift_DR* state is similar to the *Shift_IR* expect that in the former case serial data are shifted in the selected data register (see Fig. 8.7). In this case *ShiftDR* signal is set to 1, and shifting occurs on the rising edge of *ClockDR*.

If the selected data register is the boundary scan register, there are potentially many data bits that are to be shifted. If the external tester has limited buffer memory, there may be a delay in catching up with the speed of serial shifting. For this purpose, BS-1149.1 allows a pause while the external tester fetches more data for its buffer memory. This is achieved by setting *TMS* so that the controller goes into the *Pause_DR* state via *Exit1_DR* and return to *Shift_DR* via *Exit2_DR*.

While in *Pause_DR*, for the necessary amount of time that the external tester needs to fetch its next block of test data, *TMS* remains 0, which keeps the controller in the pause state. There is no need for this feature of BS-1149.1 to be used for smaller data registers such as bypass and identification register and definitely not necessary for the instruction register where instruction registers are never larger than several bits wide.

Exit1_IR (*Exit1_DR*) states. As mentioned above, the *Exit1* states are either used for transition into the pause states or for preparing for existing. In either case, no control signals are issued by the TAP controller while it is in *Exit1_IR* or *Exit1_DR* states.

Pause_IR (*Pause_DR*) states. As mentioned above, pause states allow time for an external data to fetch data from its mass storage devices. No control signals are issued in either of the two pause states. Generally, instructions are short and do not require such a pause state. Therefore, *Pause_IR* and *Exit2_IR* may be eliminated from the instruction branch of the TAP controller.

Exit2_IR (*Exit2_DR*) states. *Exit2* states of the TAP controller are auxiliary states that are used for returning to shifting or completing register loading and eventual exiting.

Update_IR (*Update_DR*) states. After an instruction is shifted in the instruction register's *capture* flip-flops, or a complete test vector is shifted in the boundary scan register's *capture* flip-flops, the TAP controller moves in the corresponding *update* state.

In the *Update_IR* state *UpdateIR* signal is issued (see Fig. 8.3), and on its rising edge, the bit pattern in the shift register chain loads into the instruction register as the current instruction. Once this is done, signals corresponding to this new instruction are activated, and a specific data register will be selected. The decoder unit of BS-1149.1 standard uses the instruction register and TAP controller signals to issue appropriate selection and clocking signals to the data registers.

The *Update_DR* state loads test data bits that have been shifted in the shift register chain of a data register into its *update* flip-flops. For a boundary scan register that is used as an input cell, this event is what is needed to make test data available for the inputs of the core logic. For an output cell, the data that have been loaded into the *update* flip-flops can now become available on the interconnects.

8.2.4   The Decoder Unit

The decoder unit is a combinational circuit that takes the existing instruction in the instruction register and signals from the TAP controller as inputs and issues signals to the appropriate data register.

8.2.5   Select and Other Units

Figure 8.2 shows two multiplexers, a flip-flop, and a tristate buffer in the architecture of boundary scan, all of which lead to the *TDO* output. The multiplexer shown near the data registers has its select inputs driven by the *Decoder Unit*, and selects the data register output specified by the current instruction. Another multiplexer whose select input is driven by the TAP controller selects a data register output or the instruction register output to go on the *TDO* of the BS-1149.1.

The flip-flop on the serial output before *TDO* is for synchronizing serial transmission of data with the test clock. Finally, the tristate buffer shown in this figure puts *TDO* in the float state when not in use.

8.3   Boundary Scan Test Instructions

Another important part of BS-1149.1 contributing to its functionality and test performance are its instructions. Boundary scan instructions are categorized into three groups. The first group is the mandatory instructions that must be implemented in any BS-1149.1 compliant test circuitry.

The second group is the optional instructions that are defined in the standard, but designers have a choice of not using them. The third group of instructions is user-defined that are there for extensibility and flexibility of the standard. What follows discusses operation of the mandatory instructions [3].

8.3.1 Mandatory Instructions

Mandatory instructions have been referred to as the mandatory module in the more recent BS-1149.1 documentations. These instructions are *Bypass*, *Sample*, *Preload*, *Extest*, and *Intest*, and the details and examples of which will be shown below.

8.3.1.1 *Bypass* Instruction

The *Bypass* instruction is used for shortening the scan path and bypassing those units that do not participate in a certain round of test. The data register of a component that is being bypassed becomes a single cell that is the bypass register. As far as the bypassed component is concerned, it only takes one clock cycle to pass through it and reach the next core. Figure 8.12 shows that reducing the length of scan chain is achieved by putting the left-hand side hardware unit in the bypass mode. This reduction causes the test data to get to the component shown on the right-hand side of this figure. Note: In the boundary scan cells shown in this figure *C* stands for *Capture* flip-flop and *U* for *Update*.

Putting a boundary scan compliant chip or core in the bypass mode can be done while it is performing its normal functions. For this purpose, the TAP controller of the chip first goes into its instruction side of the states and shifts and updates the chip's instruction register with the code for *Bypass*. It then moves into the data side of the states, and in *Capture_DR* loads a 0 into the Bypass register cell. Following this state, it moves into the *Shift_DR* state (see Fig. 8.12) and remains there for as long as it is being bypassed.

Figure 8.12 also shows the chip that is not being bypassed. As shown, the boundary scan register of the chip on the right is selected and the serial data are shifted into this data register. The shifting here also occurs in the *Shift_DR* state of its TAP controller.

8.3.1.2 *Sample* Instruction

The mandatory BS-1149.1 *sample* instruction works in the noninvasive mode and takes a snap-shot of the input interconnect values and outputs of core logic. After the sampling, the data will be shifted out through *TDO*. Figure 8.13 shows the boundary scan cells while the sample instruction is being performed and when the sampled data are being shifted out.

This instruction begins with its corresponding code loaded into the instruction register in *Update_IR* state of the TAP controller. This will cause the boundary scan register to be selected and logically placed between *TDI* and *TDO*. The TAP controller then moves to its left branch into the data register states. In the *Capture_DR* state, while the core logic is performing its normal functions, its inputs from the interconnect and outputs from the core are captured in the *capture* flip-flops of the scan cells (left-hand side diagram of Fig. 8.13). This takes place because in the *Capture_DR* state, while the *Sample* is in the instruction register, the BS-1149.1 decoder logic sets *ShiftBR* = 0 and issues *ClockBR*. (Note that *BR* is for Boundary Register-related signals from the decoder logic).

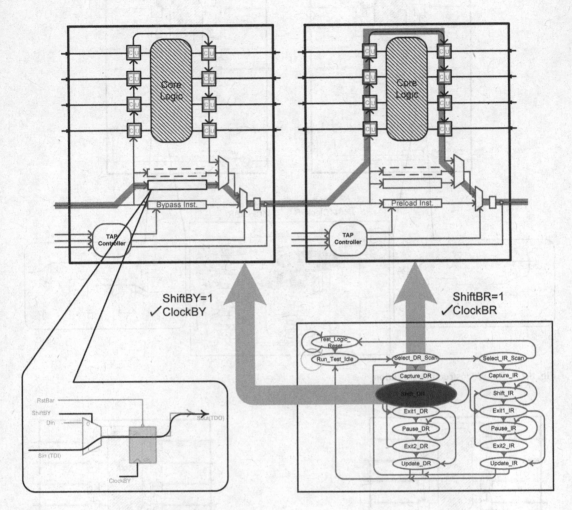

Fig. 8.12 *Bypass* instruction execution

After capturing has been done, the TAP controller moves into the *Shift_DR* state (right-hand side part of Fig. 8.13), in which *ShiftBR* is set to 1, while in this state, the sampled values will be shifted out through *TDO*.

8.3.1.3 *Preload* Instruction

Another BS-1149.1 mandatory, noninvasive instruction is the *Preload* instruction. Once again, the execution of this instruction begins by first entering the instruction branch of the TAP controller and loading the corresponding instruction bit pattern in the *Update_IR* state of the controller. The *Preload* instruction initializes the scan cells.

Figure 8.14 shows the operation of the *Preload* instruction. After this instruction is loaded in the instruction register, the TAP controller moves into the *Shift_DR* state to perform the first phase of the instruction, while in this state, *SelectBR* and *ShiftBR* are set to 1, causing the Boundary register to be selected and put in the shift mode. This is shown on the right-hand side diagram of Fig. 8.14. After the completion of shifting the necessary test data, the TAP controller is put into the *Update_ DR* state to perform the second phase of this instruction. As shown in the right-hand side diagram

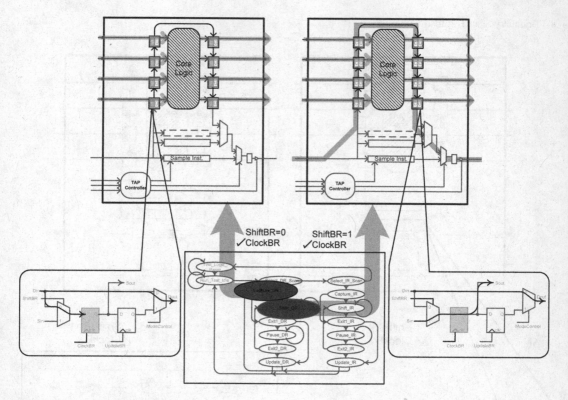

Fig. 8.13 *Sample* instruction execution

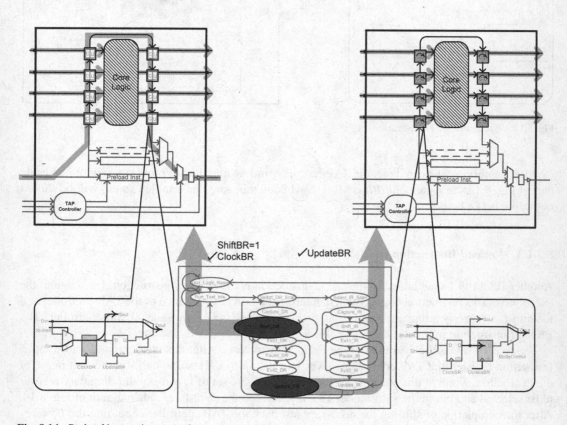

Fig. 8.14 *Preload* instruction execution

of Fig. 8.14, in the *Update_DR* state of the controller, *UpdateBR* clocks the *update* flip-flops of the boundary scan register. Data accumulated in the *update* flip-flops can be used in the other test instructions.

Note that there is no conflict between operations of *Sample* and *Preload* instructions as far as the flip-flop signaling is concerned. Therefore it is permissible to use the same opcode for both instructions.

8.3.1.4 *Extest* Instructions

An important instruction in the 1149.1 boundary scan standard is *Extest*. This instruction tests interconnections between two chips. The instruction operates in pin-permission mode, which means that it takes over the interconnects, and the chips whose interconnections are being tested cannot operate in their normal mode of operation.

The first time this instruction is being executed it must follow the complete execution of *Preload*. As discussed, *Preload* loads test data into *update* flip-flops of the boundary scan cells. Following *Preload*, as soon as *Extest* is loaded into the Instruction register, the preloaded test data become available on the output pins. After completion of the first round of *Extest* testing, new test data will be shifted as test response from the previous round of testing is being shifted out. The discussion that follows assumes that the first round of testing has already taken place.

As shown in Fig. 8.15, when the *Extest* instruction is active, interconnections between outputs of a core logic and inputs of another are being tested. For this purpose, after test data have been shifted into the *capture* flip-flops of the source core logic, its TAP controller goes into *Update_DR* state to load the test data into the *update* flip-flops (see the left-hand side diagram of Fig. 8.15). In this state, with *ModeControl* set to 1, the test data will be driving the outputs, and through the interconnections, the inputs of the next chip.

The chip receiving the test data is shown on the right-hand side diagram of Fig. 8.16. As shown, in the *Capture_DR* state of TAP controller of this chip, data from the pins are clocked into the *capture* flip-flops of the boundary scan cells. At the end of this operation, test data from the left chip outputs are now clocked into the right chip's scan cells.

In the next phase of *Extest*, TAP controller of source and destination chips are put in *Shift_DR* state. This phase is shown in Fig. 8.17. The output cells of the left chip are receiving new test data, while the input cells of the right chip are shifting out the test response. Completion of this phase prepares the output cells of the left chip for the next round of testing, for which the procedure shown in Fig. 8.15 starts again.

8.3.1.5 *Intest* Instruction

The procedure for *Intest* is similar to *Extest* for shifting in test data and shifting out response. While *Extest* is performed on two interconnecting chips, *Intest* applies test data to inputs of a chip and reads out the test response from the output of the same chip. *Intest* examines the functionality of the core logic, and is performed in pin-permission mode, which takes control inputs and outputs of core logic.

As with *Extest*, the *Preload* instruction precedes *Intest*. This provides test data in the *update* flip-flops (see the right-hand side diagram of Fig. 8.14) of the scan cells that are on the inputs of the core logic. These test data are used in the first round of testing. As soon as the *Intest* instruction is loaded into the instruction register in the *Update_IR* state, the contents of *update* flip-flops of the input cells will drive the inputs of core logic.

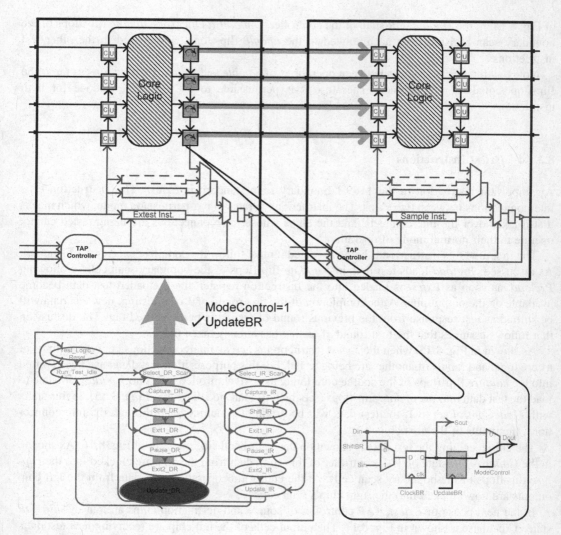

Fig. 8.15 Output cells updating test data in *Extest*

After the first round of testing, the test data inputs will be provided serially, while the test response from the previous round is shifted out. As shown on the right-hand side diagram of Fig. 8.18, data shifted in the *capture* flip-flops will be applied to the inputs of the core logic in the *Update_DR* state. For this purpose, boundary register is selected, *ModeControl* is issued, and *UpdateBR* of the corresponding flip-flops is activated.

The left-hand side diagram of Fig. 8.18 shows the output scan flip-flops during the execution of the *Intest* instruction. When the TAP controller reaches the *Capture_DR* state, the output scan register cells parallel load the outputs of core logic in their *capture* flip-flops.

After the completion of a round of test, the captured response must be shifted out and new test data shifted in. This is done when the TAP controller enters its *Shift_DR* state. Shift process in *Intest* is similar to that of Fig. 8.17 for *Extest*, except that all the inputs and output cells being shifted belong to the same chip.

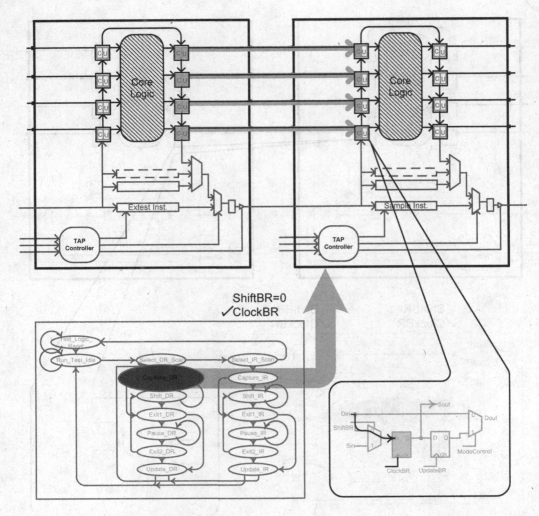

Fig. 8.16 Input cells capturing test data in *Extest*

8.4 Board Level Scan Chain Structure

The previous sections showed how a scan chain can be placed around a core logic and control its inputs and outputs for test purposes. We also discussed how *TDI* and *TDO* shift test data in and out of a scan chain, and how *TMS* makes the TAP controller on an IC to perform various test applications. Most of our discussion in a previous were centered around one or at most two (in case of *Extest*) ICs or core logics.

At the board or chip level, where there are many ICs or cores, various arrangements of scan registers can play an important role in saving test hardware and test time. This section shows some of these arrangements.

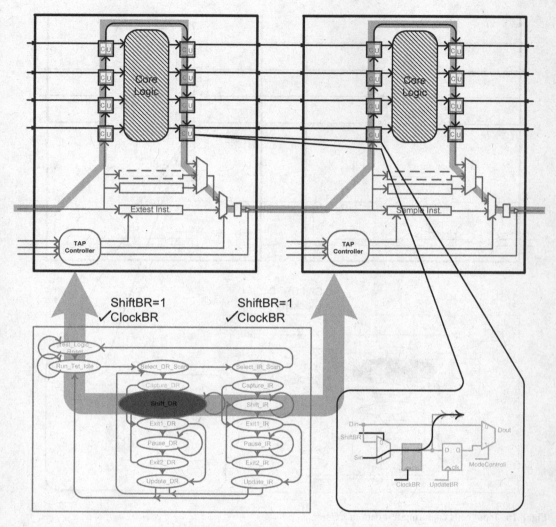

Fig. 8.17 Shifting new test data and captured response

8.4.1 One Serial Scan Chain

The "*one serial scan chain*" arrangement puts all modules on a board or chip in the same scan chain using the same TAPs. The serial data on the single *TDI* go through all the modules, and the serial data are shifted out on the only *TDO* that is available. Figure 8.19 shows this arrangement.

Since this structure uses only one TAP controller, all the boundary scan chains will be in one state. However, different instructions can be loaded, and boundary scan of each module can run its own instruction. The *one serial scan chain* arrangement has a low hardware overhead. However, because test data for a module have to travel through the entire scan, test time for this arrangement is relatively high.

8.4.2 Multiple-scan Chain with One Control Test Port

A scan architecture that has several scan chains (two in the example that follows) that are controlled by the same TAP controller is shown in Fig. 8.20. This architecture is referred to as "*multiple-scan*

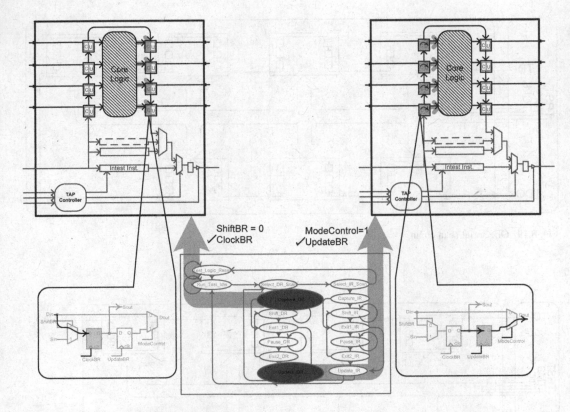

Fig. 8.18 *Intest* input and output cells

chain with one control test port". Here, since there is only one controller, like the *one serial scan chain* architecture, the different modules have to be tested at the same time and, as before, they can be using different instructions.

The advantage of *multiple-scan chain with one control test port* over the single-scan chain is that the test data can more quickly to get to the modules being tested. For every additional scan chain, a new set of *TDI* and *TDO* will be added to the test hardware.

8.4.3 Multiple-scan Chains with One TDI, TDO but Multiple TMS

Another arrangement for board or chip level scan architecture is to use multiple-scan chains with one *TDI* and *TDO*, but with multiple *TMS*. This arrangement that is shown in Fig. 8.21 partitions the scan chain into several chains with the same *TDI* and *TDO*, but different *TMS* inputs.

Although, this arrangement does not allow various scan chains to work at the same time, the test data can travel faster in the scan chain that is active. The advantage of this arrangement over that of Fig. 8.20 is that each additional chain here adds one new *TMS*, whereas in Fig. 8.20, each additional chain requires a *TDI* and a *TDO*.

8.4.4 Multiple-scan Chain, Multiple Access Port

Figure 8.22 shows groups of modules on a chip or board put into independent structures. The architecture is shown is referred to as "*Multiple-scan chain, multiple access port*". This arrangement is fast, but has a large pin and hardware overhead.

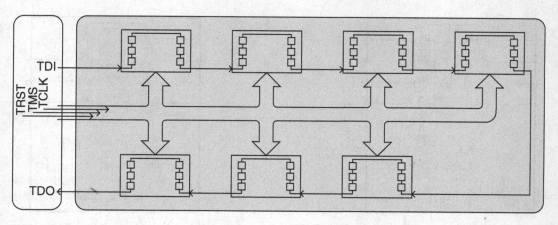

Fig. 8.19 One serial scan chain

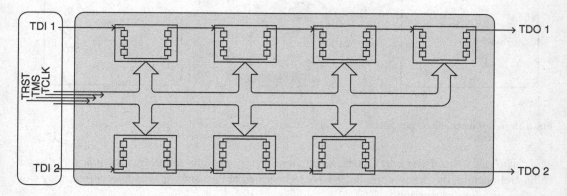

Fig. 8.20 Multiple-scan chain with one control test port

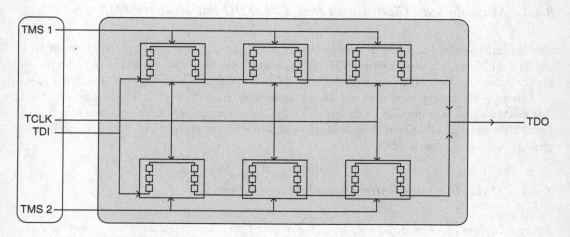

Fig. 8.21 Multiple-scan chains with one *TDI* and *TDO* but multiple *TMS*

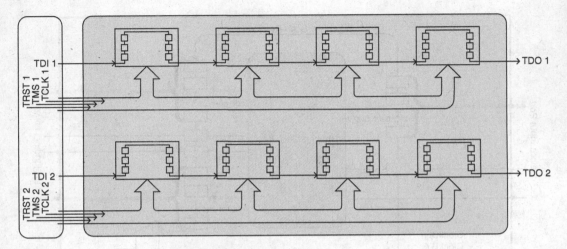

Fig. 8.22 Multiple-scan chain multiple test access port

8.5 RT Level Boundary Scan

As we did in Chap. 7, we will apply the DFT technique of this chapter to an RT level design to illustrate the applications of this DFT technique. The design we use is the Adding Machine of Chap. 2.

The purpose of this presentation is two fold. On one side, we are showing how an RT level designer uses various BS-1149.1 components to make his or her design testable. By use of Verilog code, we will show how these components are sized and configured to fit the boundary of our example circuit. The other purpose of this illustration is to show the operation of an ATE for testing a BS-1149.1 compliant circuit. This section illustrates this by use of a Verilog virtual tester operating as an ATE that controls an on-chip TAP controller and tests the circuit with boundary scan.

For demonstration purposes, we also attach another unit to our Adding Machine to be able to perform instructions like *Extest* that involve multiple modules. This fictitious module connects to the outputs of the Adding Machine, like an IO device would.

8.5.1 Inserting Boundary Scan Test Hardware for CUT

The first step in the design of boundary scan is the insertion of the necessary BS-1149.1 components in the hardware of CUT. Figure. 8.23 shows the CUT (Adding Machine) and its associated test hardware. The components shown will be discussed here.

For boundary scan test of our Adding Machine, we implement five mandatory instructions and size the boundary scan registers to cover input and output ports of our CUT. Other than sizing the instruction register, configuring the decoder, and sizing the boundary scan registers, all the other BS-1149.1 components shown in Fig. 8.23 are as defined in the standard that we discussed in Sect. 8.2.

8.5.1.1 Instruction Register

Figure 8.24 shows Verilog parameters defining five mandatory instruction codes for our design. These parameters are defined in the boundary scan decoder. According to this, the instruction register must be three bits wide.

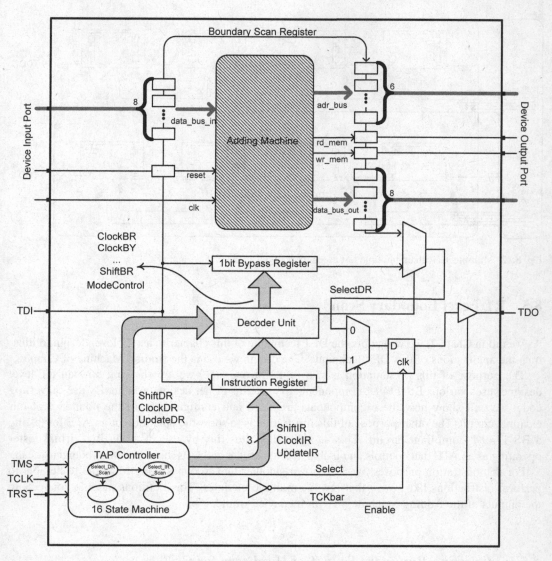

Fig. 8.23 Inserting boundary scan in our Adding Machine

```verilog
parameter [2:0]bypass_instruction = 3'b111;
parameter [2:0]intest_instruction = 3'b011;
parameter [2:0]sample_instruction = 3'b010;
parameter [2:0]preload_instruction = 3'b001;
parameter [2:0]extest_instruction = 3'b000;
```

Fig. 8.24 Instruction bit patterns

8.5.1.2 Decoder Unit

The decoder unit uses the instruction register outputs along with control signals from the TAP controller (see Fig. 8.23) to issue register control and select inputs. The Verilog code for the decoder unit has parts for the specific instruction that we support in our boundary scan implementation. Figure 8.25 shows a portion of the decoder code that handles control signals related to the *Bypass* instruction.

```
. . .
case (Instruction)
   . . .
   bypass_instruction: begin
      if (ShiftDR == 1'b1) DataSelect = Sel_BY;
      else  DataSelect = Sel_IR;
      ClockBY = (~ClockDR & TCLK);
      ShiftBY = ShiftBY;
      IRClockIR = (~ClockIR & TCLK);
      IRShiftIR = ShiftIR;
      IRUpdateIR = (UpdateIR & TCLK);
   end
```

Fig. 8.25 Bypass section of *Decoder* unit

8.5.1.3 Boundary Scan Register

The Adding Machine example has an 8-bit input bus, a *reset* input that resets its controller, and, of course, a clock (*clk*) input. On the input side, our boundary scan register will include nine scan cells for 8-bit *data_bus_in* input, and one for *reset*. The *clk* input will not be scanned.

On the output side, our CUT has a 6-bit address bus (*adr_bus*), an 8-bit *data-bus-out*, and two memory control outputs (*rd_mem* and *wr_mem*). We use sixteen boundary scan output cells for these outputs. Our boundary scan register is parameterized for its size. The input register is instantiated by use of #(9), and the output register uses another instance of the same component with #(16) for its size parameter.

8.5.1.4 Testable Design

Figure 8.26 shows the Verilog code that corresponds to the block diagram of Fig. 8.23. BS-1149.1 components and the netlist of our Adding Machine (*CPU_net CUT* (...), in the bottom half of the figure) are instantiated here.

The first line in this code generates *TCKbar* that is used in the *TDO* output flip-flop. Following this statement, the instruction register using #(3) for its size parameter is instantiated. The TAP controller is instantiated next, and is followed by the *Decoder* unit. As shown, the decoder uses generic TAP controller signals (e.g., *ShiftDR* and *ClockDR*), and using the instruction register output (i.e., Instruction) as input, it generates specific register control signals, such as *DRInShiftBR*, and *DRInClockBR* for the input side Boundary register. Signals that are generated from *ShiftDR* and *ClockDR* for the 16-bit wide output boundary scan register are *DROutShiftBR* and *DROutClockBR*.

The next instantiations after *Decoder* in Fig. 8.26 are the boundary scan input and output registers that are configured for the required number of cells. This is followed by the instance of our Adding Machine, i.e., *CPU_net CUT*. The last part of Fig. 8.26 shows instantiations for the *Bypass* register followed by the multiplexers, flip-flop, and the tristate gate that drive the *TDO* serial output.

8.5.2 *Two Module Test Case*

The complete module described above is our boundary scan Adding Machine (*BS_Adding_Machine*). To complete our test case, we take this and a nonfunctional module (*BS_Fictitious_module*) and form a model that may represent the chip we are designing that has 1149.1 boundary scan incorporated in it. The nonfunctional module we are using has a boundary scan input register that

```
module BS_Adding_Machine #(parameter in =9, parameter out =16)
                          (TDI, TMS, TCLK, Ckin, Pin, TDO,
                           Pout);
  . . .
  assign TCKbar = ~TCLK ;

  BS_IR #(3) IR (.DIN(3'b000),.SIN(TDI),.ShiftIR(SHI),
                 .ClockIR(CKI),.UpdateIR(UPI),.RstBar(RstBar),
                 .SOUT(TDOI),.DOUT(Instruction));

  TAPController TC (.TMS(TMS),.TCLK(TCLK),.RstBar(RstBar),
                    .sel(sel),.Enable(enable),.ShiftIR(ShiftIR),
                    .ClockIR(ClockIR),.UpdateIR(UpdateIR),
                    .ShiftDR(ShiftDR),.ClockDR(ClockDR),
                    .UpdateDR(UpdateDR));

  Decoder DCD (.Instruction(Instruction),.TCLK(TCLK),
               .ShiftIR(ShiftIR),.UpdateIR(UpdateIR),
               .ClockIR(ClockIR),.ShiftDR(ShiftDR),
               .UpdateDR(UpdateDR),.ClockDR(ClockDR),
               .DRInShiftBR(SHDI),.DRInUpdateBR(UPDI),
               .DRInClockBR(CKDI),.DRInTMS(TMSDI),
               .DROutShiftBR(SHDO),.DROutUpdateBR(UPDO),
               .DROutClockBR(CKDO),.DROutTMS(TMSDO),
               .IRShiftIR(SHI),.IRUpdateIR(UPI),
               .IRClockIR(CKI),.ShiftBY(SHB),
               .ClockBY(CKB),.Select_DR(Select_DR));

  BS_BSR #(in) BRi (.DIN(Pin),.SIN(TDI),.UpdateBR(UPDI),
                    .ClockBR(CKDI),.ShiftBR(SHDI),
                    .ModeControl(TMSDI),.RstBar(RstBar),
                    .SOUT(in2out),.DOUT(Cpu_in));

  BS_BSR #(out)Bro (.DIN(Cpu_out),.SIN(in2out),.UpdateBR(UPDO),
                    .ClockBR(CKDO),.ShiftBR(SHDO),
                    .ModeControl(TMSDO),.RstBar(RstBar),
                    .SOUT(TDOD),.DOUT(Pout));

  CPU_net CUT (.reset(Cpu_in[0]),.data_bus_in(Cpu_in[1:8]),
               .clk(Ckin),.adr_bus(Cpu_out[0:5]),
               .rd_mem(Cpu_out[6]),.wr_mem(Cpu_out[7]),
               .data_bus_out(Cpu_out[8:15]));

  BS_BYR BPR (.DIN(1'b0),.SIN(TDI),.ShiftBY(SHB),.ClockBY(CKB),
              .RstBar(RstBar),.SOUT(TDOB));

  MUX4_1 MX1 (.i1(1'b0),.i2(TDOI),.i3(TDOD),.i4(TDOB),
              .sel(Select_DR),.out(TDODRG));

  MUX2_1 MX2 (.i1(TDODRG),.i2(TDOI),.sel(sel),.out(TDOinit));

  D_FF TDOF (.D(TDOinit),.CLK(TCKbar),.RstBar(rstBar),.Q(TDOr));

  tristate TSO (.in(TDOr),.enable(enable),.out(TDO));
endmodule
```

Fig. 8.26 Adding boundary scan to Adding Machine

is connected to the outputs of the Adding Machine, four test ports, and a 1-bit output. This module can be regarded as an interface unit that our processor uses.

Figure 8.27 shows the complete schematic of our design. The design consists of two components, each of which has its own TAP controller. The scan chain of the complete design is formed by putting the scan chains of the two components in series. The architecture formed here is the *one serial scan*

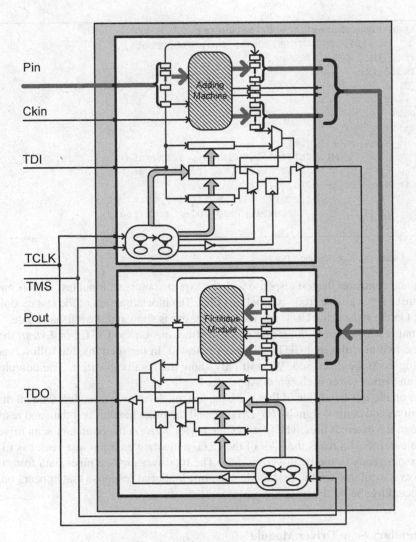

Fig. 8.27 Boundary scan testable design, consisting of two components

chain configuration discussed in Sect. 8.4.1. Figure 8.28 shows the Verilog code of this serial scan chain arrangement.

The diagram of Fig. 8.27 represents a complete board or IC with BS-1149.1 boundary scan. The design is only complete when the procedure for testing it by an ATE is also developed. The next section shows a Verilog testbench that imitates an ATE for testing this circuit.

8.5.3 Virtual Boundary Scan Tester

A virtual tester, as we discussed in Chap. 7, imitates an actual ATE. Because of the complexity of BS-1149.1 signaling, it is important for a planned test procedure to be verified together with the implemented test architecture.

The virtual tester that we discuss in this section is a Verilog testbench that instantiates the circuit of Fig. 8.27 (Verilog code of Fig. 8.27 is *system* module shown in Fig. 8.28) applies predetermined test

```
module system #(parameter in = 9, parameter out = 1)
                (TDI, TMS, TCLK, Ckin, Pin, TDO, Pout);
    input TDI, TMS, Ckin, TCLK;
    input [0:8] Pin;
    output TDO;
    output Pout;
    wire TDO1;
    wire [0:15] P1;
    BS_Adding_Machine #(9, 16)
                    M1(.TDI(TDI),.TMS(TMS),.TCLK(TCLK),.Pin(Pin),
                        .Ckin(Ckin), .TDO(TDO1),.Pout(P1));
    BS_Fictitious_Module #(16, 1)
                    M2(.TDI(TDO1),.TMS(TMS),.TCLK(TCLK),
                        .Pin(P1),.TDO(TDO),.Pout(Pout));
endmodule
```

Fig. 8.28 Serial scan chain of our test case

vectors to it, and compares the test response with the expected outputs. Input test vectors and expected output responses are available from external text files. The block diagram of the test module is shown in Fig. 8.29. On the right is the CUT, and the virtual tester is shown in the left-hand side.

This virtual tester performs *Intest* and *Extest* instructions on the CUT. For *Extest*, the interconnection of the two modules that CUT consists of is tested. In the sections that follow, various parts of this Verilog code are discussed. We will only show the details of *Intest*. The complete Verilog boundary scan virtual tester is shown in Appendix E.

As shown on the left-hand side of Fig. 8.29, the testbench has two modules. One module handles TAP instructions and control signals (*BS_Driver*), and the other handles test data and responses that are available in the external files. The first module is (*IO_Driver*) the boundary scan driver module. This module controls and reads the TAP of the CUT, and when an input test vector is to be shifted via *TDI*, it issues *ready* to the IO driver module. The IO driver module takes data from its external file and makes it available for the first module to shift in. Serial response that appears on *TDO* will also be collected by the IO driver module.

8.5.3.1 Boundary Scan Driver Module

The boundary scan driver module handles *TMS*, *TDI*, and *TDO* for *Intest* and *Extest* that are implemented here. Resetting the TAP controller, loading instructions, initializing the scan paths, and providing data that comes to it from IO driver module on *TDI* are some of the tasks of this module that will be discussed below.

Intesting procedural block. All tasks corresponding to the *Intest* instruction are enclosed in an **always** block that begins with **always@** (*intesting*) (see Fig. 8.29, left). The *Intest* instruction here only applies to the Adding Machine component of our CUT. The steps below start with resetting the machine, and in ten steps complete this instruction.

Step 1: *Resetting*. To guarantee that the TAP controller of the Adding Machine (upper block in CUT in Fig. 8.29) starts in *Test_Logic_Reset* state, *TMS* will be kept high for five consecutive clock pulses. Figure 8.30 shows a loop in the *Intest* **always** block that performs this operation.

Step 2: *First Data Set*. Before the actual *Intest* instruction is performed, the scan chain must be initialized with the first test vector. For this, the Adding Machine (core 1) for which the test data are to be loaded must be put in the *Preload* mode, and the other module in the design (fictitious

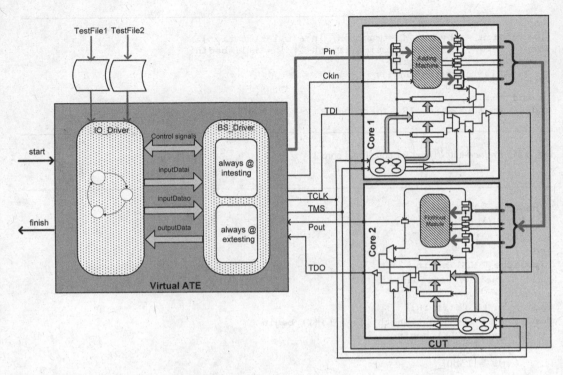

Fig. 8.29 Virtual tester testing a two-component CUT

```
. . .
for (i =0; i < 5; i = i+1) begin
    @(negedge TCLK)
        TMS = 1'b1;
end // 5 clock with TMS = 1 resets the state machine!
. . .
```

Fig. 8.30 Resetting TAP controller

module, core 2) must be bypassed. The **for** loop, shown in Fig. 8.31, shifts the codes for *Preload* and *Bypass* (two 3-bit patterns) into the instruction registers of the two cores of Fig. 8.29. After the completion of shifting the instructions, a proper sequence is generated on *TMS* to move the TAP controller to the *Capture_DR* state. At the same time, *ready* is issued to communicate with the IO driver module to inform it to provide serial data. The serial data provided as such will be placed on *TDI* while *TMS* is 0 and the controller is in the *Shift_DR* state. Figure 8.32 shows placement of indexed bits of *in_signali* on *TDI* after the falling edge of *TCLK*. Note that all the register operations in BS-1149.1 hardware are done on the rising edge of the clock. So the test-bench here provides proper data ahead of the active clock edge.

Step 3: *Loading* Intest *and* Bypass. Now that the first test vector is in the boundary scan register, *Intest* and *Bypass* instructions must be loaded in cores 1 and 2, for performing *Intest* on core 1 for the rest of input test vectors. For this purpose, using *TMS*, core 1 is put into *Shift_IR* state and a procedure similar to that shown in Fig. 8.31 is repeated.

Step 4: *Let Core Respond to Test.* Now that test data are available at the ports of core 1, we have to let it to react to the test data and generate its corresponding response. BS-1149.1 uses

```
. . .
instruction = {bypass_instruction, intest_instruction};
for ( i = 0; i < 2*instruction_length-1; i = i+1) begin
    @(negedge TCLK) begin
        TDI = instruction[i];
        TMS = 1'b0;
    end
end // stay in Shift_IR, shift n-1 first bits of instruction
. . .
```

Fig. 8.31 *Preload* and *Bypass* for core 1 and core 2

```
. . .
@(negedge TCLK) begin
    TMS = 1'b0;
    ready = 1'b1;
end
@(posedge TCLK) begin
    in_signali = inputDatai;
    ready = 1'b0;
end // go to Shift_DR;
for (i = 0; i < inputLength-1; i = i+1) begin
    @(negedge TCLK) begin
        TDI = in_signali[i];
        TMS = 1'b0;
    end
end //stay in Shift_DR, shift the inputLength-1 first bits ;
@(negedge TCLK) begin
    TDI = in_signali[inputLength-1];
    TMS = 1'b1;
end // go to Exit1_DR and shift last bit
. . .
```

Fig. 8.32 Shifting the first test vector for *Intest*

Run_Test_Idle state for this purpose. With issuing proper values on *TMS*, the TAP controller of core 1 will go in the *Run_Test_Idle* state, while the normal clock of this unit is running and it is functioning in normal mode. After waiting for a sufficient time, we will go to the state that collects the response. Fig. 8.33 shows the enabling of the system clock while TAP controller is in the idle state.

Step 5: *Capture Response.* After the Adding Machine has been given enough time to prepare its test response, we set *TMS* to 1 then to 0 (shown in Fig. 8.34) in two consecutive clocks to go from *Run_Test_Idle* state to *Capture_DR* (see state diagram in Fig. 8.10). Simultaneously we issue *ready* to tell the IO driver module that we are ready for the next input data. Figure 8.34 shows that *TMS* is set to 0 with the third falling edge of *TCLK*. This will move the TAP controller to state *Shift_DR* in which new test data will start shifting in.

Step 6: *Next Data Shift-In and Previous Response Shift-Out.* While *TMS* is 0, we remain in the *Shift_DR* state. Clocking the BS-1149.1 hardware of the Adding Machine as many times as there are output boundary scan cells (*outputlength*) will shift out the captured response. This is shown in the first **for** loop of Fig. 8.35. While this is happening *TDO* will be stored into *out_signal* indexed vector. Note in the Fig. 8.27 that the output cells of the Adding Machine are closer to *TDO* than its input cells, so the outputs will be shifted out first. The next *inputlength* clocks, while still in the *Shift_DR* state, will take input bits from the indexed *in_signali* into *TDI* and in the input scan

```
. . .
@(negedge TCLK)
   TMS = 1'b0; // go to Run_Test_Idle
@(negedge TCLK) begin
   TMS = 1'b0;
   clkenable = 1'b1;
end
for ( i =0; i < numberOfClk; i = i+1) begin
   @(negedge TCLK)
      TMS = 1'b0;
end // stay in Run_Test_Idle for numberOfClk clock cycle
clkenable = 1'b0;
. . .
```

Fig. 8.33 Applying *numOfClk* to Adding Machine in *Run_Test_Idle*

```
. . .
@(negedge TCLK) Begin
   ready = 1'b1;
   TMS = 1'b1;
end // go to Select_DR
@(negedge TCLK) begin
   in_signali = inputDatai;
   ready = 1'b0;
   TMS = 1'b0;
end // go to Capture_DR
@(negedge TCLK)
   TMS = 1'b0; // go to Shift_DR
. . .
```

Fig. 8.34 Capturing response and starting shift In–Out

```
. . .
for (i =0; i < outputLength; i=i+1) begin
   @(negedge TCLK) begin
      TDI = 1'b0;
      TMS = 1'b0;
      out_signal[i] = TDOF;
   end
end // stay in Shift_DR
for (i =0; i < inputLength-1; i=i+1) begin
   @(negedge TCLK) begin
      TDI = in_signali[i];
      TMS = 1'b ;
   end
end // stay in Shift_DR
@(negedge TCLK) begin
   TDI = in_signali[inputLength-1];
   TMS = 1'b1;
   out_signal[outputLength+1] = TDOF;
end // go to Exit1_DR
. . .
```

Fig. 8.35 Previous response shift-Out new data shift-In

cells. This is shown in the second **for** loop in Fig. 8.35. With the next clock (last part of Fig. 8.35) TMS is set to 1 to prepare for exiting (see Fig. 8.10).

Step 7: *Response Ready & Check for Last Test*. The next step is to go to *Update_DR* state by setting *TMS* = 1 and issuing *outReady* to the IO driver module for collecting the shifted out response. After receiving the *outready* signal, the IO driver module issues *lastData* if data that were shifted in were the last test data. In this case, the Verilog code of the virtual tester performing *Intest* goes to Step 8, otherwise Step 4 is taken to continue the test.

Step 8: *Let Core Respond to Last Test*. This step is similar to Step 4, and after it is completed it goes to Step 9.

Step 9: *Last Response Shift-Out*. The response due to the last test of Step 8 will be shifted out in this step. This is similar to Step 6, expect that only first **for** loop is performed.

Step 10: *Last Response and Reset*. The last step issues *outready*, sets *TMS* to go to the *Test_Logic_Reset* state, and issues *endIn* to the IO driver module to announce completion of *Intest*.

Extesting Procedural Block. Tasks performed for *Extest* are enclosed in an **always** block that begins with **always@** (*extesting*). These tasks are taken in several steps similar to those of the *Intesting* block, and are shown in Appendix E.

8.5.3.2 IO Driver Module

The IO driver module part of the boundary scan virtual tester is responsible for reading test data, and controlling the test operations. There is a memory initialization part and a sequencer that we described below.

Memory Initialization. The memory initialization part of the IO driver module reads *CPUin.txt* file into *TESTin[]* array, and *CPUout.txt* in *TESTout[]*. Test data for *Intest* will be read from *TESTin[]*, and for *Extest* from *TESTout[]*. Appendix E has the complete virtual tester code that also includes this part.

Sequencer. The sequencer part of the IO driver module implements the state diagram shown in Fig. 8.36. When *start* becomes 1, *Intesting* is issued, and the machine goes in the *intesting* state. In this state every time *ready* is seen, the next test data is placed on *inputDatai* bus. When the last data is put on *inputDatai*, *lastData* is issued. Also in the *intesting* state, when *readyOut* become 1, the data that has become available on *outputData* will be written in an output file. This continues until *endIn* value of 1 is seen, in which case the *intesting* exits and *extesting* become active. A similar signaling happens in the *extesting* state. Figure 8.37 shows the Verilog code of this part.

8.6 Boundary Scan Description Language

The boundary scan description language (BSDL) was established as part of the IEEE boundary scan standard for specifying the arrangement of BS-1149.1 hardware in a chip. BSDL provides a standard means of communication between designers, test engineers, tool developers, and silicon vendors. This standard is useful for developing test programs [3].

BSDL is not a hardware description language and does not have the ability of describing functionality of a component. BSDL specifies how boundary scan registers are used on the boundary of a chip, the features of BS-1149.1 that they support, and the boundary scan instructions that are implemented. BSDL defines a set of VHDL attributes for describing the boundary scan arrangement of a chip.

To illustrate the features and capabilities of BSDL, we will use it to describe design of the boundary scan that we did for our Adding Machine of the previous section. The complete code is shown in Fig. 8.38, and the discussion that follows references this code by line numbers.

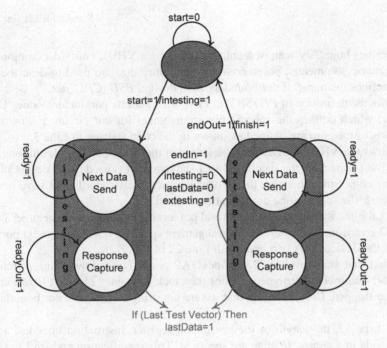

Fig. 8.36 IO driver module sequencer

```verilog
always @ (start, ready, readyOut, endIn, endOut) begin
    if (start == 1'b1) begin
        intesting = 1'b1;
        i = 0;
        lastData = 1'b0;
    end
    else if (ready == 1'b1 & intesting == 1'b1) begin
        inputDatai = TESTin[i];
        i = i+1;
        if(i == memLength1)
            lastData = 1'b1;
    end
    else if (ready == 1'b1 & extesting == 1'b1) begin
        inputDatao = TESTout[i];
        i = i+1;
        if(i == memLength2)
            lastData = 1'b1;
    end
    else if (readyOut == 1'b1 & intesting == 1'b1)
        $fwrite(f3,"%b\n",outputData);
    else if (readyOut == 1'b1 & extesting == 1'bs1)
        $fwrite(f4,"%b\n",outputData);
    else if (endIn == 1'b1)begin
        intesting = 1'b0;
        #1;
        extesting = 1'b1;
        i =0;
        lastData = 1'b0;
    end
    else if (endOut == 1'b1)begin
        extesting = 1'b0;
        finish = 1'b1;
        i = 0;
        lastData = 1'b0;
    end
end // always
```

Fig. 8.37 Verilog part of the virtual tester for operation of *Intest* and *Extest*

BSDL describes boundary scan of a chip or a core by a VHDL entity declaration (Line 1–Line 70). VHDL generic parameters, ports, constants, and attributes are used to describe the boundary scan. Line 1 defines the name of our boundary scan design, *BSD_CPU_net*.

Line 2 shows the definition of *PHYSICAL_PIN_MAP* generic parameter. Value "D1" is given to this parameter, which defines the specific pin arrangement for our circuit. Following the generic parameter, ports of the chip are defined as shown in the code starting in Line 3.

Line 13 shows the VHDL BSDL package in which the corresponding definitions and attributes are defined. Line 14 specifies the specific 1149.1 standard that is used in the design of the boundary standard of our chip. Since there are differences between the standards that have been established through the years, the specific one used must be defined.

Starting in Line 16, we are specifying several pin arrangements for our package. We are specifying D1 and D2 arrangements, where each arrangement specifies how package IO ports are mapped to the ports of the device. This part ends with Line 27 of Fig. 8.38.

The next part that starts on Line 28 defines TAP ports. This allows using different names for TAPs. The TAP statement's attribute specifies the clock frequency (1 GHz, here), and the state in which it can be stopped. Line 32 shows that we are not using test reset in our boundary scan of the Adding Machine.

Starting in Line 33, the length of the instruction register, instruction opcodes, and the default instruction opcode in *Capture_IR* state are specified. This specification ends in Line 41.

The last part that starts on Line 42 specifies what cell type each pin of the package is connected to. This part begins with specifying the length of the boundary scan register that is 25, and continues with the specification of each of the boundary scan pins. Pin specifications are concatenation of strings that are enclosed in double quotes. Each string of which specifies four values for each port. The values are for *NUM*, *CELL*, *PORT*, and *FUNCTION* fields. There is also an extra field (*SAFE*) that is used as a safe default pin value. The *NUM* field is the location of the specified port in the scan chain. Field *CELL* specifies the cell type which is one of the standard cells of the standard. Cell *BC_1* is a defined cell in the boundary scan standard. The last field is *FUNCTION* that tells the type of port, i.e., *input, output*, or bidirectional *inout*.

This concludes the presentation of BSDL. Another standard language that also considers hierarchical scan structures is HSDL, which will not be discussed here. Our purpose of presenting BSDL was merely to give an overall view of what is necessary for making chips ready for boundary scan.

8.7 Summary

This chapter treated the standard subject of boundary scan and discussed the IEEE standard like any other book on testing would do. What made this chapter different than just presenting a certain standard protocol was the use of Verilog for developing a virtual tester that interacted with the boundary scan hardware. The virtual tester demonstrated how an ATE would interact with a chip with a boundary scan. At the same time, in order to be able to perform the simulations in Verilog, we also showed hardware structures of BS-1149.1 in Verilog. Doing this we were able to discuss some of the details of hardware of this standard that would otherwise not be noticed.

The first part of the chapter discussed the IEEE Std.1149.1 standard, and Verilog was used for describing hardware components of this standard. Use of Verilog enabled us to unambiguously and clearly show the hardware structures and their interactions. By use of this language we were able to show the interaction between TAP controller and 1149.1 instructions. That is, were able to clearly show what responsibilities each part takes for control of the scan registers. In the second part, where a virtual tester was developed, interactions between the 1149.1 hardware, CUT hardware, and ATE were clarified.

```
Line 1: entity BSD_CPU_netis
-- generic parameter
Line 2: generic (PHYSICAL_PIN_MAP : string := "D1");
--logical port description
Line 3: port(
Line 4:        reset : in bit;
Line 5:        data_bus_in : in bit_vector(0 to 7);
Line 6:        clk : in bit;
Line 7:        adr_bus : out bit_vector (0 to 5);
Line 8:        rd_mem : out bit;
Line 9:        wr_mem : out bit;
Line 10:       data_bus_out : out bit;
Line 11:       TDO : out bit;
Line 12:       TMS, TDI, TCLK : in bit);
--standard use statement
Line 13: use STD_1149_1_1994.all;
--component conformance statement
Line 14: attribute COMPONENT_CONFORMANCE of BSD_CPU_net is
Line 15:       "STD_1149_1_1990";
-- device package pin mapping
Line 16: attribute PIN_MAP of BSD_CPU_net : entity is
Line 17:       PHYSICAL_PIN_MAP;
--
Line 18: constant D1 : PIN_MAP_STRING :=
Line 19: "reset: 2, data_bus_in: (3,4,5,6,7,8,9,10)," &
Line 20: "clk: 11, adr_bus: (12,13,14,15,16,17),rd_mem: 18," &
Line 21: "wr_mem: 19, data_bus_out:(20,21,22,23,24,25,26,27)," &
Line 22: "TDO: 31, TMS: 28, TDI: 30, TCLK: 29 ";
--
Line 23: constant D2 : PIN_MAP_STRING :=
Line 24: "reset: 2, data_bus_in: (7,8,9,10,11,12,13,14)," &
Line 25: "clk: 3, adr_bus: (15,16,17, 18,19,20), rd_mem: 21," &
Line 26: "wr_mem: 22, data_bus_out: (23,24,25,26,27,28,29,30)," &
Line 27: "TDO: 31, TMS: 5, TDI: 4, TCLK: 6";
-- TAP port identification
Line 28: attribute TAP_SCAN_IN of TDI  : signal is true;
Line 29: attribute TAP_SCAN_MODE  of TMS  : signal is true;
Line 30: attribute TAP_SCAN_OUT   of TDO~ : signal is true;
Line 31: attribute TAP_SCAN_CLOCK of TCLK  : signal (1.0e9,BOTH);
Line 32: -- attribute TAP_SCAN_RESET of TRST : signal is true;
--instruction register description
Line 33: attribute INSTRUCTION_LENGTH of BSD_CPU_net : entity
         is 3;
Line 34: attribute INSTRUCTION_OPCODE of BSD_CPU_net : entity is
Line 35: "BYPASS    (100, 101, 110,111)," &
Line 36: "INTEST    (011)," &
Line 37: "SAMPLE    (010)," &
Line 38: "PREALOAD (001)," &
Line 39: "EXTEST    (000)";
Line 40: attribute INSTRUCTION_CAPTURE of BSD_CPU_net : entity
Line 41:       is "001";
```

Fig. 8.38 BSDL code for boundary scan of Adding Machine

```
-- boundary scan register description
Line 42: attribute BOUNDARY_LENGTH of BSD_CPU_net : entity is 25;
Line 43: attribute BOUNDARY_REGISTER of BSD_CPU_net : entity is
Line 44: --NUM   CELL   PORT                FUNCTION SAFE
Line 45: " 0     (BC_1, reset,             input,   x)," &
Line 46: " 1     (BC_1, data_bus_in(0),    input,   x)," &
Line 47: " 2     (BC_1, data_bus_in(1),    input,   x)," &
Line 48: " 3     (BC_1, data_bus_in(2),    input,   x)," &
Line 49: " 4     (BC_1, data_bus_in(3),    input,   x)," &
Line 50: " 5     (BC_1, data_bus_in(4),    input,   x)," &
Line 51: " 6     (BC_1, data_bus_in(5),    input,   x)," &
Line 52: " 7     (BC_1, data_bus_in(6),    input,   x)," &
Line 53: " 8     (BC_1, data_bus_in(7),    input,   x)," &
Line 54: " 9     (BC_1, adr_bus(0),        output2, x)," &
Line 55: " 10    (BC_1, adr_bus(1),        output2, x)," &
Line 56: " 11    (BC_1, adr_bus(2),        output2, x)," &
Line 57: " 12    (BC_1, adr_bus(3),        output2, x)," &
Line 58: " 13    (BC_1, adr_bus(4),        output2, x)," &
Line 59: " 14    (BC_1, adr_bus(5),        output2, x)," &
Line 60: " 15    (BC_1, rd_mem,            output2, x)," &
Line 61: " 16    (BC_1, wr_mem,            output2, x)," &
Line 62: " 17    (BC_1, data_bus_out(0),   output2, x)," &
Line 63: " 18    (BC_1, data_bus_out(1),   output2, x)," &
Line 64: " 19    (BC_1, data_bus_out(2),   output2, x)," &
Line 65: " 20    (BC_1, data_bus_out(3),   output2, x)," &
Line 66: " 21    (BC_1, data_bus_out(4),   output2, x)," &
Line 67: " 22    (BC_1, data_bus_out(5),   output2, x)," &
Line 68: " 23    (BC_1, data_bus_out(6),   output2, x)," &
Line 69: " 24    (BC_1, data_bus_out(7),   output2, x)";
Line 70: end BSD_CPU_net;
```

Fig. 8.38   (continued)

References

1. Bushnell ML, Agrawal VD (2000) Essintioals of electronic testing for digital, memory & mixed-signal VLSI circuits. Kluwer, Boston
2. IEEE Standard Test Access Port and Boundary Scan Architecture (1990) *IEEE standard Board*, 345 East 74th St. New York
3. Parker KP (2000) The boundary-scan handbook, 2nd edn. Kluwer, Boston

Chapter 9
Logic Built-in Self-test

The last two chapters represented two DFT methods. Scan testing in Chap. 7 focused on testing inside a core (or the logic part), while boundary scan testing focused on interfaces between cores. This chapter presents still another DFT method, and as in Chap. 7, the focus is testing inside of a core. Unlike Chaps. 7 and 8, where our test methods heavily depended on ATEs, the DFT method presented here tries to eliminate or, at least, reduce the need for an ATE. This is done by adding testability hardware to the CUT, in a way that the added hardware tests parts or all of the CUT.

The DFT method that is our focus here is called BIST or built-in self-test. BIST is a hardware structure that produces test data, applies it to the circuit under test, collects the output response, and verifies that the output is correct [1–4]. The operation of a BIST inside a CUT is controlled by a BIST controller. BIST is a complete datapath/controller system inside a CUT such that when it engages, it tests the CUT. Effectiveness of BIST is measured by its fault coverage.

We take an approach similar to that of scan design for presenting the BIST DFT technique. Section 9.1 discusses the basics. We then discuss basics leading to BIST architectures, where several common architectures are discussed. In the last section, we use an RT level example and show how several built-in test hardware architectures can be incorporated in this complete RTL example.

9.1 BIST Basics

This section gives a general introduction to BIST, defines the main flow, terminologies, and BIST variations [5, 6].

9.1.1 Memory-based BIST

The first thing that comes to mind when we talk about an on-chip hardware that tests another hardware is to implement a small on-chip ATE with an input memory, output memory, and a controller.

As shown in Fig. 9.1, a memory-based BIST takes test data from its input memory, applies it to the CUT, reads the output, and compares it with its expected response data. The BIST controller basically does what a test program would do in a DFT method such as scan. Namely, the BIST controller decides when and what data to be applied to CUT, controls CUT's clocking, and decides when to read the expected response. As in a DFT method such as scan, we try to minimize the test

Z. Navabi, *Digital System Test and Testable Design: Using HDL Models and Architectures*,
DOI 10.1007/978-1-4419-7548-5_9, © Springer Science+Business Media, LLC 2011

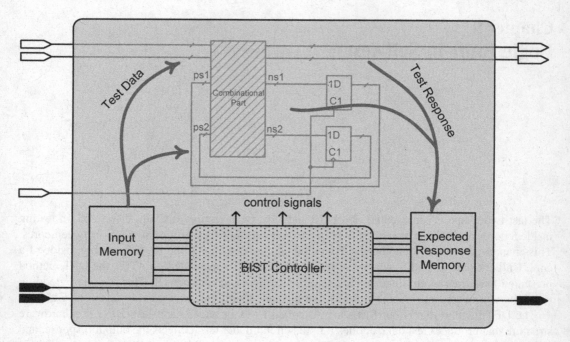

Fig. 9.1 A basic BIST

time and maximize the fault coverage. For this purpose, again like scan, we have different BIST architectures with varying effectiveness and test time.

Unlike the scan DFT method in which a *test program* (or in our HDL environments, a virtual tester) is responsible for performing the test, coming up with test data, performing the test, and analyzing the output, in BIST, the *BIST architectures* must provide all the necessary data and control circuitry. It is this important difference that requires special hardware structures for BIST handling.

9.1.1.1 Providing Test Data

Instead of the memory shown in Fig. 9.1 for the test data input, a special hardware that can produce test data must be used. In the scan method of the previous two chapters, test data resided in large storage spaces in the ATE. This virtually unlimited space could contain large test data sets generated by random and/or deterministic test generation methods. The only limitation in the size of test data for such DFT methods is the time it takes to send test data to the CUT.

BIST architectures, on the contrary, do not enjoy this luxury, and on chip memory that can be allocated for test purposes is very limited or nonexistent. This means that storing large test data on a chip is not an option for built-in testing. Instead of having a set of test data in a memory, BIST architectures generate their own. The drawback here is that we have a limited control on the data that are generated on-chip, and the coverage we could get from deterministic test vectors cannot be obtained.

Instead of a block of memory for test data, BIST architectures use various forms of test pattern generators (TPGs). A TPG is a hardware structure that generates exhaustive or pseudo-random test data. Since TPGs cannot easily be made to generate test data resulted from deterministic test generation, BIST test vectors are not as efficient in fault coverage as externally provided data.

9.1.1.2 Test Response Analysis

Instead of the memory as shown in Fig. 9.1 for response memory, a special hardware for checking CUTs response is needed. In the scan methods, test response vectors are shifted out and collected by the ATE testing the CUT. These data are compared with data in the ATE's storage memory. Unfortunately, the limited CUT's chip area does not allow the use of on-chip memories to the extent that is needed for test purpose. This problem is resolved by storing a signature, or a compressed version, of all the test responses. This way, instead of checking individual test vector responses, the BIST architecture just checks the signature obtained from test responses against the golden signature in BIST hardware.

Although using a signature eliminates the need for large on-chip memories, a new problem called aliasing is introduced. Aliasing is referred to the situation of having the same signature for different test response data sets. This problem can let faults go undetected or make faults that are distinguishable not recognized as such. The problem of aliasing can partially be dealt with by using larger or multiple signatures, and handling of which is done by the BIST hardware.

The hardware generating a signature of collected test responses is referred to as output response analyzer (ORA). An ORA is an on-chip BIST hardware, various forms of which will be discussed in a section of this chapter.

9.1.2 BIST Effectiveness

Effectiveness of a BIST is measured in terms of the stuck-at fault coverage. Although BIST architectures can detect structural, functional, and logical faults, the stuck-at coverage is just a measure of its effectiveness for all fault types it is designed to detect.

BIST effectiveness depends on its architecture, the design of its TPG and ORA, and the BIST controller. Planning an architecture, TPG, and ORA are done by simulation prior to completing the design of a BIST. Fine tuning of BIST is done by repeated use of fault simulation.

9.1.3 BIST Types

Just like a scan architecture incorporated in a circuit, a BIST architecture that resides on a chip along with CUT engages and starts testing the CUT's core logic when it is put in the test mode [3]. Various forms of engagements define the BIST types.

9.1.3.1 Offline BIST

The kind of BIST that requires the normal operation of CUT to be halted for the BIST to engage in is called an offline BIST. External system pins control the operation of BIST, and the BIST reports CUT's error conditions using external pins. Since an offline BIST has an exclusive control on test data and the hardware of a CUT, the time it takes for testing the CUT is relatively short.

9.1.3.2 Online BIST

In contrast to offline BIST, an online BIST always operates inside a CUT. Some online BISTs use the same data a CUT uses while performing its normal operation. Other online BIST structures use free time slots when certain parts of a CUT are idle to test those parts of the circuit. In general, an online BIST has to find the right time and/or the right data to engage in and test a CUT. Because of this sneak-in type of work, it takes a relatively long time for an online BIST to test an entire CUT.

9.1.3.3 Hybrid BIST

Other than the fact that a BIST controller resides on the same chip as the CUT during its normal operation, conceptually, there are very few differences between BIST and scan designs. Because of this, some of the limitations of BISTs on test data generation and output response analysis can be overcome by getting external help by means of scans. This would mean that the on-chip BIST and the off-chip ATE would share the tasks of data generation, response analysis, and handling a CUT's test program, and thus a hybrid BIST.

On the one hand, a hybrid BIST requires a simpler and less expensive ATE than a scan design. On the other hand, the fault coverage of a hybrid BIST is improved by the use of scan.

9.1.3.4 Concurrent BIST

A concurrent BIST does not produce new test data for testing a CUT and uses the same data that CUT uses for its normal operation.

9.1.4 Designing a BIST

Design of a BIST involves its architecture, TPG, ORA, and test procedure. Once these details are known, hardware implementation of the architecture, TPG, and ORA follow. The designed BIST test procedure is implemented by the BIST controller. Fault simulation plays a major role in finalizing the design of BISTs [1, 2].

9.1.4.1 Architecture Design

Architecture of a BIST defines where in the CUT test data generators are applied, and where response data are collected from. As in scan testing, the architecture is designed for minimizing test time and maximizing coverage. Decisions made for the design of BIST architecture are about the number of TPGs and ORAs, their types, how they are placed relative to the CUT, and other details of test application and response collection. Furthermore, integration of BIST with scan and boundary scan are parameters that define a BIST architecture.

Usually, there are several templates for architectures, one of which is selected for a specific CUT. The architecture of CUT, its scan design, its boundary scan, and desired test time are some of the issues based on which a preliminary BIST architecture is configured. Some typical BIST architectures will be discussed in Sect. 9.4 of this chapter.

9.1.4.2 Designing TPGs

After a general outline of a BIST architecture is known, and the number, types, and position of TPGs in the CUT are known, lengths and sequence of data generated by the TPGs must be worked out. For this preliminary decisions for the type of data that need to be generated are made, and the rest are decided by fault simulation and evaluation of fault coverage the TPGs can yield.

Selection of additional test points and adding new TPGs or extending the existing ones to cover the new test points is not an unlikely situation. Although most of these decisions start with given templates and are finalized by the use of fault simulation by trial and error, there have been works done on using intelligent methods for configuring TPGs.

Section 9.2 discusses types and parameters related to TPGs. Several common TPGs will be discussed.

9.1.4.3 Designing ORAs

Design of ORAs is very similar to that of TPGs, and fault simulation is used for configuring them. As TPGs, fault simulation is used in a trial and error process to configure ORAs for better coverage of faults.

In designing a BIST, new ORAs are added, or existing ones are extended to give more observability to a CUT and to reduce ORA aliasing. Some research works have been done on making intelligent decisions for ORA design and configuration. The topic of ORAs is discussed in Sect. 9.3.

9.1.4.4 BIST Procedure

In scan and boundary scan designs, an ATE runs a test program to execute the test procedure designed for the specific design of scan or boundary scan. In BIST, the same is done by the BIST controller.

After the BIST design is complete for a CUT, TPGs, ORAs, the way TPGs are clocked, the number of clocks the circuit under test receives, the time or times at which ORAs are collected, and finally the signature(s) the ORAs must produce for a good CUT will be known. All these can be verified and adjusted by use of fault simulation.

The design of a BIST is completed by designing a BIST controller that implements the test procedure BIST is executing. Figure 9.2 shows a block diagram of a BIST controller with signals controlling events that cause the testing of the CUT. Output signals shown here describe some of the functionalities performed by a BIST controller.

A BIST controller has a state machine that sequences through various phases of CUT testing. In addition, one of several counters of a BIST counts the number of times test data is applied to a CUT, and perhaps one keeps a count of when response has to be collected. Functions controlled by the BIST controller that relate to test data generation include initializing the TPGs, clocking them, selecting their outputs to apply to CUT, and reconfiguring them for different test data production. Functions related to ORAs include initializations, reconfigurations, stopping and starting them, and reading their values. Finally, functions related to CUT control include clocking the CUT certain number of times for preparation of data.

As shown in Fig. 9.2, BIST hardware also includes a part that we refer to as BIST compare. This part contains the good circuit responses that have been obtained during simulation. Comparators in this part compare expected ORA responses with those that are built-in in this part. This part is controlled by the BIST controller, which tells it when to perform the required comparisons. The BIST compare unit reports its compare results to the controller.

Fig. 9.2 BIST controller

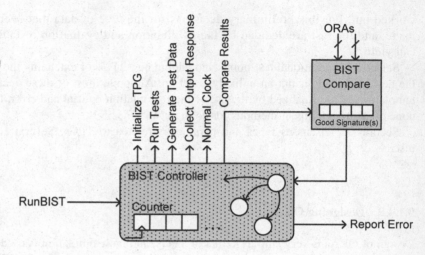

9.2 Test Pattern Generation

A major part of any BIST hardware is its test pattern generation scheme. This is the part that contributes to providing access to various parts of a CUT and the overall fault coverage. In certain situations, TPGs are based on functionality of a part of a CUT, other times they may be providing random test data for a sequential or combinational part. For example, testing an ALU requires the test pattern at the function-select inputs to be based on the functionality of the part, whereas testing a given ALU function may require random or pseudo random test patterns.

Testing memories, logic units, state machines, bus control logic parts, and various parts of a system may require different types of TPGs. Since data generation for a BIST must be done on-chip, we are limited by the amount of hardware that can be allocated to this part of a BIST. This section discusses TPGs, their hardware, implementation, and their applications.

9.2.1 Engaging TPGs

In the normal mode of operation, a TPG is idle and its outputs are not used. In the test mode, the TPG connected to the part being tested is engaged and drives the CUT input. Figure 9.3 shows two ways of engaging a TPG. Figure 9.3a shows a multiplexer allowing normal data or TPG output to drive the inputs of a CUT, and Fig. 9.3b shows a TPG with a transparent mode that allows normal data to pass through it when not in the test mode.

9.2.2 Exhaustive Counters

An n-bit binary counter provides 2^n consecutive test vectors. The Verilog code of an n-bit up/down counter is shown in Fig. 9.4. The counter is cascadable and has carry-in (cin) and carry-out ($cout$) input and output for this purpose. The counter has parallel load inputs for starting it at any stage or for reducing its count sequence.

Using counters for exhaustive testing of logic units with a large number of inputs is very inefficient in terms of the number of test vectors. However, by partitioning combinational circuits into cones with inputs, all of which drive a given output, the number of inputs of each partition will be

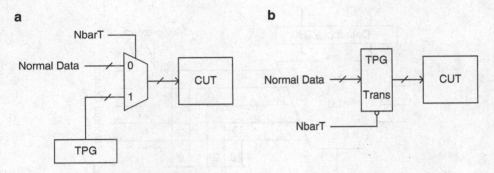

Fig. 9.3 Engaging a TPG (**a**) multiplexed, (**b**) transparent mode

```
module counter #(parameter length = 8)(input [length-1:0] d_in,
                                        output cout, input clk,
                                        ld, u_d, en, cin, output
                                        reg [length-1:0] q);
    always @ (posedge clk) begin
      if (en) begin
        if (ld) q <= d_in;
        else if (cin) begin
          if (u_d) q <= q + 1;
          else q <= q - 1;
        end
      end
    end
    assign cout = &{q, cin};
endmodule
```

Fig. 9.4 Up/down counter Verilog code

limited. In such a situation, exhaustive testing becomes more practical if applied to each logic cone independently.

Counters are also useful for exhaustively selecting all functions of an ALU, each of which to be tested by data coming from other TPGs having more limited test vectors. Figure 9.5 shows a 5-bit counter connected to the function inputs of a 16-bit ALU. The counter selects every one of the 32 functions of the ALU, while for each function, other TPGs provide test data for ALU data inputs.

9.2.3 Ring Counters

An n-bit ring counter has n states, and each state has only one bit set at 1 and the rest at 0. Starting with a 1 in a bit position, the next count sequence of a ring counter moves the 1 by one bit position. This continues until the 1 reaches the last counter bit, at which point it rotates back to the opposite end. For obvious reasons, a ring counter is also called a one-hot or a walking-1 circuit.

A ring counter can be implemented by a shift-register that is initialized with a 1 in one bit position. This implementation is shown in Fig. 9.6. Alternatively, a $\log_2 n$ binary counter driving a decoder can implement the same functionality.

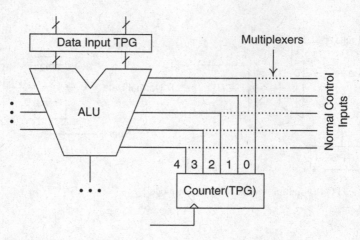

Fig. 9.5 Counter testing every ALU function

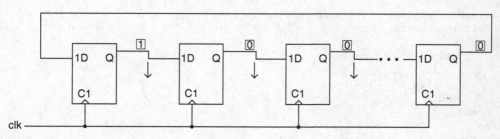

Fig. 9.6 Ring counter

A ring counter is useful for activating one-hot bus select inputs. Other scenarios such a counter can be useful for include memory testing and testing data path control inputs.

9.2.4 Twisted Ring Counter

A twisted ring counter is also called a marching-1 circuit. Unlike the ring counter in which a 1 walks from one end of the counter to the other end, in a twisted ring counter, 1s start marching from one end to the other end of the counter. A twisted ring counter is also called a Johnson counter. An n-bit twisted ring counter has $2n$ states. Figure 9.7 shows the count sequence of a 4-bit twisted ring counter.

As shown in Fig. 9.8, a twisted ring counter can be implemented by placing an inverter in the feedback of the shift-register implementation of a ring counter. A marching-1 circuit is particularly useful for memory testing.

Inserting an XOR gate at the input of the left-most shift-register flip-flop makes a configurable ring counter that can either work as a regular ring counter or a twisted one, i.e., walking-1 or marching-1. Figure 9.9 shows such a counter with asynchronous resetting mechanism (*init*) and a *twisted* mode select input. When this input is 1, the circuit works as a twisted ring counter.

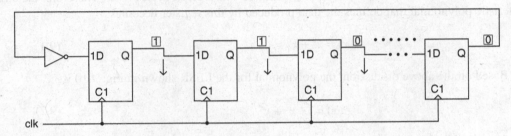

Fig. 9.7 Twisted ring counter sequence

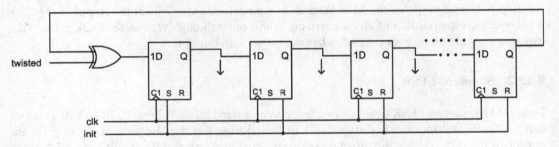

Fig. 9.8 Twisted ring counter

Fig. 9.9 Configurable ring counter

9.2.5 Linear Feedback Shift Register

The binary counter that we discussed earlier has 2^n states for n flip-flops, but has the problem that the rate of change of an upper-order bit is twice as slow as its most immediate lower-order bit. This problem becomes worse when comparing bits that are distanced farther. The problem caused by this situation is that the CUT inputs that are connected to the upper counter bits are not exercised as much as those connected to the lower order bits. Furthermore, capacitive or bridging effects of pins connected to upper order bits may go undetected, because of slow relative change of such pins.

On the contrary, the ring counters we discussed earlier did not have this problem. For an n-bit ring counter, the rate of change of each bit is the clock frequency divided by n.

However, important problems with the ring counters are the limited number of distinct outputs and uniformity of the data they produce.

A linear feedback shift register (LFSR) can solve some of the problems we face with the counters discussed above. When an n-bit binary counter produces 2^n unique input vectors, an LFSR can be made to generate up to $2^n - 1$ unique pseudo-random test vectors [5].

An LFSR consists of a series of flip-flops wired as a shift-register with feedbacks through XOR gates. The XOR gates are modulo-2 adders, and the flip-flops are considered as delay elements [7].

9.2.5.1 LFSR Characteristic Equation

Data produced by an LFSR are based on what is referred to as its characteristic equation, which is defined by the way its feedback is formed. For example, consider the single feedback circuit of Fig. 9.6. We number these flip-flops from left to right as 1 to n. The data at the output of flip-flop n at time t_n was also here n clock cycles back at time t_0.

We represent the time component coefficient of each data by x, thus x^n becomes the coefficient of data at time n, and x^0 becomes the coefficient of data n clock cycles back. Therefore, the characteristic polynomial that defines the data produced by this register becomes:

$$P(x) = x^0 + x^n = 1 + x^n. \tag{9.1}$$

Based on the above discussion, the polynomial for the LFSR shown in Fig. 9.10 is:

$$P(x) = 1 + x^2 + x^3. \tag{9.2}$$

As shown in this figure, in addition to the feedback from the right-most flip-flop (number 3) to the left-most flip-flop, there is also a feedback that adds the output of flip-flop 2 to data coming into flip-flop 3. Considering that the XOR behaves as a modulo-2 adder, data at the output of flip-flop 3 become the superposition of all data arriving at this point through various feedback paths. This explains the addition of x^2 in E.q. 9.2, when compared with E.q. 9.1.

9.2.5.2 Standard LFSR

Figure 9.11 shows an LFSR type that is referred to as a standard or external-XOR LFSR. This circuit has feedbacks from flip-flop stages back to the left-most flip-flop. Because the XOR gates are outside of the shift-register, this structure is also referred to as external-XOR LFSR [8, 9].

The structure in this figure has XOR gates (modulo-2 adders) in the feedback path and switches that turn feedback contributions on or off. An XOR gate with a switch input in the off (0) position is effectively removed from the feedback path. Equation 9.3 is the general form of

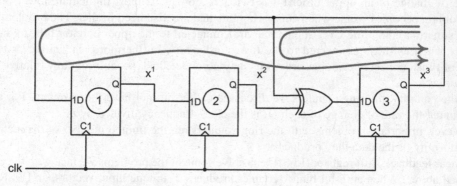

Fig. 9.10 LFSR with third degree polynomial

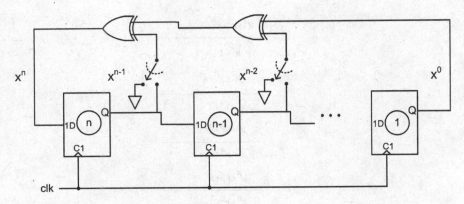

Fig. 9.11 Standard LFSR

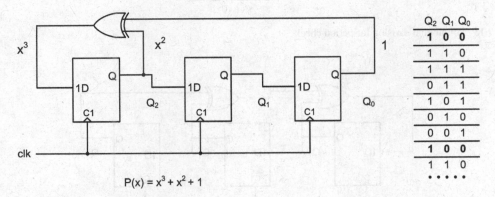

$$P(x) = x^3 + x^2 + 1$$

Fig. 9.12 A third degree standard LFSR

the polynomial for this LFSR. The h parameters represent the switch positions, where h_n and h_o are always 1.

$$P(x) = x^n + h_{(n-1)}x^{(n-1)} + h_{(n-2)}x^{(n-2)} + \ldots + 1. \qquad (9.3)$$

Figure 9.12 shows an LFSR of this type and its corresponding polynomial. Starting with any non-zero initial value, this LFSR cycles through all 3-bit combinations. An example is shown in this figure.

9.2.5.3 Period of LFSR

Polynomials can be divided and multiplied in modulo-2 system. If polynomial of an LFSR divides the polynomial, $x^T + 1$ (remainder is 0), and T is the smallest positive number for which this is true, then T is the period of the polynomial. Figure 9.13 shows dividing $x^7 + 1$ by $x^3 + x^2 + 1$, which is the polynomial of LFSR of Fig. 9.12. Since this division has no remainder, the period of this LFSR is 7. This can be verified by the sequences shown in Fig. 9.12 (the 100 pattern repeats after 7 clock cycles).

Figure 9.14 shows another LFSR with a different polynomial. The sequences this LFSR cycles through are also shown. The smallest positive integer, T, with which $x^T + 1$ can be formed such that it can be divided by the polynomial of Fig. 9.14 is 4. Thus the period of this LFSR is 4, which is also verified by the sequences shown. In an n-stage LFSR, $T = 2^n - 1$, then the LFSR with this

$$x^4+x^3+x^2+1$$

$$x^3+x^2+1 \overline{)\;\; x^7+1}$$

$$x^7+x^6+x^4$$

$$x^6+x^4+1$$

$$x^6+x^5+x^3$$

$$x^5+x^4+x^3+1$$

$$x^5+x^4+x^2$$

$$x^5+x^2+1$$

$$x^5+x^2+1$$

$$0$$

Fig. 9.13 Polynomial division for period check

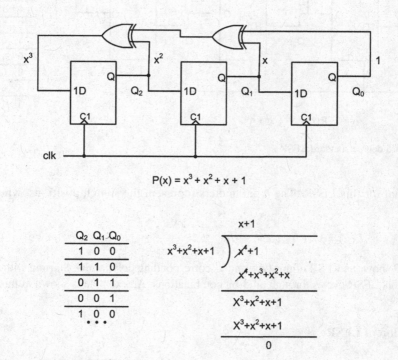

$$P(x) = x^3 + x^2 + x + 1$$

Q_2	Q_1	Q_0
1	0	0
1	1	0
0	1	1
0	0	1
1	0	0

$\cdots$

$$x+1$$

$$x^3+x^2+x+1 \overline{)\;\; x^4+1}$$

$$x^4+x^3+x^2+x$$

$$x^3+x^2+x+1$$

$$x^3+x^2+x+1$$

$$0$$

Fig. 9.14 Third degree polynomial with $T = 4$

period produces all possible n-bit combinations except 0. This kind of LFSR is called maximum-length LFSR. A maximum-length LFSR has a good randomness in terms of frequency of 1s and 0s for each bit, and for this reason is a good source of test data for data inputs of combinational circuits. The characteristic polynomial of a maximum-length LFSR is called a primitive polynomial. Table 9.1 lists primitive polynomials of degree 1 to n. For a given shift-register length, more than one maximum-length polynomial may exist. This table just lists one.

Table 9.1 Primitive polynomials

Bits	Feedback polynomial	Period
n		2^n-1
2	x^2+x+1	3
3	x^3+x^2+1	7
4	x^4+x^3+1	15
5	x^5+x^2+1	31
6	x^6+x^5+1	63
7	x^7+x^6+1	127
8	$x^8+x^6+x^5+x^4+1$	255
9	x^9+x^5+1	511
10	$x^{10}+x^7+1$	1,023
11	$x^{11}+x^9+1$	2,047
12	$x^{12}+x^{11}+x^{10}+x^4+1$	4,095
13	$x^{13}+x^{12}+x^{11}+x^8+1$	8,191
14	$x^{14}+x^{13}+x^{12}+x^2+1$	16,383
15	$x^{15}+x^{14}+1$	32,767
16	$x^{16}+x^{14}+x^{13}+x^{11}+1$	65,535
17	$x^{17}+x^{14}+1$	131,071
18	$x^{18}+x^{11}+1$	262,143
19	$x^{19}+x^{18}+x^{17}+x^{14}+1$	524,287

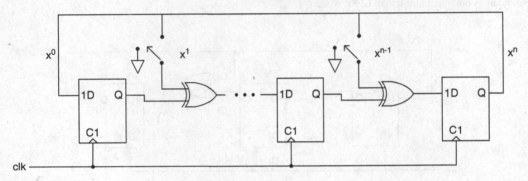

Fig. 9.15 Modular LFSR

9.2.5.4 Modular LFSR

Another LFSR type that uses XOR gates internal to its register is called a modular or internal-XOR LFSR. The modular LFSR is also referred to as a type-2 LFSR. The standard LFSR is type-1. Figure 9.15 shows a generic form of a modular LFSR [9].

Equation 9.3 is rewritten as in E.q. 9.4 to order the terms according to Fig. 9.15. The h parameters express the position of the switches.

$$P(x) = 1 + h_1 x^1 + \ldots + h_{(n-1)} x^{(n-1)} + x^n. \tag{9.4}$$

A switch that is put in the off position (0) effectively removes the XOR that it is an input of. A 4-stage type-2 LFSR is shown in Fig. 9.16. According to Table 9.1, this is not a maximum-length LFSR and since the polynomial of this LFSR $(x^4 + x^2 + x + 1)$ divides $x^7 + 1$, the period of this LFSR is 7. As shown in Fig. 9.16, the sequence 0011 repeats after seven clock cycles.

As a maximum-length modular LFSR, consider the circuit shown in Fig. 9.17. The polynomial for this circuit is $x^3 + x^2 + 1$, which divides $x^7 + 1$, thus its period is 7.

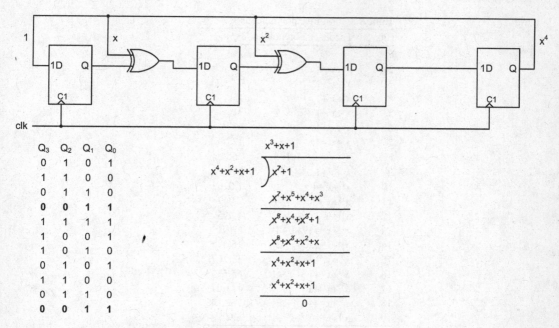

Fig. 9.16 A nonmaximum-length LFSR type2

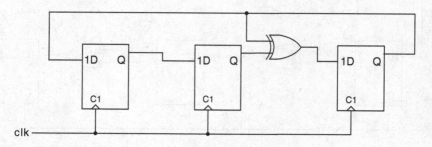

Fig. 9.17 3-Stage maximum-length LFSR

9.2.5.5 LFSR with Serial Input

The sequence of test data an LFSR produces as a TPG can further be influenced by injecting serial input into it. Among various applications, this feature is also useful for using an LFSR as a signature generator, altering test data sequences, and in hybrid BISTs for injecting external data. Figure 9.18 shows addition of a serial input to the LFSR of Fig. 9.17.

Figure 9.18 also shows test sequences that this LFSR generates as serial data are shifted in. Initially, the LFSR contains all 0s, and the serial input is 1. After the first clock cycle, register contents become 100. In nine consecutive clocks, serial data bits 110011101 (left bit, first) are applied to the serial input of the LFSR. As clocks are applied, the sequence 000100101 (from left to right) appears on the serial output of the LFSR. After nine clock cycles, the LFSR contains 001, as shown.

Serial data output and register contents after application of serial data inputs can be calculated by modulo-2 division of input polynomial by the LFSR polynomial. The input polynomial is formed by using the first data bit value for the coefficient of the highest order term (x), and the last bit value for the lowest order term (x^0). Thus, the serial input data polynomial is:

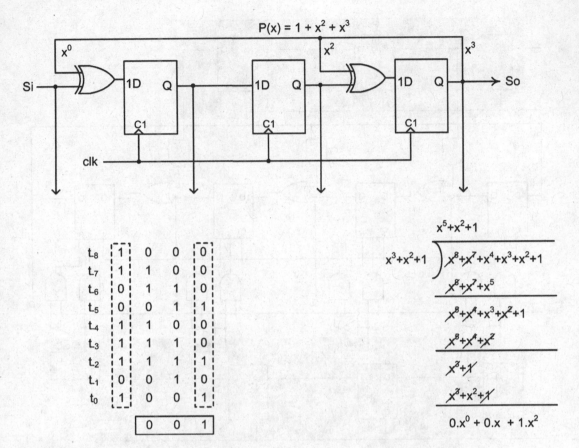

Fig. 9.18 LFSR with serial input

$$G(x) = x^8 + x^7 + x^4 + x^3 + x^2 + 1.$$

As shown in Fig. 9.18 dividing this polynomial by $P(x)$, the LFSR polynomial results in a polynomial for the quotient and another for the remainder, as shown below:

$$Q(x) = x^5 + x^2 + 1, \quad R(x) = x^2.$$

Like the input sequence polynomial $G(x)$, the quotient $Q(x)$ defines bit values in given time units. Setting the reference time at t_0 after completion of clocking the 9-bit input sequence, the above $Q(x)$ indicates that the serial output is 1 at t_{-5}, t_{-2}, and t_0.

The order of the remainder polynomial is the same as that of the LFSR, i.e., highest order term is on the right. $R(x)$ defines the contents of the LFSR register after application of serial data. In Fig. 9.18, 001 for $R(x)$ corresponds to the LFSR flip-flops from left to right.

9.2.5.6 Configurable LFSR

LFSRs can be configured for their polynomials and their initial values (LFSR seed). Such configurations are useful for changing test data generation schemes during BIST operation, for hybrid BISTs, or for when LFSRs are used for ORAs. The latter topic will be discussed in Sect. 9.3. Nevertheless, configuration schemes that apply to TPGs as well as ORAs are discussed here.

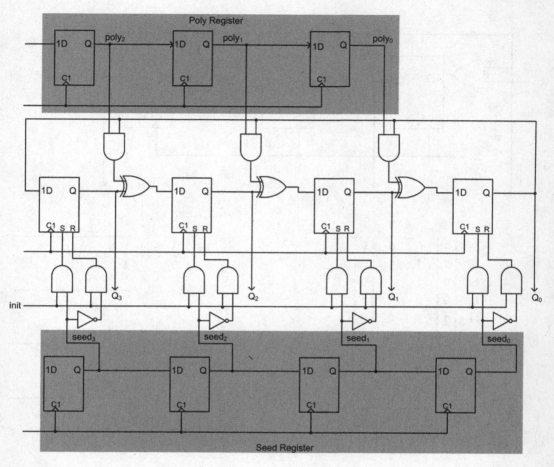

Fig. 9.19 A configurable LFSR

Configuring an LFSR polynomial can be achieved by a vector of AND gates in the feedback paths on XOR inputs. LFSR seed can be loaded into it asynchronously by use of asynchronous flip-flop set and clear inputs, or synchronously, by loading parallel data into the register. Figure 9.19 shows a 4-stage LFSR that can be configured for its polynomial and its seed. As shown, the *seed* register defines seed, and *poly* the polynomial. Depending on the values (1 or 0) in the *poly* register, feedbacks to the XOR inputs are turned on or off, which determines the LFSR polynomial. Initializing the LFSR with a seed takes place when the *init* input becomes 1. Here, we have flip-flops with asynchronous active high S (set) and R (reset) inputs. If a *seed* bit value is 1, with *init* = 1 the *sc* input becomes 10 causing the flip-flop to set to 1, and if a *seed* bit value is 0, *sc* becomes 01 causing it to reset to 0. When *init* is 0, both S and R inputs are inactive (0). Values for *poly* and *seed* can be externally provided, or serially shifted into a CUT via scan registers. Figure 9.19 shows the latter.

Figure 9.20 shows a parameterized configurable LFSR Verilog code. The code shown corresponds to the part in the middle of Fig. 9.19. As shown, the LFSR parallel output is provided on Q n-bit vector. The AND and XOR logic gates feeding the LFSR flip-flops in Fig. 9.19 are implemented by the expression on the right-hand side of the assignment to Q inside the for loop.

```
module LFSR #(parameter n = 8) (input clk, init, en,
                                input [n-1:0] seed,
                                input [n-1:0] poly,
                                output reg [n-1:0] Q);
    integer i;
    always @(posedge clk, posedge init) begin
        if (init == 1'b1) Q <= seed;
        else if (en == 1'b1) begin
            Q[n-1] <= Q[0];
            for (i=0; i<n-1 ; i=i+1 ) begin
                Q[i] <= (Q[0] & poly[i] ) ^ Q[i+1];
            end //for
        end
    end
endmodule
```

Fig. 9.20 Configurable LFSR Verilog code

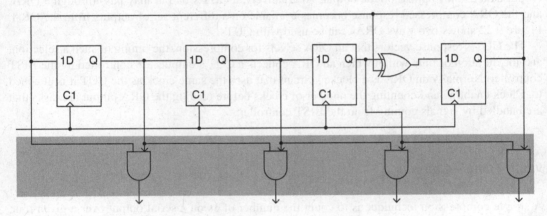

Fig. 9.21 Weighted LFSR

9.2.5.7 Weighted LFSR

Randomness of LFSRs makes probability of each bit being 1 or 0, 50%. Although this is a desirable situation for most fault detections, there are hard-to-detects faults that may require a higher probability of occurrence of certain logic values on the circuit inputs. For such cases, we can add a combinational logic to the output of an LFSR to add a weight to change a given logic value [5, 10]. Figure 9.21 shows a maximum-length 4-stage LFSR with an AND vector that produces parallel outputs with 0.25 for the probability of occurrence of 1. For 0.75 probability, a vector of OR gates should be used.

Nonuniform weights can be obtained by the design of combinational circuits specifically made to produce different probabilities for each test vector bit. Furthermore, a programmable register structure such as that for a configurable LFSR can be used for programming the weights of test data.

In the above sections, we discussed various forms of LFSRs. These hardware structures are used for parallel or serial test data generation. An LFSR can be used in its entirety, or just certain bits can be used to feed a CUT's inputs. With additional hardware, LFSRs can be made configurable in output weights and/or sequences they generate.

9.3 Output Response Analysis

On the one hand, TPGs described in Sect. 9.2 generate test data for on-chip BIST. ORAs, on the other hand, compress responses from a CUT and make them available for the BIST to analyze. We have to have an ORA for the BIST to just check a short signature of all the responses, instead of checking every response of the CUT for the test data that is applied to it.

As with TPGs, ORAs must be brief in use of hardware to fit on the same chip as the rest of the BIST hardware, and of course, the CUT itself. ORAs must be efficient in compressing data so that different sets of test response vectors compress to different values, i.e., have low aliasing. In this section, we discuss various hardware structures for compressing a CUT's response to input test data.

9.3.1 Engaging ORAs

As part of BIST architecture, an ORA attaches to a CUT's output and collects and compresses output data as they appear on the output. Alternatively, a CUT's output may pass through a ORA, and the ORA compressed response becomes available on a different set of outputs from the ORA. Figure 9.22 shows two ways ORAs can be used with CUTs.

The CUT's response vectors that an ORA selects for compression, the timing of such a selection, and the duration of time within which an ORA collects a CUT's outputs are controlled by the BIST controller. Normally an ORA is a clocked circuit that uses the same clock as the CUT it is attached to. Clock enabling and counting the number of clocks before reading the ORA output are tasks that are handled by signals coming from the BIST controller.

9.3.2 One's Counter

A simple compression technique is to count the number of 1s on a serial output over a given time interval. This is decided by the BIST procedure and implemented by the BIST controller. Figure 9.23 shows a One's counter that collects a signature from stream of bits coming from the serial output of a CUT [11].

As shown here, the BIST controller controls the time interval in which the number of 1s on the CUT's serial output are to be counted. This is done by the controller issuing the enable input of the

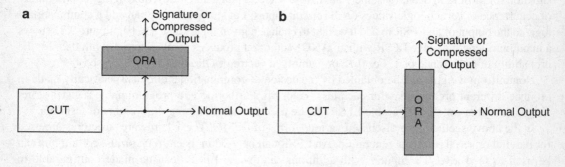

Fig. 9.22 Engaging ORAs

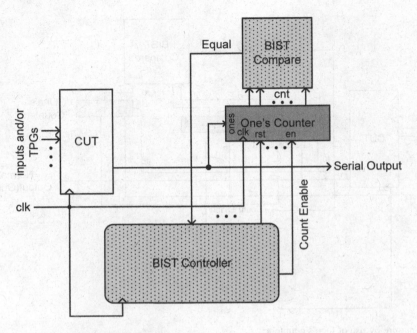

Fig. 9.23 Using One's counter ORA on a serial output

```
module onescounter #(parameter n = 1)
                    (input clk, rst, en, ones,
                     output reg [n - 1:0] cnt);
    always @(posedge clk, posedge rst) begin
        if (rst) cnt <= 0;
        else if (en) begin
            if (ones) cnt <= cnt + 1;
        end
    end
endmodule
```

Fig. 9.24 One's counter Verilog code

counter. At the end of this time interval, the output of the One's counter is read by the BIST compare circuitry, and is compared with its predetermined correct count. The BIST compare reports the compare results to the BIST controller. In case of the One's counter circuit, the ORA's signature is simply the number of 1s. The signature may be more complex for other ORAs discussed in the following subsections.

Note in this block diagram that the CUT, BIST controller, and the ORA use the same clock signal. This means that every serial output bit that is synchronized with the clock is participating in formation of the signature.

Figure 9.24 shows Verilog code for an n-bit One's counter. Note here that the enable input and the input on which 1's are being counted act the same. The counter has a reset input (*rst*) that is also controlled by the BIST controller at the start of a BIST session.

One's counters, and other serial data ORAs, can also be used for vector outputs of CUTs. A simple extension of the signature collection of Fig. 9.23 to a CUT's vector output is to use multiple One's counters. Figure 9.25 shows the use of $i + 1$ One's counters for creating a signature for the data that appears on CUT's vector output, *Outbus[i:0]*.

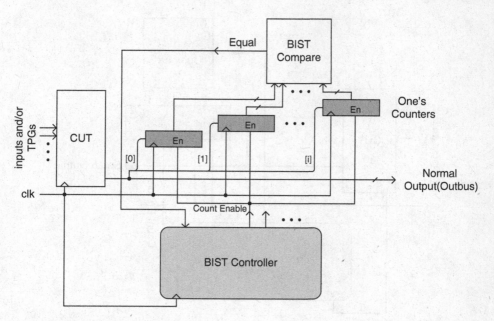

Fig. 9.25 Signature by use of One's counters

As shown in this figure, there is a One's counter for every line of the vector of the CUT shown in the figure. The number of 1s counted on each line is collected in the line's corresponding counter. The outputs of all counters form the signature that is given to the BIST hardware for checking with the correct signature. As before, controlling when the countings begin and when they end is done by the BIST controller and is decided during simulation by the procedure we select for built-in testing.

Another technique for collecting a signature from a multi-bit bus by this serial ORA is to multiplex vector lines and select a different bus line with each clock. Figure 9.26 shows this mechanism. For collecting a signature from *Outbus[i:0]*, a modulo-($\log_2(i+1)$) counter that is controlled by the BIST controller is used. This counter selects the bus line to be sampled. The counter output goes to the select input of an multiplexer that selects one of its $i + 1$ data inputs. With every count of the select-counter, a line of *Outbus* is selected to feed the One's counting input of the One's counter. When the total number of collected samples reaches the predetermined count, the BIST controller reads its compare output and determines whether an error has occurred.

This method of collecting a signature from a multi-bit bus reduces the size of the signature, but increases the possibility of aliasing.

9.3.3 *Transition Counter*

Another serial ORA is the transition counter. This ORA connects to a serial output of a CUT and counts the number of transitions that occur on the output. Figure 9.27 shows the circuitry needed to make a transition counter from a standard counter that has two enable inputs [12].

The flip-flop in this figure holds the old value of the serial input coming from the CUT. Determined by the XOR gate, if the old value and the new value of serial CUT data are different, i.e., a transition has occurred, then the counter is enabled and counting occurs. As before, the other enable input of the counter (*en1*) is controlled by the BIST controller. Figure 9.28 shows Verilog code for an *n*-bit transition counter.

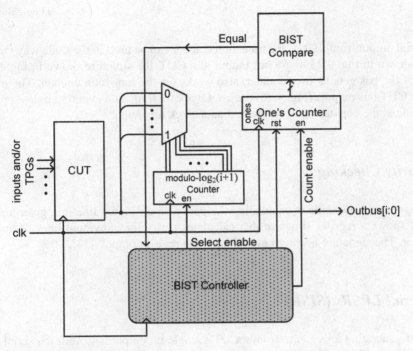

Fig. 9.26 Multiplexing bus lines for serial ORA

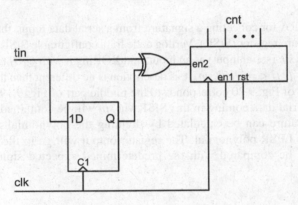

Fig. 9.27 Transition counter

```
module TransitionCounter #(parameter n = 8)
                          (input clk, rst, en, tin,
                           output reg [n - 1:0] cnt);
   reg q;
   wire en2;
   always @(posedge clk, posedge rst) begin
      if (rst) cnt <= 0;
      else if (en) begin
         if (en2) cnt <= cnt + 1;
      end
   end
   always @(posedge clk) q <= tin;
   assign en2 = tin ^ q;
endmodule
```

Fig. 9.28 Transition counter Verilog code

For a serial output from a CUT, the transition counter can be used in the same way One's counter is used, as shown in Fig. 9.23. For a bus output of a CUT, the structure shown in Fig. 9.25, where every line of the bus gets its own counter, also works for the transition counter. The multiplexing solution for CUT bus outputs (Fig. 9.26) does not work for transition counting unless provisions for keeping a line at the counter input for more than a clock are made.

9.3.4 Parity Checking

A simple signature generation is calculating the parity on an output line in a given time interval. Figure 9.29 shows a register structure that calculates parity for individual lines of an n-bit CUT output vector. The signature becomes available on the register output [13].

9.3.5 Serial LFSRs (SISR)

A method of generating a signature from a CUT's single-bit output is to shift the serial output into an LFSR with serial input [14]. The structure used for this purpose and the relation between input, data, serial output data, and contents of the register are the same as those discussed in Sect. 9.2.5.5 and shown in Fig. 9.18.

When used as an ORA for collecting a signature from a serial data input, this structure is referred to as serial input signature register (SISR). Verilog code for a configurable SISR is shown in Fig. 9.30. Except for inclusion of *sin* (serial input) as an input, and XORing it with $Q[0]$ to assign to the register left most bit (i.e., $Q[n-[1]] <= Q[0]^\wedge sin$), this description is no different than that shown in Fig. 9.20. Recall that description of Fig. 9.20 corresponds to the middle part of Fig. 9.19.

The signature of serial data coming in this SISR via *sin* will be contained in the register parallel outputs. This signature can be calculated by dividing the polynomial corresponding to the input sequence by the LFSR polynomial. The register output will go to the BIST compare component, where it will be compared with the predetermined expected signature for the CUT's golden model.

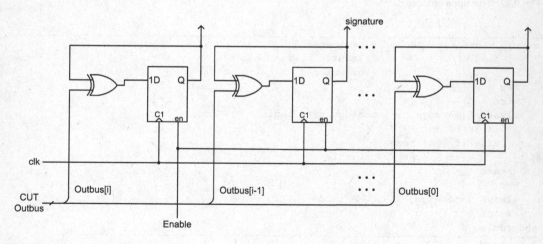

Fig. 9.29 Parity generation

```
module LFSR #(parameter n = 8)
              (input clk, init, en, sin, input [n - 1:0]poly,
               seed, output reg [n - 1:0] Q;
   integer i;
   always @(posedge clk, posedge init) begin
      if (init == 1'b1) Q <= seed;
      else if (en == 1'b1) begin
         Q[n - 1] <= Q[0] ^ sin;
         for (i = 0; i < n - 1; i = i + 1) begin
            Q[i] <= (Q[0] & poly[i]) ^ Q[i + 1];
         end //for
      end
   end
endmodule
```

Fig. 9.30 Configurable SISR

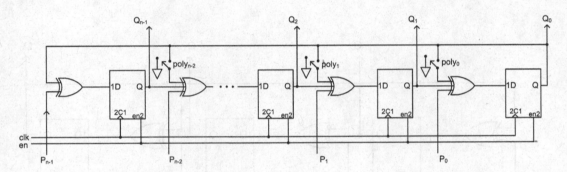

Fig. 9.31 An *n*-bit MISR

SISRs can be used as serial ORAs in any of the three configurations discussed in Sect. 9.3.2. In certain BIST architectures, serial output from a SISR right most bit is used as pseudo-random serial test data for other parts of a CUT. In such cases, the SISR is used both as an ORA and a TPG. When used for serial test data generation, this structure is referred to as shift-register sequence generator (SRSG).

9.3.6 Parallel Signature Analysis

ORA structures discussed so far in this section were all based on single serial inputs. Of course, we have shown how these structures can be used for collecting signatures from multi-bit busses by repeating them the required number of times. Another ORA that is inherently parallel and is a better fit for calculating signatures from bus outputs of CUTs is a multiple input signature register (MISR). As with other ORAs, MISRs are also considered as data compressors.

Structure of an MISR is basically the same as an LFSR except for the addition of XOR gates at the inputs of register flip-flops for bringing in parallel data [15, 16]. Figure 9.31 shows an *n*-stage MISR. Parallel data inputs to be compressed (*P[n-1:0]*) are inputs of the XOR gates at the flip-flop inputs.

Switches at the XOR inputs determine whether the feedback exists or not, i.e., removes the corresponding polynomial term if set to 0. Therefore, the XOR gates are always there, some have three inputs for feedback, and others are just 2-input gates when feedback is off. For control by the BIST controller, the MISR shown has an enable input that tells it when to participate the incoming parallel data in the formation of the signature.

```
module MISR #(parameter n = 8 )
              (input clk, rst, en, input [n - 1:0] poly, seed, P,
               output reg [n - 1:0] Q);
    integer i;
    always @(posedge clk, posedge rst) begin
        if (rst == 1'b1)
            Q <= seed;
        else if (en == 1'b1) begin
            Q[n - 1] <= (Q [ 0 ] & poly [n - 1]) ^ P[n - 1];
            for (i = 0; i < n - 1; i = i + 1) begin
                Q[i] <= (Q[0] & poly[i]) ^ P[i] ^ Q [i + 1];
            end//for
        end
    end
endmodule
```

Fig. 9.32 Generic MISR Verilog code

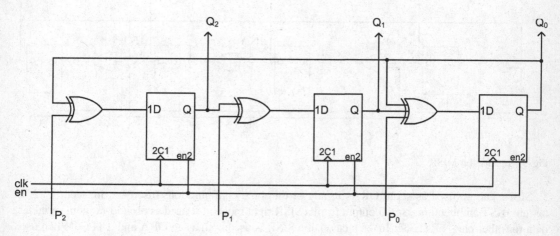

Fig. 9.33 A 3-state MISR

Figure 9.32 shows Verilog code that corresponds to this structure. As shown, 1s and 0s in *poly* vector input determine the MISR polynomial, and 1s and 0s in *seed* form the initial value (*seed*) of the register.

Figure 9.33 shows a 3-stage MISR that is based on the LFSR of Fig. 9.18. The Verilog code of Fig. 9.32 turns into this MISR by using $n = 3$, *poly* = 011, and a *seed* for the initial value. The actual *poly* is 1011, where the left-most bit is always a 1 (coefficient of x^0). The signature appears at the *Q* outputs.

Figure 9.34 shows using a MISR for signature analysis of a CUT output. As with other ORAs, the BIST controller controls the MISR enable input to define the time interval in which CUT outputs are taken and affect the signature. At the start of a BIST session, the controller issues *rst* to initialize the MISR with a *seed*. For avoiding aliasing, a BIST session may consist of several rounds of test data application and signature analysis. At the beginning of each round, the MISR receives a new seed and/or a new poly. The number of clocks for each round are counted by the BIST controller and the corresponding signature is collected. When the BIST session terminates all collected signatures must match the predetermined expected signatures. Alternatively, a larger signature can be used to save test time, but obviously required more hardware for the MISR and BIST compare.

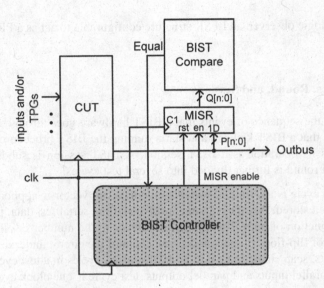

Fig. 9.34 Using an MISR at a CUT output

9.4 BIST Architectures

As discussed thus far in this chapter, a BIST in a CUT consists of TPGs, ORAs, comparators, and a BIST controller that controls operation and timing of these units. BIST architectures define various arrangements of such units within a CUT. Because BISTs cannot use deterministic test data, and have to rely on pseudo-random tests, fault coverage not obtainable by TPGs is compensated for by architectures gaining observability and controllability of a CUT. This section presents templates for BIST architectures. It is important to note that details of a BIST heavily depend on the CUT's architecture, and the templates presented here must be configured for the specific CUTs.

9.4.1 BIST-related Terminologies

Some of the terminologies related to BIST are presented in this section. We first define various TPG and ORA types and then terminologies related to BIST test sessions.

9.4.1.1 TPGs and ORAs

Various TPGs and ORAs were discussed in Sects. 9.2 and 9.3. Those based on LFSRs are particularly important in BIST architectures. Some of the ways LFSRs are used are as follows:

LFSR. Linear feedback shift register, used for test pattern generation. The term LFSR is often used as a general term to refer to all LFSR based TPGs and ORAs.

MISR. Multiple input signature register, used for output response analysis.

SISR. Serial input signature register, a signature register with a single serial input. Used as an ORA.

SISA. Serial input signature analyzer, basically the same as SISR.

PRPG. Pseudo random pattern generator, an LFSR-based TPG with parallel outputs.

SRSG. Pseudo random sequence generator, an LFSR-based TPG with only one serial output. A parallel SRSG is equivalent to PRPG.

BILBO. Built-in logic observer, an LFSR structure configurable to act as a PRPG, an MISR, and a standard register.

9.4.1.2 Test Cycles, Round, and Sessions

We have referred to the sequence of events that a BIST hardware goes through as the BIST procedure. From the time that a BIST kicks in and starts running the BIST procedure until the time that the BIST completes its operation is a BIST session. A BIST session is subdivided into several rounds of test, and a round is further divided into several test cycles.

Test cycle. A test cycle is the time in which a complete test vector is applied to a CUT, and its test response has been stored. A test cycle may involve shifting serial test data, parallel loading the test data, or a combination of both. The largest contributor to the number of clock cycles in a test cycle is the number of flip-flops in the longest scan chain of the circuit under test. For a CUT with 10000 flip-flops in its, scan chain, there will be 10,000 clock cycles in a test cycle. For a combinational circuit with parallel inputs and parallel outputs, test cycle is one clock cycle.

Test round. A round of test consists of a number of test cycles between collections of ORA signatures. At the start of a test round, TPGs and ORAs are initialized and possibly configured for certain polynomials. After application of a certain number of test vectors, ORA signatures is collected or saved, and another round of test may begin. The number of test cycles in a round of test is the same as the number of test vectors applied in that round. A ball-park number for the number of test vectors for a large CUT consists of multiple cores is about one million. This translates to one million test cycles per round of test.

Test session. A signature is obtained at the end of a test round. For reducing aliasing, a test session may involve more than just one round of test. In a test session with r rounds of test, the BIST controller collects all r signatures and then compares them with the expected signatures. In effect, having multiple rounds of test in a test session increases the length of the signature. Between two rounds of test, TPGs and MISRs can be reseeded and/or programmed for new polynomials. A typical test session includes only one test round. The number of test rounds in a session is in the signal digit range.

For 10,000 clock cycles per test cycle, 1,000,000 test cycles per test rounds, and three test rounds for a BIST session, the total test time is 5 min. BIST architectures try to reduce this.

9.4.2 A Centralized and Separate Board-level BIST Architecture (CSBL)

CSBL is a simple BIST Architecture that encloses a CUT, but does not affect its internal hardware [3].

9.4.2.1 CSBL Hardware

Figure 9.35 shows CSBL architecture for a CUT that is represented by its Huffman model. As shown, a PRPG is used for TPG, the output of which is multiplexed to be used as the CUT's input in test mode ($NbarT = 1$). The m bits of output of the CUT are sampled by a multiplexer to be used as the serial input of a SISR. The output sampling process is the same structure we discussed in Sect. 9.3.2 for collecting a signature from multi-bit output busses.

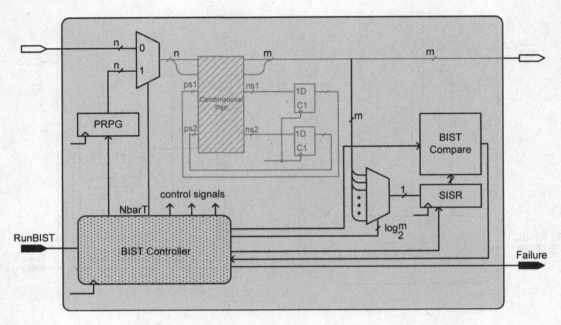

Fig. 9.35 CSBL BIST architecture

9.4.2.2 CSBL Test Process

CSBL test process begins with the external *RunBIST* signal telling the BIST to take over the operation of the CUT. The BIST controller issues *NbarT* and runs its session for a specified number of clocks. During this time, the PRPG and SISR are enabled, and with each clock a new random number is generated by PRPG and applied to the CUT's input. At the same time, a bit of the m outputs of CUT is sampled and clocked into the SISR. After the BIST session is completed, the BIST-Compare compares the SISR parallel output with the expected signature and issues its error signal if a discrepancy is found.

9.4.2.3 CSBL Features

As shown in Fig. 9.35, this method does not involve CUT's internal registers. Therefore, it is not suited for general finite-state controllers and circuits with a lot of feedbacks. This BIST architecture is most useful for pipeline circuits with a limited feedback.

9.4.3 Built-in Evaluation and Self-test (BEST)

The built-in evaluation and self-test (BEST) BIST architecture can be considered as the chip version of CSBL that was primarily used for board level testing [17]. BEST hardware can be separated from that of the CUT it is used for, or it can be integrated in it. In the former case, TPGs and ORAs used by BIST are separate entities that are only used in testing. The other alternative is for TPG and ORA to combine with internal hardware of CUT.

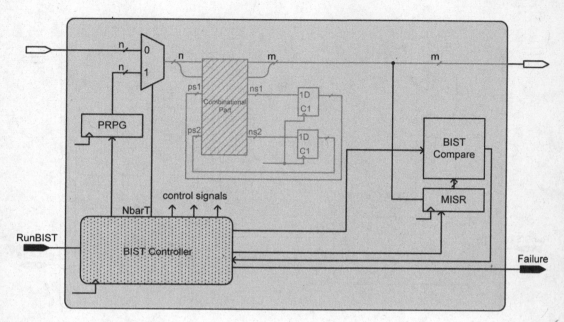

Fig. 9.36 Separate BEST BIST architecture

Figure 9.36 shows a BEST BIST architecture of the separate type. Other than the MISR used for ORA, this architecture is no different than CSBL discussed earlier. Testing procedure for this architecture is the same as that discussed for CSBL. The only difference is clocking MISR instead of SISR.

Controllability and observability of a CUT with BEST and CSBL BIST architectures is limited to the CUT's primary inputs and primary outputs. Because of this, good fault coverage can only be expected if long BIST sessions are run. In which case, we still need to run extensive fault simulations to determine the duration of the BIST session, as well as PRPG and MISR polynomials. If needed for a better fault coverage, multiple rounds of testing with MISR reseeding can be used. As previously discussed, MISR reseeding reduces the chance of signature aliasing for long test sessions.

9.4.4 Random Test Socket (RTS)

The random test socket (RTS) BIST architecture alleviates the low controllability and observability of CUT in CSBL and BEST architectures. RTS shifts pseudo random test patterns in the feedback registers of the circuit being tested.

9.4.4.1 RTS Hardware

The application of this BIST method to a CUT can be explained by considering the CUT's Huffman model. As shown in Fig. 9.37, like the BEST architecture, RTS uses a PRPG and a MISR at CUT's primary inputs and primary outputs. In addition to this, RTS has a SRSG that generates pseudo random scan inputs, and a SISA that generates a signature from scanned outputs from the internal CUT's feedback registers.

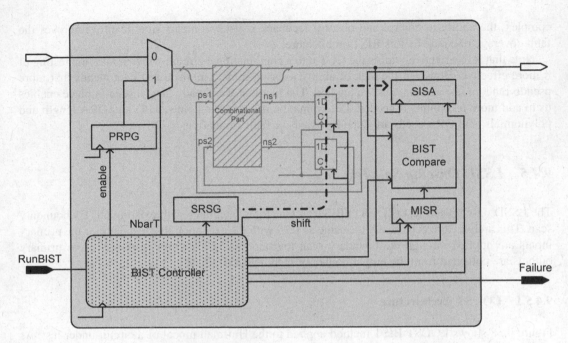

Fig. 9.37 RTS (random test socket) BIST architecture

9.4.4.2 RTS Test Process

The test process of RTS BIST is basically the same as scan testing, with the exception that test data generation and response collection are done internally instead of being external scanned-in and scanned-out.

After *RunBIST* is issued, the BIST controller puts the feedback registers in shift mode, and applies the shift clock simultaneous with applying clock to SRSG. While the SRSG is generating serial pseudo random test data, these data are shifted in the CUT's feedback register. After clocking as many times as the number of feedback flip-flops, these flip-flops come out of the shift mode, and the BIST controller puts them in parallel mode. Note that while in the shift mode, the SISA collects data shifted out from the previous test application. When shifting ends, the SISA clock is also disabled.

While the feedback registers are in parallel mode, the BIST controller asserts *NbarT* to apply PRPG output to the CUT's primary inputs. In this mode, PRPG and MISR clocks are also enabled. With the application of the CUT's clock, circuit's primary outputs are clocked into the MISR, and its pseudo primary outputs (state outputs) are clocked into its feedback registers. Clocking PRPG generates a new input vector for the next round of test, while clocking MISR collects the output of the current test. Pseudo outputs of the CUT collected in the feedback registers will be accumulated in SISR when the next scan data are shifted.

At the end of each round of test, the BIST controller is responsible for comparing SISA and MISR outputs with the expected signatures.

9.4.4.3 RTS Features and Improvements

RTS BIST controller uses a counter and a state machine for keeping track of the number of shifting, and for controlling register clocking. Although the shift process makes the controller somewhat complex, the ability to control and observe feedback register contents significantly improves the fault coverage obtained by this BIST architecture.

Note that RTS effectively turns the CUT into a combinational circuit, which makes the testing of it more effective. However, the lack of ability to bring in deterministic test data means that more pseudo random test vectors must be applied. The fact that we already have a scan in place enables us to add more test points if needed. Determination of extra test points, TPG and ORA length and polynomials, and ORA seeds are done by extensive fault simulations.

9.4.5 LSSD On-chip Self Test

The LSSD on-chip self test (LOCST) BIST architectures integrates with existing CUT's boundary scan. This architecture is basically the same as RTS with the exception that test data for the primary inputs are applied through the boundary scan registers, and CUT's test responses from primary outputs are collected from the output boundary scan registers [3, 18].

9.4.5.1 LOCST Architecture

Figure 9.38 shows LOCST BIST method applied to the Huffman model of a circuit under test. As shown, the LOCST BIST architecture puts primary input and output boundary scan registers in series with the feedback register scan chain. This forms a single chain for feeding in test data and collecting test response. A SRSG with a serial output connects to the beginning of the chain at the primary input side and provides test data for the CUT's primary inputs and pseudo primary inputs. The SISR that connects to the end of the scan chain collects test response from circuit's primary and pseudo primary outputs.

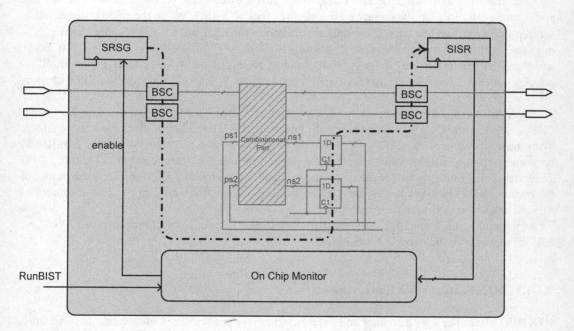

Fig. 9.38 LOCST BIST architecture

9.4.5.2 LOCST Test Process

A LOCST BIST test session begins when the external *RunBIST* signal is issued. At this time, for every test cycle, the controller puts the boundary scan and feedback registers in shift mode and clocks them as many times as there are flip-flops (total boundary-scan and feedback). Simultaneous with this, the SRSG is clocked to produce serial test data for the scan for the current test, and SISR is clocked to absorb serial response from the scan, from the previous test cycle. When clocking complete feedback registers, output boundary scan registers are put in parallel mode and are clocked once. This captures the test response to be shifted out when shifting in the next test.

The BIST controller repeats the above test cycle for as many times as it has been programmed to do so for the given test session. Recall that determination of the number of test cycles in a test session is done by extensive fault simulation for obtaining a desired fault coverage.

9.4.5.3 LOCST Features

As with the RTS architecture, LOCST treats the CUT as a combinational circuit that gives it a better controllability and observability.

LOCST takes advantage of the existing boundary scan cells, but this increases test time since serial data have to be clocked in for the primary inputs. Inclusion of multiple rounds of testing in a given BIST session is possible by reconfiguring and/or reseeding the LFSRs.

9.4.6 Self-testing Using MISR and SRSG

The most widely used BIST in industry today is self-testing using MISR and PRPG (SRSG) (STUMPS) [19]. This BIST architecture solves the problem of long internal scan chain of a CUT (as in RTS and LOCST) by splitting the internal scan into several individually accessible scan chains. Preferably, the scan chains should be of equal or close lengths, but there is no penalty if this is not done.

9.4.6.1 STUMPS Structure

Figure 9.39 shows a generic form for the application of STUMPS BIST method to the Huffman model of our CUT. As shown three separate scan chains are used here. The serial input of each scan chain is driven by a bit of the parallel output of a PRPG. This result in pseudo-random serial test data shifted into the scan chain. The serial outputs of the internal CUT scan chains drive the bits of the parallel input of an MISR.

The block diagram shown in Fig. 9.39 only deals with pseudo inputs and pseudo outputs, and it does not specify how primary input test data are applied and how primary output test response values are read. STUMPS leaves these issues to be decided by the individual implementations.

One possible implementation is to include boundary scan cells as we did with LOCST BIST architecture. However, the boundary scan here is to be used by an ATE to scan in test data into primary input cells and scan out contents of primary output cells. Extending this option one step further, the ATE can also provide LFSR configuration data and seeds for the STUMPS PRPG and MISR. This configuration, shown in Fig. 9.40, is an example of a hybrid BIST.

Another complete BIST implementation based on STUMPS is to use the mechanism we used for RTS for providing primary input test data and collecting a signature of primary output test responses. In this configuration, a PRPG provides input test data, and a separate MISR collects the outputs. At the end of a round of testing, both MISRs (that of STUMPS and that of outputs) will be checked by the BIST compare hardware.

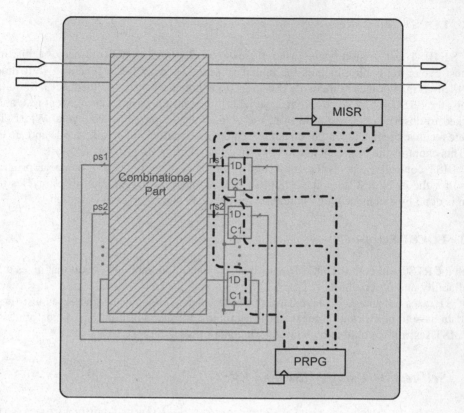

Fig. 9.39 The generic form of STUMPS

9.4.6.2 STUMPS Test Process

Depending on implementation, STUMPS test process is similar to RTS or LOCST. The only difference is in the number of clocks that is needed in each test cycle to shift pseudo random serial data into the CUT's internal registers. Obviously, fewer clock cycles are needed because several registers are receiving test data in parallel.

9.4.6.3 STUMPS Features

In board level testing, various STUMPS scan registers correspond to internal scan of individual chips. At the chip level, STUMPS registers are segments of the chip's scan register. This implementation is similar to multiple scan configuration discussed in Chap. 7. The biggest advantage of STUMPS is that it uses significantly less clock cycles per test cycle.

9.4.7 Concurrent BIST

All architectures discussed earlier apply to sequential or combinational circuits and are off-line. An architecture that is somewhat different than the above architectures is CBIST (concurrent BIST). This architecture only applies to combinational circuits and performs its testing while the CUT is also performing its normal functions.

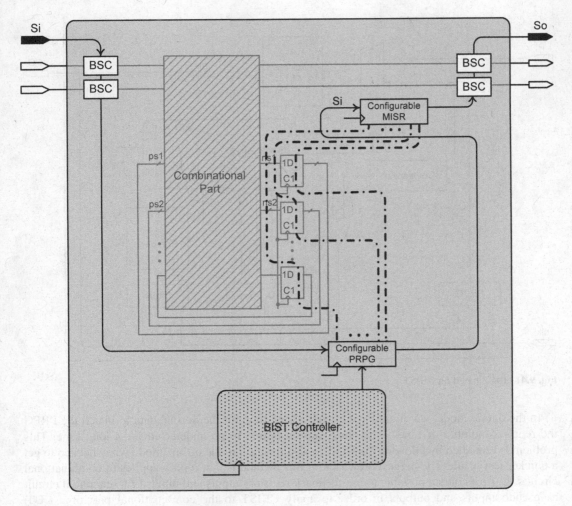

Fig. 9.40 Hybrid BIST based on STUMPS

9.4.7.1 CBIST Structure and Operation

Figure 9.41 shows a BIST system that incorporates a CBIST for online testing. This BIST operates in online and offline modes. When in the online mode, *NbarT* that is issued by the BIST controller is 0. In this case, the CUT receives its normal input from its primary inputs. A comparator compares the incoming inputs with the contents of a PRPG, and when a match is found the PRPG, a match counter, and the output MISR are clocked. The counter keeps track of the number of clocks in the test session and reports it to the BIST controller. The MISR collects the output signature only when an input match is found.

While system is running, at the end of a BIST session that is determined by the match counter, the BIST controller reads the result of BIST Compare that compares the MISR output with the expected signature after the given count. The BIST controller reports a failure if for the set of test data that PRPG has generated, the MISR signature does not match the expected one.

This BIST also runs in offline mode. In this case, the BIST controller asserts *NbarT*, which enables the PRPG, match counter, and the MISR with every clock. While this happens, *NbarT* causes the PRPG output to feed the CUT and ignores the primary inputs.

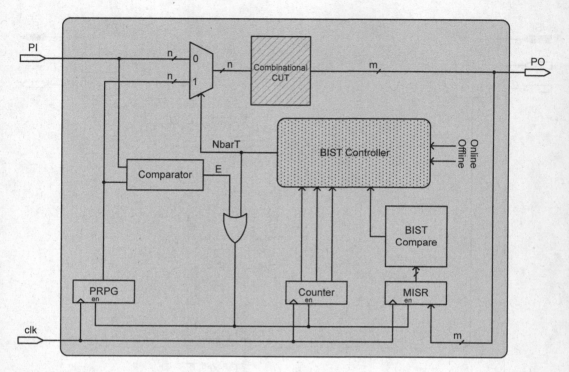

Fig. 9.41 BIST based on CBIST

In the online mode, we might have to wait a long time for the normal data to match the PRPG the required number of times. This may cause a failure to go undetected for a long time. This problem is remedied by allowing the BIST to also function in the offline mode when needed to get a quicker test result. The biggest drawback of this method is that it only applies to combinational circuits. It is, of course, possible to use feedback register outputs and inputs of a sequential circuit as pseudo inputs and outputs in order to apply CBIST to the combinational part of a CUT. However, the large number of bits of primary inputs added to the pseudo primary inputs makes normal input combinations harder to match the PRPG output, which makes the online test process to take a long time.

9.4.8 BILBO

A BIST architecture that only defines the structure of scan registers is built-in logic block observer (BILBO). BILBO combines an LFSR TPG, a MISR ORA, and a scan register (shift-register) with the internal register of CUT [20, 21].

9.4.8.1 BILBO Architecture

Figure 9.42 shows BILBO architecture for the internal register of a CUT. This structure replaces the feedback register in the Huffman model of Fig. 9.1. The register has an n-bit $P[n-1:0]$ parallel input and $Q[n-1:0]$ parallel output. In the standard Huffman model, these signals becomes next-state and present-state signals, respectively. In addition, the structure shown in Fig. 9.42 has a serial input and a serial output (Si and So).

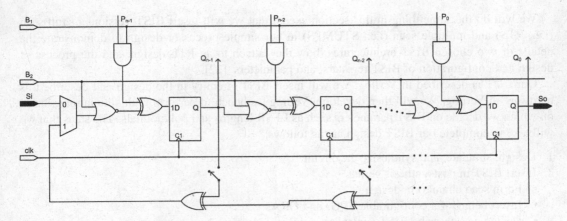

Fig. 9.42 BILBO register structure

Two mode control inputs for the BILBO register are B_2 and B_1. When $B_2B_1 = 00$, the register turns into a shift-register and shifts the complement of data coming in via Si. $B_2B_1 = 01$ is the MISR mode, where the P parallel input will be compressed into the existing contents of the register. $B_2B_1 = 10$ resets the BILBO flip-flops. Finally, $B_2B_1 = 11$ puts the register in normal parallel mode.

9.4.8.2 BILBO Test Process

A test cycle in BILBO begins with $B_2B_1 = 00$, which shifts test data into the register, and at the same time shifts out the register contents captured in the previous test cycle. After this, B_2B_1 is set to 01, which causes the output response that has been made available on the P inputs of the register to be compressed by the MISR function of BILBO. To exit the test mode, B_2 is set to 1. BILBO can be regarded more as a combination of full scan and BIST. Since BILBO is only concerned with the internal registers, a complete BIST based on BILBO requires a mechanism for applying test vectors to the CUT's primary inputs and collecting its primary outputs. In this respect BILBO is similar to STUMPS, and similar solutions to those discussed in Sect. 9.4.6 can be used for primary inputs and primary outputs.

9.4.9 Enhancing Coverage

The architectures discussed earlier are just some of the many possibilities that we have for making a CUT self testable. In an actual design, combination of techniques discussed here is used for better fault coverage and shorter test time. It is also important to note that different parts of a digital system may require different BIST architectures, and there may not be a single solution, even for the same design.

Some of the ways BIST architectures can be enhanced include insertion of test points, reseeding, running multiple rounds of test, and using configurable LFSRs. All such enhancements must be verified by extensive fault simulations before finalizing the design of a BIST.

9.5 RT Level BIST Design

Recall in Chap. 7 that we used a simple processor to demonstrate full scan and multiple scan methods. We inserted the scans and developed a virtual tester that showed how an ATE would handle the testing of an scan-inserted design. We used full scan and multiple scan designs.

We will do the same thing in this section except that we will insert BIST versions of full scan (i.e., RTS) and multiple scan (i.e., STUMPS) in our simple processor design to demonstrate the details of two chosen BIST architectures, how they attach to an RTL design, and the process of design and configuration of BIST registers, and parameters [22].

Our CUT is described in Verilog. We will insert BIST circuitry in the design and describe this hardware in Verilog. We will then develop a Verilog testbench to perform fault simulations, based on which we decide on BIST parameters such as LFSR lengths and polynomials. The steps that we will take to complete our BIST design are as follows:

1. Design, simulate, and synthesize the circuit
2. Insert BIST in postsynthesis netlist
   - Form scan chain(s) in design
   - Insert behavioral configurable TPGs and ORAs
   - Design and describe BIST controller
   - Form CUT in netlist format and behavioral BIST hardware
3. Configure the BIST
   - Come up with several sets of polynomials for the LFSRs (configurations and number of test cycles)
   - Develop a testbench to calculate good signatures for each configuration
   - Develop a testbench to evaluate each configuration based on fault coverage
   - Choose the right configuration
4. Incorporate the selected configuration in the hardware of LFSRs and BIST controller

We will exercise the above steps for RTS and STUMPS BIST designs for our adding machine. We start with RTS for which all the above steps will be shown in detail.

9.5.1 CUT Design, Simulation, and Synthesis

As mentioned, the first step in any DFT method is to complete the design of the circuit to be tested in an HDL. The next step is to synthesize it and generate a netlist to be used for evaluation of the DFT method.

Our design is the Adding Machine that was first discussed in Chap. 2. This circuit has been synthesized to our standard netlist format, and portions of the netlist have been shown in the previous chapters, including Chap. 7, where we inserted several scans in it. Figure 9.43 shows the block diagram of the Adding Machine, based on which its RT level Verilog code has been developed.

9.5.2 RTS BIST Insertion

As shown in Fig. 9.37, RTS BIST architecture adds a PRPG to the primary input of the circuit being tested, an MISR to its primary output, and a SRSG and a SISA to the beginning and end of the chain made of its internal state flip-flops. Figure 9.44 shows implementation of this BIST architecture in the circuit of Fig. 9.43.

9.5.2.1 Scan Insertion in netlist

The scan for our design is shown by a heavy dotted line in Fig. 9.44. The scan serial input (Si) is the left-most bit of AC, and its output (So) is the least significant bit of control flip-flops. This step

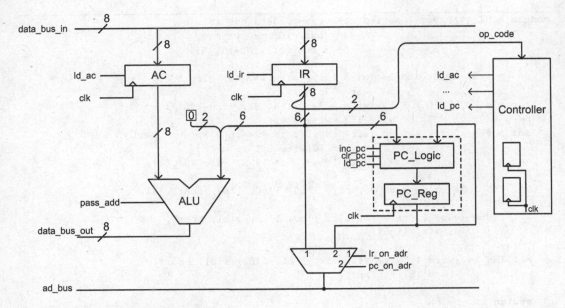

Fig. 9.43 CUT for BIST insertion

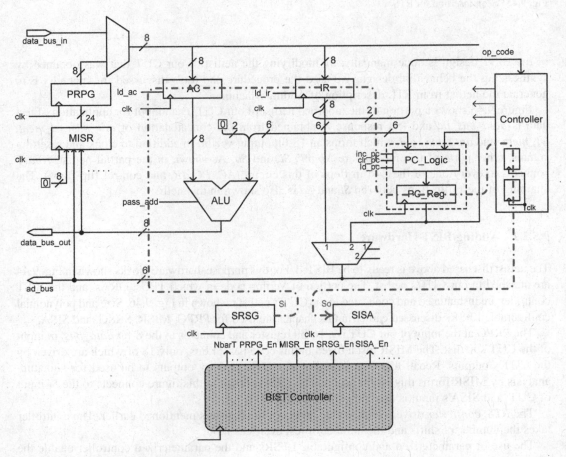

Fig. 9.44 Adding Machine with RTS

```
module AddingCPU_ScanInserted(clk, reset, data_bus_in,
                              {adr_bus, rd_mem, wr_mem,
                               data_bus_out}, NbarT, Si, So);
   . . .
   and_n #(3, 0, 0) AND_95 (wire_191, {wire_142_15, wire_5_22,
                            wire_135_4});
   xor_n #(2, 0, 0) XOR_19 (wire_187, {wire_191, wire_190});
   notg #(0, 0) NOT_38 (wire_193, wire_8_0);
   and_n #(3, 0, 0) AND_96 (wire_194, {data_bus_in_0_2,
                            rd_mem_10,wire_8_1});
   and_n #(2, 0, 0) AND_97 (wire_195, {wire_10_8, wire_193_0});
   . . .
   dff INS_1 (wire_6, wire_96, clk, 1'b0, 1'b0, 1'b1, NbarT, Si,
             1'b0);
   dff INS_2 (wire_2, wire_98, clk, 1'b0, 1'b0, 1'b1, NbarT,
             wire_6, 1'b0);
   . . .
   dff INS_24 (wire_4, wire_249, clk, 1'b0, 1'b0, 1'b1, NbarT,
             wire_18, 1'b0);

   assign So = wire_4;
endmodule
```

Fig. 9.45 Scan insertion for RTS

of the BIST design is done manually, by modifying the netlist of our CUT that was obtained by synthesizing the behavioral design. We used the procedure and tools discussed in Appendix E to generate the netlist from RTL description of Adding Machine.

Figure 9.45 shows a portion of the modified netlist of our CUT. Data input for application of test data is *data_bus_in*, and test response is obtained from the concatenation of *adr_bus*, *rd_mem*, *wr_mem*, and *data_bus_out*, which forms an 18-bit output vector. In addition to clock and resetting signals, other ports of this module are *NbarT*, *Si*, and *So*. As shown in the partial netlist, *NbarT* connects to every one of the 24 flip-flops of this circuit (*AC*, *PC*, *IR*, and control flip-flops). The chaining of these flip-flops between *Si* and *So* is also shown in this netlist.

9.5.2.2 Adding BIST Hardware

The netlist discussed above is ready to be BISTed. For this purpose, hardware blocks shown in Fig. 9.44 are attached to our CUT's netlist. The portion of Verilog code in which TPGs, ORAs, and the BIST controller are instantiated and connected to the CUT's netlist is shown in Fig. 9.46. Size and polynomial configurable LFSRs discussed earlier in this chapter are used for PRPG, MISR, SRSG, and SISA.

The PRPG at the input of our CUT is a 16-bit register and connects to the 8-bit *data_bus_in* input of the CUT's netlist. The MISR instantiated in this code has 24 bits, only 18 of which are driven by the CUT's outputs. Recall that concatenation of CUT's primary outputs to be used for signature analysis by MISR forms this 18-bit vector. The SRSG shown in this figure connects to the *Si* input of CUT, and SISA's input is driven by *So* from the CUT.

The *RTS_controller* drives clock enable inputs of the registers mentioned earlier. The controller takes the number of shifts and the number of test cycles as input.

The use of parameterized and configurable LFSRs and the parameterized controller enable the test engineer to adjust these components for obtaining a good fault coverage.

```
module AddingCPU_RTSArchitecture ();
   . . .

   parameter PRPG_Size = 16;
   parameter SRSG_Size = 16;
   parameter MISR_Size = 24;
   parameter SISA_Size = 16;

   AddingCPU_ScanInserted AddingCPU (clk, 1'b0, PRPG_Out[7:0],
                                     PO, NbarT, Si, So);
   LFSR #(PRPG_Size) PRPG (clk, internalRst, PRPG_En, PRPG_Poly,
                      PRPG_Seed, PRPG_Out);
   MISR #(MISR_Size) MISR_1 (clk, internalRst, MISR_En,
                         MISR_Poly, MISR_Seed,
                         {8'b00000000, PO}, MISR_Out);
   SRSG #(SRSG_Size) SRSG_1 (clk, internalRst, SRSG_En,
                         SRSG_Poly, SRSG_Seed, Si);
   SISA #(SISA_Size) SISA_1 (clk, internalRst, SISA_En,
                         SISA_Poly, SISA_Seed, So,
                         SISA_Out);
   RTS_Controller #(Shift_Cnt, numOfTstCycl) RTS_Controller_1
                  (clk, masterRst, NbarT, internalRst, PRPG_En,
                   SRSG_En, SISA_En, MISR_En, done);
   . . .
endmodule
```

Fig. 9.46 Instantiating BIST components

9.5.2.3   Design of the BIST Controller

The BIST controller, the outline of which is shown in Fig. 9.47, has a sequencer and two counters. The sequencer performs a complete test session after the *rstIn* is set to 0. The test session here has one round of testing that includes *numOfTstCycl* (number of test cycles). In each test cycle, the number of bits specified by the *ShiftSize* parameter are shifted in and out of the internal CUT's flip-flops. The first counter shown in Fig. 9.47 keeps track of the number of shifts, and the second counter counts the test cycles.

The sequencer steps through the test procedures and relies on the two counters for knowing when to enable the LFSRs (PRPG, MISR, SRSG, SISA) and when to stop testing. The sequencer issues *PRPG_En*, *SRSG_En*, *SISA_En*, and *MISR_En* to control the LFSRs. When a test session is complete it issues the *done* signal.

Figure 9.48 shows the details of sequencing of the sequencer through its six states. Note that the register part of the sequencer was included in Fig. 9.47, and the code in Fig. 9.48 is for the combinational part of the sequencer.

In the *Reset* state, the BIST controller is working in normal mode. After *rstIn* is deasserted, the next state *GenData* is taken. In this state, clock enable input of PRPG is enabled for it to generate a new pseudo random data for the CUT's parallel output. The shift counter is reset here for the counting to begin in the next state.

In the *ShiftData* state, the shift counter (the middle bracket in Fig. 9.47) is enabled to keep track of the number of shifts being done. Meanwhile, *SRSG_En* and *SISA_En* are issued. The former generates a pseudo-random serial data for *Si* of the CUT's internal scan, and the latter collects the scan output (*So*) and puts it in *SISA* signature. Setting *NbarT* to 1 causes the internal flip-flops to work in the shift mode.

```
module RTS_Controller #(parameter ShiftSize = 1,
                        numOfTstCycl = 50)
                       (clk, rstIn, NbarT, rstOut, PRPG_En,
                        SG_En, SISA_En, MISR_En, done);
   input clk, rstIn;
   output reg NbarT, rstOut, PRPG_En;
   output reg SRSG_En, SISA_En, MISR_En;
   output reg done;
   reg [2:0] present_state, next_state;
   reg [5:0]shtCount; //vector size should be log2 of ShiftSize
   reg shtCount_Rst, shtCount_En;
   reg [15:0]testVectorCount; //Its size should be log2 of
                              //numOfTstCycl
   reg testCount_Rst, testCount_En;
   . . .
  //Sequencer
  always @(posedge clk, posedge rstIn)
     if(rstIn) present_state <= Reset;
     else          present_state <= next_state;
  always @(present_state or shtCount) begin :Combinatorial
     . . .
  end
  //Counting number of bits shifted into scan chain
  always @(posedge clk) begin
     if(shtCount_Rst)  shtCount <= 0;
     else if(shtCount_En)
        shtCount <= shtCount + 1;
  end
  //Counting number of applied test vectors
  always @(posedge clk) begin
     if(testCount_Rst)  testVectorCount <= 0;
     else if(testCount_En)
        testVectorCount <= testVectorCount + 1;
  end
endmodule
```

Fig. 9.47 BIST controller processes

After the proper number of shifts, *ShiftSize* will be reached, and the sequencer's next state becomes *NormalMode*. In this state, the CUT is put in the normal mode, which causes its internal flip-flops to work in parallel mode. Therefore, when clock arrives, the internal state of the CUT will be captured in the scan register flip-flops. These data will be shifted out during the next test cycle when we are back in the *ShiftData* state.

After the *NormalMode*, the sequencer enters the *GenSignature* state. In this state, the MISR that is connected to the 18 output lines is enabled, which causes it to collect the data on these lines into its signature. In this state, a check is made to see if proper number of test cycles have been completed. This is done by checking the output of *TestVectorCount* counter (the third bracket in Fig. 9.47).

If proper number of test vectors have been applied (number of test cycles), a round of test is complete. Since our BIST session here has only one round, completion of this round means the completion of the test session, which causes the sequence to go in the *Exit* state. In this state, *done* is issued for one clock duration. The logic outside of BIST controller compares signatures, when the *done* signal is issued.

```
module RTS_Controller #(parameter ShiftSize = 1,
                        numOfTstCycl = 50)
                       (clk, rstIn, NbarT, rstOut, PRPG_En,
                        SRSG_En, SISA_En, MISR_En, done);
    . . .
    always @(present_state or shtCount) begin : Combinatorial
        NbarT = 1'b0; rstOut = 1'b0; MISR_En = 1'b0;
        PRPG_En = 1'b0; SRSG_En = 1'b0; done = 1'b0;
        SISA_En = 1'b0; shtCount_Rst = 1'b0; shtCount_En = 1'b0;
        testCount_Rst = 1'b0; testCount_En = 1'b0;
        case (present_state)
            `Reset : begin
                next_state = `GenData;
                rstOut = 1'b1;
                NbarT = 1'b1;
                testCount_Rst = 1'b1;
            end
            `GenData : begin
                next_state = `ShiftData;
                PRPG_En = 1'b1;
                shtCount_Rst = 1'b1;
            end
            `ShiftData : begin
                next_state = (shtCount < ShiftSize - 1) ?
                              `ShiftData : `NormalMode;
                shtCount_En = 1'b1;
                SRSG_En = 1'b1;
                SISA_En = 1'b1;
                NbarT = 1'b1;
            end
            `NormalMode : begin
                next_state = `GenSignature;
                NbarT = 1'b0;
            end
            `GenSignature : begin
                next_state = (testVectorCount < numOfTstCycl - 1) ?
                              `GenData : `Exit;
                testCount_En = 1'b1;
                MISR_En = 1'b1;
            end
            `Exit : begin
                next_state = `Exit;
                done = 1'b1;
            end
            default : next_state = `Reset;
        endcase
    end
    . . .
endmodule
```

Fig. 9.48 Sequencing BIST control states

9.5.2.4 BISTed CUT Model

Verilog descriptions for the CUT netlist, and behavioral and configurable descriptions for all components instantiated in Fig. 9.46, are now available. The next step is examining these components and properly configuring them.

9.5.3 *Configuring the RTS BIST*

Configuring the RTS BIST of Fig. 9.46 is done by a Verilog testbench that involves extensive fault simulations for obtaining a good fault coverage for the least amount of test time. Polynomials, seeds, final signatures, and the number of test cycles for reaching these signatures will be decided after this configurations phase.

The testbench instantiates the CUT and its BIST circuitry, provides clocking for these components, and in a procedural **initial** statement performs the tasks of BIST evaluation and configuration. The outline of this testbench is shown in Fig. 9.49.

More details of the **initial** statement of Fig. 9.49 are shown in Fig. 9.50. As shown here, LFSR seeds are assigned values. These seeds clock into the LFSRs before a BIST session begins. There are two **while** loops in the procedural part of the configuration testbench. The first **while** loop calculates good signatures for several given configurations, and the second loop performs fault simulation to select the best of the given configurations.

9.5.3.1 Acceptable Configurations

We define our BIST configuration as a set of values for polynomials of the four LFSRs of our BIST. For other BIST parameters such as the seeds and number of test cycles, fixed numbers are used in our testbench.

Rather than automating the step of generating several configurations to choose from, we have manually generated them. For this purpose, we have generated a text file, each line of which has four bit patterns that correspond to the polynomials for PRPG, SRSG, MISR, and SISA, respectively.

```verilog
module AddingCPU_RTSArchitecture ();
    parameter PRPG_Size = 16;
    parameter SRSG_Size = 16;
    parameter MISR_Size = 24;
    parameter SISA_Size = 16;
    parameter Shift_Cnt = 1; //Scan_Size = 24
    parameter numOfTstCycl = 100;
    parameter numOfConfig = 1;
    . . .
    AddingCPU_ScanInserted AddingCPU (. . .);
    LFSR #(PRPG_Size) PRPG(. . .);
    MISR #(MISR_Size) MISR_1(. . .);
    SRSG #(SRSG_Size) SRSG_1(. . .);
    SISA #(SISA_Size) SISA_1(. . .);
    RTS_Controller #(Shift_Cnt, numOfTstCycl)
                 RTS_Controller_1(. . .);

    always #5 clk = !clk;

    initial begin
        . . .
    end
endmodule
```

Fig. 9.49 BIST configuration testbench

```
module AddingCPU_RTSArchitecture ();
    . . .
    initial begin
        . . .
        PRPG_Seed = 12;
        SRSG_Seed = 5;
        MISR_Seed = 13;
        SISA_Seed = 24;
        //Generate Dictionary of Good Signatures
        //for Various Configurations
        while (!$feof(cfgFile)) begin
            . . .
        end
        // Fault Simulation
        $FaultCollapsing ( AddingCPU, "AddingCPU.flt");
        while (!$feof(cfgFile)) begin
            . . .
            while(!$feof(faultFile)) begin
                . . .
            end // "while(!$feof(faultFile))"
        end // "while (!$feof(cfgFile)) "
        $stop;
    end
endmodule
```

Fig. 9.50 Main task in the procedural statement of configuration testbench

9.5.3.2 Good Signatures

The next task after developing a set of acceptable configurations is to generate good circuit signatures for each configuration. The part of the procedural statement in our configuration testbench that performs this task (first bracket of Fig. 9.50) is shown in Fig. 9.51.

The Verilog code shown here reads a configuration set from the *Configuration.txt* file, extracts four binary vectors from it, and assigns them to the LFSR polynomial inputs. It then issues the BIST *masterRst* (master reset), and waits for the BIST controller to complete by issuing its *done* output. When *done* is issued, the **while** loop in Fig. 9.51 writes signatures from *MISR* and *SISA* in a signature file. Generating good circuit signatures continues for every available configuration set in the configuration text file.

9.5.3.3 Evaluating Configurations by Simulation

The rest of the code of the **initial** statement, the outline of which we showed in Fig. 9.50, is shown in Fig. 9.52. This part that corresponds to the second bracket in Fig. 9.50 is responsible for evaluating every proposed configuration of the polynomials by performing fault simulation. At the start of this part of the testbench, **$FaultCollapsing** is called to generate list of faults for the *AddingCPU* instances of our CUT.

The outer **while** loop in Fig. 9.52 loops for every available configuration in the configuration file (file with logical name *cfgFile*). For every configuration, good circuit signatures for MISR and SISA are read from the signature file, and the polynomials of the configurations are assigned to their corresponding LFSRs.

```
module AddingCPU_RTSArchitecture ();
   . . .
   initial begin
      sigFile = $fopen ("Signature.txt", "w");
      . . .
      // Generate Dictionary of Good Signatures
      // for Various Configurations
      cfgFile = $fopen ("Configuration.txt", "r");
      i = 0;
      while (!$feof(cfgFile)) begin
         i = i + 1;
         //Apply Configurations
         status = $fscanf(cfgFile, ."%b %b %b %b\n", PRPG_Poly,
                          SRSG_Poly, MISR_Poly, SISA_Poly );
         masterRst = 1'b1; #1 masterRst = 1'b0;

         //Wait for good signature
         @(posedge done);
         $fwrite( sigFile, "%b %b\n", MISR_Out, SISA_Out );
      end
      $fclose(sigFile);
      #1;
      // End Dictionary of Good Signatures

      // Fault Simulation for every configuration
      . . .
      $stop;
   end
endmodule
```

Fig. 9.51 Calculating good signatures

The inner **while** loop in Fig. 9.52 is responsible for performing fault simulation for every fault generated by the **$FaultCollapsing** PLI function. As shown in this loop, after injecting a fault in the netlist of CUT, the BIST controller is reset and started by turning *masterRST* on and off. We then wait for the BIST controller to issue its done signal. When this happens, *MISR* and *SISA* signature outputs for the injected fault become available on the outputs of these registers.

As shown in the inner loop of Fig. 9.52, *MISR_Out* and *SISA_Out* are compared with good signatures read from the good signature file that was prepared by the code of Fig. 9.51. A mismatch in either signature indicates that the injected fault has been detected. In this case, the number of detected faults for the given BIST configuration is incremented to be used for calculation of fault coverage when all faults have been simulated.

After all faults in the fault list are considered, the inner loop exits, and fault coverage, configuration polynomials, and other BIST parameters are reported. The test designer chooses the right configuration of parameters of the LFSRs based on test time and desired coverage.

9.5.4 Incorporating Configurations in BIST

The results of the above evaluations are polynomials for the LFSRs, seeds, the number of test cycles, and good MISR and SISA signatures. For completing the design of the BIST, these parameters must be incorporated into various components of the BIST hardware.

```
module AddingCPU_RTSArchitecture ();
   . . .
   initial begin
      . . .
      //starting "Fault Simulation"
      $FaultCollapsing(AddingCPU, "AddingCPU.flt");
      i = 0;
      while (!$feof(cfgFile)) begin
         i = i + 1;
         //extract golden signature
         status = $fscanf( sigFile, "%b %b\n",
                           Golden_MISR_Out, Golden_SISA_Out);
         //Apply Configurations
         status = $fscanf(cfgFile, "%b %b %b %b \n",
                          PRPG_Poly, SRSG_Poly, MISR_Poly,
                          SISA_Poly);
         #1;
         faultFile = $fopen ("AddingCPU.flt", "r");
         numOfFaults = 0; numOfDetected = 0;

         while(!$feof(faultFile)) begin
            status = $fscanf(faultFile,"%s s@%b\n",wireName,
                             stuckAtVal);
            numOfFaults = numOfFaults + 1;
            $InjectFault(wireName, stuckAtVal);
            masterRst = 1'b1; #1 masterRst = 1'b0;

            @( posedge done ); //Wait for signature
            //compare
            if({MISR_Out, SISA_Out}!=
               {Golden_MISR_Out, Golden_SISA_Out})
               numOfDetected = numOfDetected + 1;
            $RemoveFault(wireName);

         end   // "while( !$feof(faultFile))"

         coverage = numOfDetected * 100.0  /  numOfFaults;
         $fwrite(resultFile, "%b %b %b %b %d %d   %f\n",
                 PRPG_Poly, SRSG_Poly, MISR_Poly, SISA_Poly,
                 numOfTstCycl, numOfTstCycl * Shift_Cnt,
                 coverage );
      end //"while ( !$feof(cfgFile) ) "
      $stop;
   end
endmodule
```

Fig. 9.52 Fault simulation for every configuration

LFSR polynomials and seeds are directly incorporated in the corresponding LFSRs. The polynomials determine feedback XOR structures, while LFSR seeds determine their reset states. Extra LFSR hardware is needed if a BIST session is determined to have multiple rounds that use different polynomials and/or seeds.

Once the number of test cycles has been decided, this number will be hardwired in the counter in the BIST controller that is used for keeping track of this parameter. Another counter and extra circuitry may be required if a BIST session requires multiple rounds of testing.

Finally good circuit signatures must also be incorporated in a BIST circuitry. The BIST compare component, which we used in describing our BIST block diagrams, will be made to compare the signatures obtained from various ORAs with bit pattern that are hardwired into them. This part of the BIST requires a simple combinational logic circuit.

9.5.5 Design of STUMPS

The next RTL BIST design we consider here is inclusion of STUMPS in the Adding Machine example. Figure 9.53 shows TPGs and ORAs used for our STUMPS implementation. This implementation splits the internal registers of our design into three segments of equal lengths. A 12-bit PRPG (*PRPG2*) is used for feeding the serial inputs of the three scan registers. The three serial scan outputs are captured in *MISR2* that is also a 12-bit register. Using a larger register for output signatures reduces the possibility of aliasing.

Another PRPG (*PRPG1*) is used for the circuit's primary inputs, and the primary outputs are compressed by *MISR1* signature register. For test purposes, *PRPG1* applies data to the CPU data

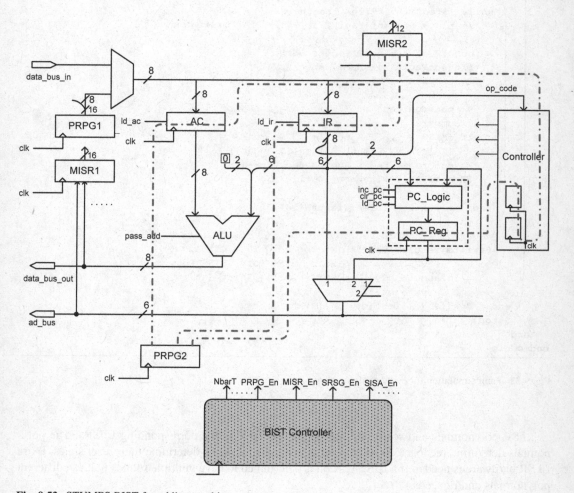

Fig. 9.53 STUMPS BIST for adding machine

bus, *PRPG2* applies serial data to internal registers, *MISR1* collects parallel output, and *PRPG2* collects serial data appearing on three scan chains.

9.5.5.1 Inserting Scan Registers

Figure 9.54 shows partial netlist of the Adding Machine after synthesis (generated by *NetGen* explained in Appendix E). The ports of this module have three serial inputs and three serial outputs for the scan chains. The flip-flops shown here indicate that flip-flops *INS_1* to *INS_8* form one chain, *INS_1* to *INS_16* the second chain, and *INS_17* to *INS_24* form the third scan chain. The third scan chain includes six bits of the PC and two control flip-flops.

```
module AddingCPU_MultiScanInserted(clk, reset, data_bus_in,
                            {adr_bus, rd_mem, wr_mem,
                            data_bus_out}, NbarT, ir_Si,
                            ac_Si, pc_Si, ir_So, ac_So,
                            cntrl_So);

. . .
and_n #(3, 0, 0) AND_95 (wire_191, {wire_142_15, wire_5_22,
                       wire_135_4});
xor_n #(2, 0, 0) XOR_19 (wire_187, {wire_191, wire_190});
notg #(0, 0) NOT_38 (wire_193, wire_8_0);
and_n #(3, 0, 0) AND_96(wire_194, {data_bus_in_0_2,
                     rd_mem_10, wire_8_1});
and_n #(2, 0, 0) AND_97 (wire_195, {wire_10_8, wire_193_0});
. . .
. . .
dff INS_1(wire_6, wire_96, clk, 1'b0, 1'b0, 1'b1, NbarT,
        ir_Si, 1'b0);
. . .
dff INS_8(wire_26, wire_173, clk, 1'b0, 1'b0, 1'b1, NbarT,
        wire_54, 1'b0);
dff INS_9(wire_130, wire_180, clk, 1'b0, 1'b0, 1'b1, NbarT,
        ac_Si, 1'b0);
. . .
dff INS_16(wire_52, wire_208, clk, 1'b0, 1'b0, 1'b1, NbarT,
        wire_50, 1'b0);

dff INS_17(wire_65, wire_211, clk, 1'b0, 1'b0, 1'b1, NbarT,
        pc_Si, 1'b0);
. . .
dff INS_24(wire_4, wire_249, clk, 1'b0, 1'b0, 1'b1, NbarT,
        wire_18, 1'b0);

assign ir_So = wire_26;
assign ac_So = wire_52;
assign cntrl_So = wire_4;
endmodule
```

Fig. 9.54 Insertion of three scan chains for STUMPS

9.5.5.2 Adding BIST Components

Figure 9.55 shows instantiation of the Adding Machine along with BIST related LFSRs and the BIST controller. This partial code corresponds to the BISTed CUT shown in Fig. 9.44. Because this Verilog description is for evaluation of the BIST, the multiplexer that would be required at the *data_bus_in* primary input for the actual hardware is not included here.

One thing to note here is the distribution of *PRPG2* output bits 8, 4, and 0 to the three scan chains. This is quite arbitrary, and is done for more randomness in data shifted into the registers. The rest of this Verilog code (including LFSR and controller descriptions) follows closely what was done for RTS in Fig. 9.46. The difference in STUMPS controller is that it requires fewer clocks for shifting data into the scan registers.

9.5.5.3 STUMPS Configuration

The testbench for evaluation of STUMPS BIST parameters is almost identical to that of RST discussed in Sect. 9.5.3.

```verilog
module AddingCPU_STUMPSArchitecture ();
    parameter PRPG1_Size = 16;
    parameter PRPG2_Size = 12;
    parameter MISR1_Size = 16;
    parameter MISR2_Size = 12;

    AddingCPU_MultiScanInserted AddingCPU
        (clk, 1'b0, PRPG1_Out[7:0], PO, NbarT, ir_Si,
         ac_Si, pc_Si, ir_So, ac_So, cntrl_So);
    LFSR #(PRPG1_Size) PRPG_1
        (clk, internalRst, PRPG1_En, PRPG1_Poly, PRPG1_Seed,
         PRPG1_Out);
    LFSR #(PRPG2_Size) PRPG_2
        (clk, internalRst, PRPG2_En, PRPG2_Poly, PRPG2_Seed,
         PRPG2_Out);
    MISR #(MISR1_Size) MISR_1
        (clk, internalRst, MISR1_En, MISR1_Poly, MISR1_Seed,
         PO, MISR1_Out );
    MISR #(MISR2_Size) MISR_2
        (clk, internalRst, MISR2_En, MISR2_Poly, MISR2_Seed,
         {3'b0, cntrl_So, 3'b0, ac_So, 3'b0, ir_So}, MISR2_Out);

    assign {pc_Si,ac_Si,ir_Si} =
           {PRPG2_Out[8],PRPG2_Out[4],PRPG2_Out[0]};

    STUMPS_Controller
           #(Shift_Cnt, numOfRounds)
           STUMPS_Controller_1(clk, masterRst, NbarT, InternalRst,
                           PRPG1_En, PRPG2_En, MISR1_En,
                           MISR2_En, done);
    . . .
endmodule
```

Fig. 9.55 Adding STUMPS BIST components

9.5.6 RTS and STUMPS Results

The results we obtained by inserting RTS and STUMPS in the Adding Machine example was that for basically the same amount of hardware, STUMPS BIST sessions are almost three times as fast. Coverage we obtained for RTS was about 71%, and for STUMPS it was 74% for the same test time.

9.6 Summary

In this chapter, we covered BIST from an RTL designer's point of view. In the first parts of the chapter, various forms of TPGs and ORAs are discussed, and how they are realized in an RT level HDL environment are shown. Section 9.4 covered various BIST architectures. In this part, we used the generic sequential circuit model so that application of BIST architectures to various other digital designs could be easily understood. On the one hand, our discussion of BIST procedures in this section did not get into clock level details, and it mainly stood at the task level, e.g., loading test data, shifting serial response, etc. On the other hand, Sect. 9.5 where exact RT level descriptions of the CUT, BIST controller, and TPGs and ORAs were discussed, the clock level details of all BIST tasks become clear.

Actually Sect. 9.5 puts together all the materials discussed in the earlier sections of this chapter. There is a twofold benefit in presentation of HDL models and discussing test procedures in an RT level language. One side of this is in clarification of component bindings and timings of the events for the purpose of education, and the other benefit is that this presentation brings test concepts one step closer to home for the RTL designers using HDLs in their designs. The evaluation testbenches and BIST controllers discussed here are typical of what is needed in real designs and can be used as templates in such cases.

References

1. Agrawal VD, Kime CR, Saluja KK (1993) A tutorial on built-in self-test, part 1: principles. IEEE Des Test Comput 10(1):73–82
2. Agrawal VD, Kime CR, Saluja KK (1993) A tutorial on built-in self-test, part 2: applications. IEEE Des Test Comput 10(2):69–77
3. Abramovici M, Breuer MA, Friedman AD (1994) Digital systems testing and testable design. IEEE Press, Piscataway, NJ (revised printing)
4. McCluskey EJ (1986) Logic design principles: with emphasis on testable semiconductor circuits. Prentice Hall, Englewood Cliffs, NJ
5. Rajski J, Tyszer J (1998) Arithmetic built-in self-test for embedded systems. Prentice-Hall, Upper Saddle River, NJ
6. Agrawal VD, Dauer R, Jain SK, Kalvonjian HA, Lee CF, McGregor KB, Pashan MA, Stroud CE, Suen L-C (1987) BIST at your fingertips handbook. AT&T, June 1987
7. Bardell PH, McAnney WH, Savir J (1987) Built-in test for VLSI: pseudorandom techniques. Wiley, New York
8. C Dufaza, Cambon G (1991) LFSR-based deterministic and pseudo-random test pattern generator structures. In: Proceedings of the European test conference, pp 27–34, Apr. 1991
9. Golomb SW (1982) Shift register sequences. Aegean Park Press, Laguna Hills, CA
10. Waicukauski JA, Lindbloom E, Eichelberger EB, Forlenza OP (1989) WRP: A method for generating weighted random test patterns. IBM J Res Dev 33(2):149–161
11. Savir J, McAnney WH (1985) On the masking probability with ones count and transition count. In: Proceedings international conference on computer-aided design, pp 111–113, November 1985
12. Hayes JP (1976) Transition count testing of combinational logic circuits. IEEE Trans Comput 25(6):613–620
13. Das SR, Sudarma M, Assaf MH, Petriu EM, Jone W-B, Chakrabarty K, Sahinoglu M (2003) Parity bit signature in response data compaction and built-in self-testing of VLSI circuits with nonexhaustive test sets. IEEE Trans Instrum Meas 52(5):1363–1380
14. Peterson WW, Weldon EJ Jr (1972) Error-correcting codes. MIT Press, Cambridge, MA

15. Hassan SZ, McCluskey EJ (1984) Increased fault coverage through multiple signatures, in Digest of Papers. Fault-Tolerant Computing Symposium, pp 354–359, June 1984
16. Williams TW, Daehn W, Gruetzner M, Starke CW (1987) Aliasing errors in signature analysis registers. IEEE Des Test Comput 4(4):39–45
17. Perkins CC, Sangani S, Stopper H, Valitski W (1980) Design for in-situ chip testing with a compact tester. In: Proceedings of international test conference, pp 29–41, November 1980
18. Eichelberger EB, Lindbloom E (1983) Random-pattern coverage enhancement and diagnosis for LSSD logic self-test. IBM J Res Dev 27(3):265–272, March 1983
19. Bardell PH, McAnney WH (1982) Self-testing of multiple logic modules. In: Proceedings of international test conference, pp 200–204, November 1982
20. K¨onemann B, Mucha J, Zwiehoff G (1979) Built-in logic block observation techniques. Proceedings of international test conference, pp 37–41, October 1979
21. K¨onemann B, Mucha J, Zwiehoff G (1980) Built-in test for complex digital circuits. IEEE J Solid-State Circuits 15(3):315–318
22. Roy S, Guner G, Cheng K-T (2000) Efficient test mode selection and insertion for RTL-BIST. In: Proceedings of international test conference, pp 263–272, October 2000

Chapter 10
Test Compression

Test application time is one of the main sources of complexity in testing IP cores. Test data volume is also a major problem encountered in the testing of SOCs. High volume of test data is not only exceeding the memory and I/O channel capacity of ATE, but it is also leading to high testing time that impacts the cost of test [1]. Test cost depends on test data volume, the maximum scan chain length, and the required time for transferring test data from ATE to CUT. This amount of time depends on ATE channel capacity and test data bandwidth. Test data bandwidth is the rate at which test vectors can be scanned in and test response scanned out [2]. For a specified ATE channel capacity and bandwidth, reducing test time is possible by test data compression, changing the scan chain structure, and using logic BIST. Changing the structure of internal scan chains is not possible for an existing IP core. Logic BIST requires redesigning IP cores, the extensive cost associated with which may not be justified. So, the only option that remains for reducing test time is to reduce test data by compression [1].

This chapter begins with a general discussion of compression and how it relates to an ATE and the circuit it is testing. The section that follows this general discussion presents various compression techniques and the corresponding algorithms. Code-based and scan-based techniques are presented here. We then focus our attention on decompression hardware for on-chip decompression.

10.1 Test Data Compression

Unlike compaction methods that reduce the number of test vectors, compression methods do not alter the number of test vectors. Instead, they reduce the number of bits per test vector. Unlike compaction, compression requires decompression. Figure 10.1 shows three different ways of applying test vectors from an ATE to a circuit under test (CUT). In its simplest form, test data can be applied to the CUT without any compression or compaction. This is shown in Fig. 10.1 by the dashed path designated by circled number 1. Another alternative is the test data that is compacted can also be applied to the CUT directly as shown by the dotted line designated by circled number 2. The dashed-dotted lines (number 3) show the third alternative depicted in Fig. 10.1. This is the path that test data should travel if it is stored in a compressed format. In this case, the compressed test data should be decompressed using on-chip decompression hardware before it is applied to the CUT.

As shown in Fig. 10.1, decompression hardware consists of some synchronization logics and a decoder. Cyclical scan chains, shift registers, and counters are some of the components used in the design of the decoder logic. The on-chip decompression hardware shown here has a synchronization logic, *Synchronizer 1*, that serially receives test data bits from the ATE at the ATE's speed. The data are saved in the synchronization logic buffer and are made available to the decoder when the decoder requests them.

Z. Navabi, *Digital System Test and Testable Design: Using HDL Models and Architectures*,
DOI 10.1007/978-1-4419-7548-5_10, © Springer Science+Business Media, LLC 2011

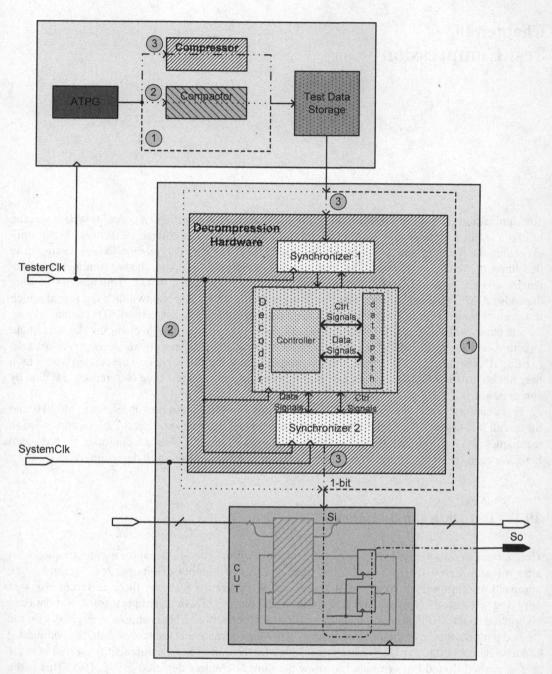

Fig. 10.1 Test application alternatives

From the incoming data, the decoder forms decoded words of size L_{sym} (symbol length). Based on the compression method, the L_{sym} bits wide decompressed data are either completely generated after a fixed number of tester clock cycles, or generated in an uneven distributed fashion over several tester clock periods. Decoding for Dictionary-based and Huffman methods is of the former type, and decoding of the Run-length is an example of the latter type. In either case, the application of decoded data to the CUT uses the system clock. The synchronization logic on the output side of the decoder, *Synchronizer 2*, of Fig. 10.1 synchronizes the decoded data with the system clock. Details of decoder and synchronization logic are discussed in Sect. 10.3.1.

Compressing test data is either an actual compression in which test data vectors are reduced to fewer bits, or it is virtual in which case the same test data is used multiple times, which makes it look like it has been compressed. In this chapter, several compression methods are described first, that is then followed by the decompression hardware of each of the methods.

10.2 Compression Methods

Test data compression methods explained in this section are categorized as code-based and scan-based schemes. Code-based schemes [1, 3–8] are performed after the test generation process using a software program. The input of the compressor is a set of randomly or deterministically generated test patterns and its output is compressed test data, compressed data decompressed by on-chip decompression hardware and shifted into the internal scan or boundary scan chain of a CUT. Scan-based schemes are based on broadcasting a value into multiple scan chains or applying the same set of test data to multiple CUTs [8–16]. In some cases, a simple encoder is needed to analyze the generated test pattern and encode it in order to make it ready to be applied to multiple-input scan chains. In such cases, a simple decompression hardware is also required to decode the data and apply it to the CUT scan chains. Generally, in scan-based compression methods reducing data is done by sharing the same data multiple times, therefore compression and decompression are not as solidly defined as in code-based methods. In the rest of this section, various code-based and scan-based compression methods are discussed.

10.2.1 Code-based Schemes

Code-based schemes are used to compress test cubes or test vectors. Test cubes are the test vectors with several don't care bits that are generated by deterministic ATPG tools. Although, don't care bits are usually filled by ATPG tools randomly, some ATPG tools may leave some of them unspecified. In compressing test cubes, these don't care bits should be filled with binary values. They can be filled in a way that a better compression is achieved. As an example, in Run-length coding, unspecified bits are filled by 0 to get larger length of 0s.

To encode a data in code-based schemes, it is partitioned into blocks called symbols. Each symbol is then converted to a codeword using a compression algorithm. Several considerations are taken into account for partitioning a data into symbols. Considerations are based on the way of applying the data to the CUT, the number of scan chains of a CUT and the length of each scan chain. Figures 10.2 and 10.3 show two ways of partitioning test data into symbols. In each figure, the box on the left represents test patterns, and box(es) on the right is (are) the scan chain(s).

The first method is used when the CUT has a single scan chain with mk scan elements. The symbols are taken from consecutive bits of the test vectors. Figure 10.2 shows an example of this case for a scan chain of size mk. Partitioning the test vector to obtain symbols is illustrated above the test pattern. The dashed arrows show the corresponding data of each scan element after the mk^{th} clock cycle.

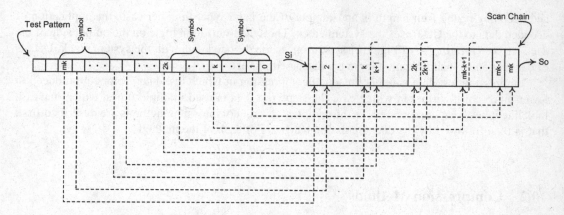

Fig. 10.2 Corresponding value of a scan chain after *mk* clock cycles

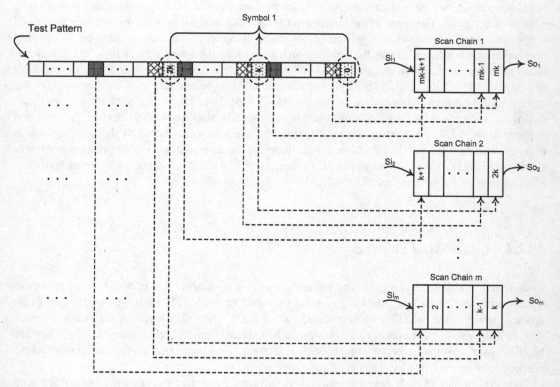

Fig. 10.3 Corresponding value of *m* scan chains after *k* clock cycles

Figure 10.3 shows the case of having a CUT with *m* scan chains, where the length of each scan chain is *k*, and *m* bits of data are shifted into the scan chains in parallel. After *k* clock cycles, data reaches the end of the chains, and the serial outputs of the chains become available. In this case, partitioning test data to create symbols that are to be encoded differs from the case of having a CUT with a single scan chain. Here, the symbols are taken from every *k*th bit of the test vector. There are such *k* *m*-bit symbols. Each *m*-bit symbol is encoded (compressed), sent to the decompression unit, decoded and then applied to *m* scan chains in parallel.

Table 10.1 Four categories of code-based compression methods

Symbol length (L_{sym})	Codeword length (L_{cw})	Example
Fix	Fix	Dictionary code
Fix	Variable	Huffman code [1, 6, 18]
Variable	Fix	Run_length code [8]
Variable	Variable	Golombs code [3]

Code-based schemes can be classified into four groups, depending on whether the symbols and codewords have fixed or variable size [17]. Based on this assumption, four types of encoding exist that are summarized in Table 10.1. An example method of each category is also named in this table.

Code-based schemes can be lossy or lossless. In lossy methods, in each encoding/decoding process, several bits are destroyed or lost. So, the original data is not reconstructed after decoding the codewords. Therefore, compressing test data in this way may decrease the fault coverage. In a lossless compression, encoding test vectors are done such that the original data can be uniquely reconstructed from the codewords. Since in lossless methods the fault coverage does not change, these methods are more often used for compressing test data.

In the rest of this section, test data compression algorithms that are shown in Table 10.1 are explained. All of these methods are lossless. Decoders for these methods are discussed in the next section.

10.2.1.1 Huffman Codes

Huffman [18] coding is a lossless compression method. It compresses data by replacing each fixed length symbol by the corresponding variable length codeword [6]. The idea is to make the codewords that occur most frequently to have a smaller number of bits, and those that occur least frequently to have a larger number of bits. Huffman uses a table of occurrence frequency of each symbol to build up an optimal representation of each symbol as a binary string. As mentioned earlier, compression is accomplished by giving frequent patterns short codewords and infrequent patterns long codewords. It is proven that Huffman code provides the shortest average codeword length among all uniquely decodable variable length codes. A Huffman code is obtained by constructing a Huffman tree. The path from the root to each leaf gives the codeword for the binary string corresponding to the leaf.

Procedure. The procedure for constructing the Huffman tree of a given test set with a given occurrence frequency of each symbol is described here.

1. Sort the symbols in a descending order of their frequencies.
2. Consider each nodes as a leaf node of a tree.
3. While there is more than one node, merge two with the smallest frequency to form a node whose frequency is the sum of the frequency of the merged nodes.
4. Assign a 0 (or a 1) to the left branch of all nodes and the other logic value to the other branch.

The following example shows how Huffman code is used to compress a set of test data.

Example 1: Huffman encoding. Figure 10.4a shows a stream of a test pattern in which all unspecified bits are filled randomly. This pattern is divided into 4-bit symbols. The corresponding pattern of each symbol and its occurrence frequency are depicted in Fig. 10.4b.

Figure 10.5a shows the Huffman tree of this pattern. Figure 10.5b shows the corresponding codeword of each symbol.

a

0 0 1 0 0 1 0 0 0 0 0 1 0 1 0 1 0 1 0 0 0 1 1 0 0 0 1 0 0 1 1 0

0 1 0 0 0 0 1 0 0 0 1 0 1 0 0 0 1 0 1 0 0 1 0 1 0 1 1 0 0 1 0 1

0 1 1 0 0 0 1 0 0 0 1 0 0 0 1 0 0 1 0 0 0 0 1 0 0 1 0 0 0 1 1 0

0 0 1 0 0 1 0 0 0 1 1 0 0 1 0 0 0 0 1 0 0 1 0 0 0 1 0 0 0 1 1 0

b

Symbol	Pattern	Frequency
S_0	0 0 1 0	10
S_1	0 1 0 0	9
S_2	0 1 1 0	7
S_3	0 1 0 1	3
S_4	0 0 0 1	1
S_5	1 0 0 0	1
S_6	1 0 1 0	1

Fig. 10.4 (a) Stream of a test pattern (b) Symbols and corresponding occurrence frequency

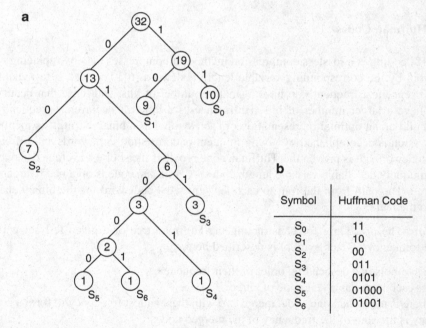

Fig. 10.5 (a) Huffman tree of test pattern (b) Symbols and corresponding codewords

The compression ratio is defined as follow:

$$\text{compression ratio} = \frac{\text{size of the pattern before compression}}{\text{size of the pattern after compression}}$$

So the compression ratio of the Huffman code can be calculated as:

$$\text{compression ratio} = \frac{n \times L_{\text{sym}}}{\sum_{i=1}^{n} L_{\text{cw}_i} \times Freq_i}$$

In the above equation, n is the number of symbols and codewords. Note that in the Huffman coding symbol length is fixed. L_{cw_i} is the length of the codeword corresponding to the ith symbol, and finally $Freq_i$ is the occurrence frequency of the ith symbol. According to this definition, the compression ratio of the Huffman code discussed in Example 1 is calculated as:

$$\text{compression ratio} = \frac{32 \times 4}{10 \times 2 + 9 \times 2 + 7 \times 2 + 3 \times 3 + 1 \times 4 + 1 \times 5 + 1 \times 5} = 1.71$$

As shown in Fig. 10.5a, Huffman code is prefix-free. It means that there is no codeword which is the prefix of another codeword. This simplifies the decoding process because the decoder can recognize the end of each codeword uniquely without any lookahead. However, a Huffman decoder is complex and the number of states increases exponentially with the number of bits of a symbol. So, some variations of Huffman coding that have simpler decoders than the original Huffman are introduced. Although these coding schemes are not as efficient as Huffman, their decoders have fewer states, and hence simpler to build on-chip. Selective Huffman is a variation of Huffman that is explained here.

Selective Huffman. Selective Huffman [6] coding scheme is a variation of Huffman coding, in which only the more frequent symbols are encoded. In this method, an extra bit is added as the first bit of each codeword, indicating whether the data is encoded. If the first bit of the codeword is 0, it means that the symbol is not encoded and transmitted as is. On the other hand, if the first bit of the codeword is 1, it indicates that the symbol is encoded. As the number of codewords that should be encoded is fewer than Huffman, it is obvious that the decoder of this method has fewer states than Huffman.

Example 2: Selective Huffman encoding. Assume that the pattern that is shown in Fig. 10.4a is to be encoded using Selective Huffman coding scheme. As shown in Fig. 10.4b, S_0, S_1, and S_2 are the most frequent symbols. Therefore, only these symbols are encoded and other symbols are sent with a 0 as their starting bit indicating that they are not encoded. The Huffman tree of this method is shown in Fig. 10.6.

10.2.1.2   Dictionary-based Codes

Dictionary-based [5] methods are lossless compression methods that operate by searching for matches between the text to be compressed and a set of strings in a dictionary. In this method,

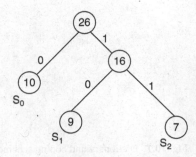

Fig. 10.6 Huffman tree based on Selective Huffman coding

strings of the symbols are selected to establish a dictionary. The basic idea is taking advantage of the number of commonly occurring sequences. For encoding, a symbol is replaced by its index to the dictionary. The dictionary may be either static or dynamic [7]. A static dictionary does not change during the coding process. In other words, the dictionary containing a set of predefined strings is built before starting the coding process. In methods using dynamic dictionaries, the dictionary has some strings at the beginning, and its contents change during the coding process, based on the data that has been already encoded. LZ77 is an example of this kind.

An English dictionary is a simple example of a static dictionary. A word in the input text is encoded as an index to the dictionary if it appears in the dictionary. Otherwise, it is encoded as the size of the word followed by the word itself. In order to distinguish between the index and the raw word, a flag bit needs to be added to each codeword. If the flag bit is 0, it indicates that the word does not exist in the dictionary. In this case, the codeword is composed of the flag bit, the size of the word, and the word itself. If the flag bit is 1, it implies that the codeword exists in the dictionary and the following bits indicate the index of the codeword in the dictionary.

Dictionary-based methods can be used for compressing test data. One of the variations of Dictionary-based codes is encoding fixed length symbols to fixed length codewords [7]. This method uses a static dictionary with *index* rows and L_{sym} columns, which should be selected appropriately. In this method, each symbol in the test pattern is encoded as an index to the dictionary if it appears in the dictionary. Otherwise, it is not encoded. A flag bit is added to each codeword indicating whether or not it is encoded. If the flag bit is 0, it indicates that the symbol does not exist in the dictionary, and the following L_{sym} bits specify the symbol. If the flag bit is 1, it implies that the symbol exists in the dictionary and the following L_{index} bits indicate the index of the codeword in the dictionary. Therefore, the codeword length is fixed and is as shown below. Figure 10.7 shows a dictionary and how a symbol is coded by that.

$$L_{cw} = \begin{cases} 1 + L_{sym} & \text{if flag} = 0 \\ 1 + L_{index} & \text{if flag} = 1 \end{cases}$$

As shown, S_0 is in the ith row of the dictionary. So, it is encoded with flag value 1 followed by L_{index} bits representing i, which is the position of S_0 in the dictionary. To encode a symbol, S_1, that does not exist in the dictionary, it is encoded as a 1-bit 0 for the flag, followed by the L_{sym}-bit original data, S_1.

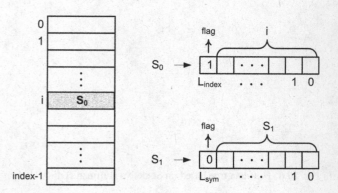

Fig. 10.7 Dictionary and coding scheme

Procedure. The procedure of encoding a given test set with using fixed to fixed Dictionary-based method is described here:

1. Construct the dictionary.
2. Partition the test set into symbols in an appropriate way for encoding.
3. Search each symbol in the dictionary.
   (a) If it is compatible with an entry of the dictionary, encode it as a 1 followed by corresponding L_{index}-bit *index* of that symbol in the dictionary.
   (b) If it is not compatible with an entry of the dictionary, encode it as a 0 followed by the L_{sym}-bit symbol.

As shown in the above procedure, the first step is constructing a dictionary, and for this purpose, the size of the dictionary should be specified. Actually, a dictionary is a memory having *index* number of words, the length of which is L_{sym}-bit. Each word of the dictionary is shifted into the scan chain(s) of the CUT after being decoded.

In the rest of this section, encoding test data using a Dictionary-based method is explained with the use of two examples. In the first example, it is assumed that the CUT has a single scan chain, and in the second example the CUT has multiple scan chains. The manner in which the test pattern is partitioned, constructing the dictionary, and encoding test data are presented in each case. As we show in the two following examples, the number of scan chains that should be loaded at the same time affects the word length of the dictionary and how the test pattern is partitioned into symbols.

Example 3: Dictionary-based method for CUT with single scan chain. Let T_D be the given test set consisting of six 12-bit test cubes, as shown in Fig. 10.8a.

Suppose that the CUT that uses these test cubes has a single scan chain. Therefore, 1-bit test data should be shifted into the scan chain in each system clock cycle. The test pattern can be partitioned into

a

1	0	1	x	1	x	1	0	0	x	0	0
x	x	1	0	1	x	1	0	x	0	0	0
x	1	0	0	1	x	0	1	1	0	1	1
0	1	1	1	1	1	1	0	1	0	0	1
x	0	1	x	1	0	x	1	1	1	1	1
0	1	x	x	x	x	x	1	1	0	1	1

b

L_{sym}

index

0	1	0	1	0	1	0	1	0	0	0	0	0
1	0	1	0	0	1	0	0	1	1	0	1	1
2	0	1	1	1	1	1	1	0	1	0	1	1

c

Symbol	Codeword
1 0 1 x 1 x 1 0 0 x 0 0	1 0 0
x x 1 0 1 x 1 0 x 0 0 0	1 0 0
x 1 0 0 1 x 0 1 1 0 1 1	1 0 1
0 1 1 1 1 1 1 0 1 0 0 1	1 1 0
x 0 1 x 1 0 x 1 1 1 1 1	0 0 0 1 0 1 0 0 1 1 1 1
0 1 x x x x x 1 1 0 1 1	1 0 1

Fig. 10.8 (a) Test cubes T_D. (b) Constructed dictionary (c) Corresponding codeword of each symbol

fixed size symbols as in Fig. 10.2. The simplest way of encoding is to consider each test cube as a symbol. In the encoding process, each test cube is searched in the dictionary. If it is found, it is encoded as a 1 followed by the index of its position in the dictionary. Otherwise, it is encoded as a 0 followed by the test cube whose unspecified bits are filled by 0 in this example. As the first step, a dictionary should be constructed. In this case, we have a dictionary with *index* number of rows and L_{sym} columns, as shown in Fig. 10.8b, i.e., three 12-bit words. Figure 10.8c shows the corresponding codeword of each symbol.

The total number of bits to shift into the scan is as there are bits in all codewords, i.e., $(1 + 2) \times 5 + (1 + 12)$, which is 28. To further reduce test application time, multiple scans can be used. This is shown in Fig. 10.3 and is illustrated in the following example.

Example 4: Dictionary-based method for CUT with multiple scan chains. Let the CUT have three scan chains, and let T_D of Fig. 10.8a be used as the test data. In this case, each bit of an entry of the dictionary is to be applied to a separate scan chain. Therefore, the size of each word of the dictionary should be 3. Figure 10.9a shows the dictionary that contains two 3-bit entries. The test pattern is partitioned into 3-bit symbols the way it is shown in Fig. 10.3. This means that, each test vector is divided into four 3-bit subvectors. The symbols are taken from every fourth bit of the test vector as is shown in Fig. 10.9b. So, each test vector is divided into four 3-bit symbols. The symbols and their corresponding codewords are shown in Fig. 10.9c.

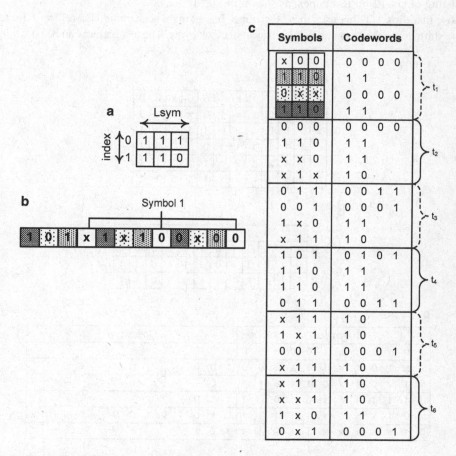

Fig. 10.9 (a) Dictionary (b) Partitioning test set (c) Symbols and their codewords

10.2.1.3 Run-length Codes

The conventional Run-length [3, 8] code is a variable to fixed coding scheme. *Runs* of data are sequences in which the same data value is repeated. In this method, a test pattern is partitioned into variable length symbols that consist of runs of consecutive 0s or 1s. Each symbol is encoded as the length of runs of consecutive 0s or 1s. Here, the runs of consecutive 0s are considered. Run-length coding is efficient for data with long sequences of equal symbols. Because of the correlation of consecutive test vectors, encoding the difference vector set is more efficient. If $T_D = \{t_1, t_2, ..., t_n\}$ is the test set we are starting with, the difference vector set, T_{diff}, is defined as: $T_{diff} = \{t_1, t_1 \oplus t_2, ..., t_{n-1} \oplus t_n\}$. Run-length code encodes variable-length symbols to fixed-length codewords. Suppose that the length of the codeword is L_{cw}. Each codeword represents the length of runs of consecutive 0s. Therefore, the maximum number of consecutive 0s can be $2^{L_{cw}} - 1$. This way, test data will be partitioned into symbols containing $2^{L_{cw}} - 1$ runs of 0s or less than $2^{L_{cw}} - 1$ runs of 0s followed by a 1. The procedure of Run-length coding is discussed here.

Procedure. To start encoding T_D or T_{diff}, a fully specified set with runs of 0s followed by a single 1 should be generated. The procedure is described here:

1. All don't care bits of T_{diff} are mapped to 0 to obtain a fully specified test set before compression.
2. Select the codeword size, L_{cw} which specifies the maximum size of runs of 0s in each symbol.

Once L_{cw} is determined, the data should be partitioned. To do so, we start from the beginning of the sequence and add each bit to the symbol until we reach a 1, or the number of consecutive 0s exceeds $2^{L_{cw}} - 1$. Then, each symbol is encoded as the length of runs of 0s. The following example illustrates this.

Example 5: Run-length coding. This example shows how to encode a test data using Run-length code. Figure 10.10a shows the stream of a difference test set, and Fig. 10.10b shows its encoding.

Here, we are assuming the codeword length is 3, which gives a maximum of seven 0s in a run. Starting from the left, a partition is marked when a 1 is reached, and the number of 0s before that are recorded.

10.2.1.4 Golomb Codes

Golomb codes [3] map variable-length runs of 0s in difference vectors to variable length codewords. Using Golomb code, each codeword has two parts: a group prefix and a tail. To encode data, each set of data should be partitioned into groups of size m. The runs of 0s in T_{diff} are mapped to group of size m. The number of such groups is determined by the length of the longest runs of 0s in T_{diff}. The set of run lengths $\{km, km+1, ..., km+m-1\}$ forms group G_k. If m is chosen to be a power of 2 (2^t), t-bit sequence, tail, uniquely identifies each member within the group.

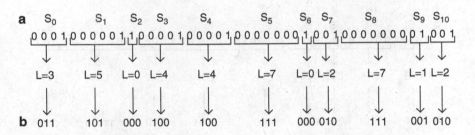

Fig. 10.10 (a) The stream of difference test set (b) The corresponding encoded stream

Procedure. As in Run-length coding, a fully specified test set with long runs of 0s followed by a single 1 should be used as the starting point. The procedure is described here:

1. All don't care bits of T_{diff} are mapped to 0 to obtain a fully specified test set before compression.
2. Select parameter m for the group size.
3. Partition data in a way that each symbol contains runs of 0s followed by a 1.
4. Using the number of 0s in a symbol and parameter m, determine the group number (k) of the symbol. Group number k becomes: (number of 0s) modulo-m.
5. For a symbol that belongs to group G_k, the corresponding codeword has a group prefix in the form of (k) ones, followed by a zero, and tail which identifies the position of the symbol in the group (between 0 and m).

Figure 10.11 shows the encoding of variable runs of 0s based on Golomb coding for $m = 4$. Example 6 shows how to encode a test set using Golomb coding.

Example 6: Golomb coding. Figure 10.12a shows the stream of a difference test set, and Fig. 10.12b shows the encoded test set.

In this example, the group size is equal to 4, so the tail of each codeword is 2-bit. Data is partitioned with a stream of 0s ending with a 1, as shown in Fig. 10.12a, which also shows the number of runs of 0s in each group. Then, each symbol is encoded based on the runs of 0s according to Fig. 10.11. S_0 has 3 runs of 0s which makes it part of group G_0. So, it has 1-bit prefix which is 0, and

Group	Run_length	Group Prefix	Tail	Codeword
G_0	0	0	00	000
	1		01	001
	2		10	010
	3		11	011
G_1	4	10	00	1000
	5		01	1001
	6		10	1010
	7		11	1011
G_2	8	110	00	11000
	9		01	11001
	10		10	11010
	11		11	11011
⋮	⋮	⋮	⋮	⋮

Fig. 10.11 Golomb encoding with $m = 4$

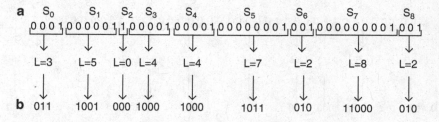

Fig. 10.12 (**a**) Stream of difference test set (**b**) Corresponding encoded stream

2-bit tail that is 11 indicating its position within group G_0. S_1 has 5 runs of 0, which makes it part of group G_1 with a prefix of 10, and its tail is 01 that is its position in group G_1. The remaining symbols are encoded similarly.

10.2.2 Scan-based Schemes

Chapter 7 discussed scan design for gaining controllability and observability into a CUT. As we showed in that chapter, test application time of a scan system is proportional to the length of the scan chain. Although multiple scan chains reduce test application time by concurrently shifting data into several scan chains, they introduce the overhead of extra input and output ports for each parallel scan. In this section, scan-based compression methods are discussed. Generally, in these methods reducing data is done by sharing the same data multiple times, therefore compression and decompression are not as solidly defined as in code-based methods.

10.2.2.1 Broadcast Scan

A scan-based compression method is the broadcast scan [11, 17, 19] that uses one tester channel to load multiple independent scan chains. When test data is generated by an ATPG, some bits are unspecified. These don't care bits can be valued in a way to detect a fault of another CUT, and thus use the test vector for testing more than one CUT. As scan chains of CUTs are independent, using a test vector to test multiple CUTs does not change the fault coverage of each CUT.

The main idea in this scan-based compression method is to share test sets among multiple CUTs and broadcast the same test set to all CUTs [13]. Normally, the majority of test vectors of a test set are those that are randomly generated, and the remaining test vectors for detecting the last 20% to 40% faults are generated by deterministic methods. If we are to merge test sets of several CUTs, random test can be shared and we need not recreate them for each CUT independently. On the other hand, deterministic tests, that are made for each CUT independently, have many don't care values that can be adjusted to detect undetected faults of multiple CUTs. Those test vectors that cannot be shared and cannot be adjusted to benefit other CUTs become part of the test set for all CUTs. Alternatively, such vectors are just dropped from the common test set, which cause some faults to go undetected.

Virtual Circuit. The sharing discussed above can be accomplished by generating tests for a virtual circuit [13] that is formed by tying together the inputs of all CUTs that are to share test sets. The virtual circuit is for test generation purposes only, and does not physically exist.

With several CUTs having different number of inputs, there are many configurations for tying their inputs together, and the way inputs are connected affects the number of generated test patterns. In the virtual circuit [13] shown in Fig. 10.13, inputs of the CUTs are tied together according to their

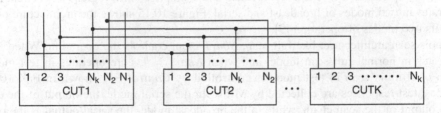

Fig. 10.13 Virtual circuit

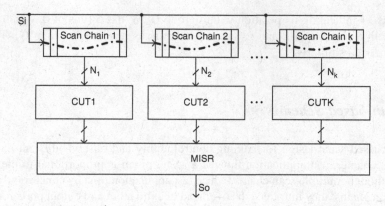

Fig. 10.14 Scan chain configuration of a virtual circuit

ordering in the CUT. Those CUTs that have fewer inputs than some, just do not use the shared inputs.

The scan chain configuration of this connection method is shown in Fig. 10.14 [13]. Since all circuits receive the same test set, we can use a single line to broadcast test patterns to all circuits. The configuration of the scan chain depends on the way the virtual circuit is formed.

As shown in Fig. 10.14, in this configuration all CUTs receive the same test patterns through the scan input. Unless we use as many serial outputs as there are CUTs that are receiving data in parallel, this compression method requires a MISR for collecting circuit responses.

10.2.2.2 Illinois Scan

Using broadcast scan for multiple scan chains that drive the same circuit may result in a reduced fault coverage because same scan cells always hold identical values [9]. The next scan compression technique is called Illinois scan, which is actually a parallel serial full scan (PSFS) [9]. As in the broadcast scan, this method has a parallel mode that reduces the test application time by dividing the scan chain into multiple partitions and shifting-in the same vector to each scan chain through a single scan-in input [9, 12, 15]. The outputs of the scan chains are observed through a MISR.

The drawback of the broadcast method is that the test vector overlap forced by this method can cause shifting-in tests that can benefit only one CUT and are useless for the others. Considering each CUT might need its own special tests that others cannot use, leads to many unnecessary shared test vectors. Avoiding this, results in many faults in the CUTs that will be left undetected. In order to solve this problem, Illinois scan technique preserves the single scan chain structure with its serial mode. This is implemented by extra multiplexers and a simple control logic [9]. As a result Illinois scan operates in two modes of broadcast and serial. Figure 10.15 shows the architecture of Illinois scan and its two modes of operation [9].

In Illinois scan architecture, like full scan, *NbarT* input controls the operation. When *NbarT* = 0, it operates in normal full scan mode, and when *NbarT* = 1, core operates in test mode. The operation of Illinois scan in test mode is controlled by *Control FF*, shown in Fig. 10.15a. In both modes, test responses are collected by MISR. In the serial mode, the output of the circuit is the serial output of the scan chain, while in the broadcast mode, the serial output is the output of the MISR.

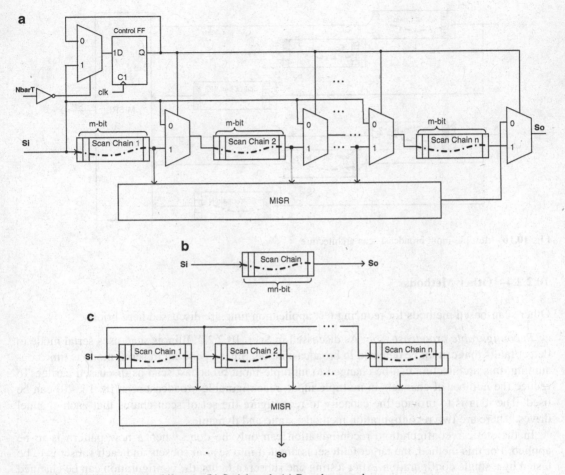

Fig. 10.15 (a) Illinois scan architecture (b) Serial mode (c) Broadcast mode

10.2.2.3 Multiple-input Broadcast Scan

As discussed in the above subsections, in broadcast scan and broadcast mode of Illinois scan, a tester channel is used to drive all scan chains (see Figs. 10.14 and 10.15). The idea of Illinois scan is to share the same scan-in pin among multiple scan chains in broadcast mode. For a test pattern, if any scan chain requires a logic value different from the logic values on the other scan chains in the same cycle, that test pattern must be serially scanned in. The Illinois scan architecture can be optimized by using multiple-input broadcast scan [12, 13, 20] that uses more than one channel. If two scan chains must be independently controlled to detect a fault, they can be assigned to different channels. Furthermore, using multiple channels enables us to use shorter scan chains that cause better fault coverage by placing fewer constraints on the ATPG. Figure 10.16 shows Multiple-input broadcast scan architecture.

The architecture shown in Fig. 10.16 shows two channels each with three scan chains. Shaded patterns used in the scan cells indicate same bit values that are shifted in the scan cells. An example of using this architecture was shown in Sect. 10.2.1.2. In Example 4, this method is used in conjunction with Dictionary-based method to further reduce the test time.

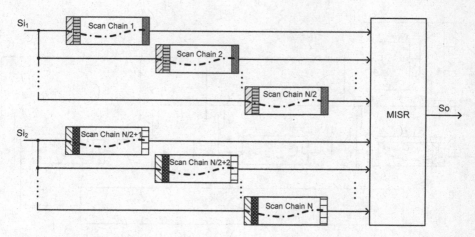

Fig. 10.16 Multiple-input broadcast scan architecture

10.2.2.4 Other Methods

Other scan-based methods for reducing test application time are discussed here briefly.

Reconfigurable broadcast scan. As discussed in Sect. 10.2.2.2, Illinois scan uses serial mode to detect faults that cannot be detected in broadcast mode. Since operating in serial mode is time consuming, this architecture can be changed to multiple-input broadcast scan as discussed earlier. To reduce the number of channels in multiple-input, reconfigurable broadcast scan [8, 13, 20] can be used. The idea is to provide the capacity to reconfigure the set of scan chains that each channel drives. There are two reconfiguration methods: static and dynamic.

In the static reconfiguration, reconfiguration can only be done when a new pattern is to be applied. For this method, the target fault set is divided into several subsets and each subset is to be tested by a single configuration. After testing one subset of faults, the configuration can be changed to test another subset [20].

In dynamic reconfiguration, the configuration can be changed while scanning in a pattern. This provides more reconfiguration flexibility. The disadvantage of dynamic reconfiguration is that it needs more control information for reconfiguration at the right time because reconfiguration occurs on a per-shift basis [20].

A reconfiguration method uses multiplexers before scan chains for selectively shifting the appropriate data into each scan chain [13, 20]. This method can be used to reconfigure scan chains statically or dynamically by changing the selector signals of the multiplexers per-test or per-shift, respectively. Figure 10.17a shows an overview of this architecture for four 4-bit scan chains. There are two serial inputs, Si_1 and Si_2. Figures 10.17b and 10.17c show static reconfiguration of this architecture in which the selector signal of the multiplexers can only be changed after shifting a test pattern into the scan chain completely [20]. Figure 10.17d shows a dynamic reconfiguration of this architecture [20]. In this case, the selector signals of multiplexers get four different values during shifting a test pattern into the scan chains. In the architecture of Fig. 10.17, multiplexers give the scan chains the choice of using either one of scan inputs. Multiple chains can use the same input causing them to receive the same data.

In static configuration, there are only as many configurations as there are independent serial inputs. This is shown for our example in Fig. 10.17b, c, where same pattern shadings indicate same data being shifted.

There is more flexibility in dynamic reconfiguration because reconfiguration is possible per-shift. In Fig. 10.17d, the value of the *select* signal of the multiplexer can be changed per-shift.

Therefore, in four clock cycles, it can be configured four different ways. In the first clock cycle, in which *select* = 1, Si_1 is shifted into scan chains 1 and 3, and Si_2 is shifted into scan chains 2 and 4. In the second clock cycle, as *select* = 1 again, Si_1 is shifted into scan chains 1 and 3, and Si_2 are shifted into scan chains 2 and 4. In the third clock cycle, as *select* = 0, Si_1 is shifted into scan chains 1 and 2, and Si_2 are shifted into scan chains 3 and 4. In the forth clock cycle, as *select* = 0 again, Si_1 is shifted into scan chains 1 and 2, and Si_2 is shifted into scan chains 3 and 4. In the figure, scan cells that receive their inputs from the same source use the same shading patterns.

LFSR-based methods. LFSR-based compression method [16] is based on LFSR reseeding. An LFSR seed configures it for generating its specific pattern. The idea here is to compute a set of seeds that lead to the generation of a given deterministic test cube after being expanded by the LFSR.

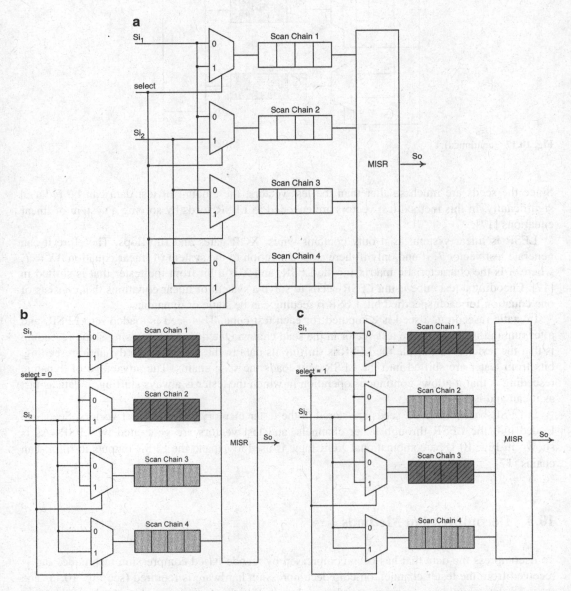

Fig. 10.17 (**a**) Reconfigurable broadcast scan architecture. (**b**) Static reconfiguration: *select* = 0 (**c**) Static reconfiguration: *select* = 1. (**d**) Dynamic reconfiguration: *select* = 0011(during four shift-in operations)

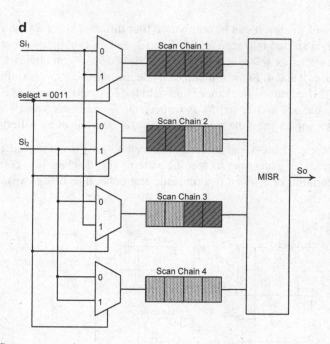

Fig. 10.17 (continued)

Since the seeds are much smaller than the test vectors, the amount of test data can be reduced significantly. In this method test vectors are encoded as LFSR seeds by solving a system of linear equations [17].

LFSR is linear system, as it only contains wires, XOR gates and flip-flops. Therefore it can generate test vector T, if and only if there exist a solution to the system of linear equation $AX = T$, where A is the characteristic matrix for the LFSR, and X is a bit from the tester that is shifted in [17]. Encoding a test cube using LFSR needs to solve a system of linear equations that consists of one equation for each specified bit. LFSR reseeding can be static or dynamic.

In static reseeding, a seed is computed for each test cube. This seed is loaded into LFSR, and after running it produces the test vector in the scan chains. One drawback of using static reseeding is that the tester is idle while the LFSR is shifting its data in the scan cells. In dynamic reseeding, bits from tester are shifted into the LFSR as it loads the scan chains. The advantage of dynamic reseeding is that it allows continuous operation in which the tester is always shifting in data as fast as it can and is never idle.

In LFSR-based methods, seeds are stored in the tester memory instead of test vectors. Seeds are loaded into the LFSR through tester channels, and test vectors are generated by LFSR. As is shown in Fig. 10.18, a combinational XOR logic is used to expand the LFSR outputs to fill n scan chains [17]

10.3   Decompression Methods

To decompress the data that has been compressed by a code-based compression technique, and is received from the tester channel, on-chip decompression hardware is required (see Fig. 10.1). The received data should be decompressed before being shifted into the scan chain(s). In such cases, explicit decoder and decompression hardware are required. On the other hand, in most scan-based

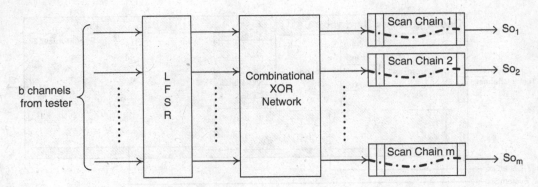

Fig. 10.18 Compressing using LFSR reseeding

compression techniques, the use of the same data by multiple chains makes it look like data has been compressed. For such methods, that we refer to as virtual compression methods, no specific decompression hardware is required.

In testing on-chip cores, test vectors of each core should be applied to scan inputs and its internal scan chain, and test responses should be observed on core output and shifted out from internal scan chain. If the data received is compressed by a code-based technique, hardware structures to decompress test data are needed before applying data to the CUT. Generally, decompression hardware consists of decoder and synchronization logics.

This section discusses decompression hardware architectures. In addition, cyclical scan chain (CSC) [21] that is used for generating the original test set from the set of difference vectors is discussed. Decompression hardware for Selective Huffman is described in detail, including its decoder and synchronization logic. For other code-based compressing methods, the architecture of decoder is discussed. For the sake of completeness, we also mention scan-based decompression methods.

10.3.1 Decompression Hardware Architecture

In this section, the overall structure of an on-chip decompression hardware is discussed and an example is presented. Figure 10.19 shows the architecture of decompression hardware in which the single internal scan chain of the CUT is used for applying test data. This general structure applies to most code-based techniques.

In every tester clock cycle, *TesterClk*, decompression unit receives 1-bit serial data through the tester channel. Since it may take the decoder (shown as *Decoder* in Fig. 10.19) more than one clock cycle to decode a bit and be ready for the new data, a synchronization logic (shown as *Synchronizer 1* in Fig. 10.19) is needed. *Decoder* informs *Synchronizer 1* that it is ready to receive a new bit of data by asserting its *Ready* signal.

After collecting enough number of bits, the decoder generates the original data whose size is that of the symbol of the compressed data, i.e., L_{sym}. Based on the compression method, the L_{sym}-bit original data is either completely generated after a fixed number of tester clock cycles and put on *ParOut* output of *Decoder*, or it is generated in an uneven distributed fashion over several tester clock periods bit-by-bit and put on *SerOut* output signal. In either case, the decoded data should be applied to the CUT scan chain using the system clock (*SystemClk*).

Since tester and system clocks are usually different, the decoded data should be synchronized with the system clock. The synchronization logic (*Synchronizer 2*) on the output side of the decoder of Fig. 10.19 is responsible for this task. With this arrangement, we are avoiding the tester being idle while shifting the decoded data into the CUT scan chain.

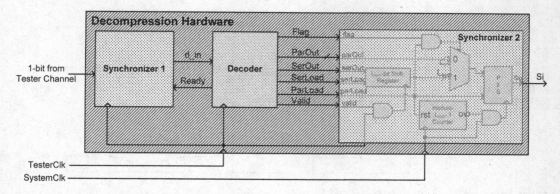

Fig. 10.19 Block diagram of decompression hardware

Fig. 10.20 Architecture of *Synchronizer 2*:
Synchronizing parallel data

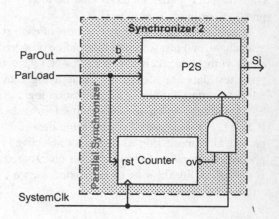

When *Decoder* is ready to send data for the scan chains, it informs and loads data into *Synchronizer 2* using *Valid*, *Flag*, and either of the load signals (*ParLoad* or *SerLoad*). Data for the synchronizer becomes available on *SerOut* or *ParOut*, depending on the compression technique used. The compression technique also influences the architecture of *Synchronizer 2*. What follows discusses synchronizing the generated data that is generated by the decoder. We look at two cases of synchronizer's input data becoming available after a fixed number of tester clocks, and data sent to the synchronizer serially over several clock periods. In both cases, the architecture of *Synchronizer 2* that is considered is for a fixed decoded data size.

In the first case, we assume that the decoder generates b-bit original data after a fixed number of tester clock cycles. The application of the decoded data to the CUT scan chain should be done with the system clock. Figure 10.20 shows the corresponding hardware. This hardware has a parallel to serial converter (*P2S* in Fig. 10.20), and control hardware for control of *P2S* clocking and shifting. The data input of this hardware is *ParOut*, which is the b-bit output of the decoder that is loaded into *P2S* when *ParLoad* is asserted. After loading the decoded data in to *P2S*, it is shifted into the CUT scan chain in the next b system clock cycles. Since the clock of the system is faster than the tester clock, a data block is shifted into the CUT scan chain before decoding a new codeword. This sets a minimum for the size of the codeword.

For the case that the decoded data is generated bit-by-bit, an additional shift register is needed to collect serial data before it is serialized again with the system clock. Figure 10.21 shows the

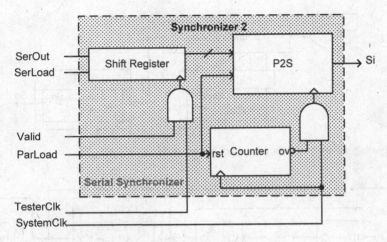

Fig. 10.21 Architecture of *Synchronizer 2*: Synchronizing serial data

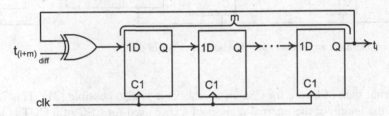

Fig. 10.22 Architecture of a CSC

Synchonizer 2 hardware for this case. As shown, the parallel input of *P2S* is coming from the shift register instead of the decoder, as was done in the previous case.

10.3.2 Cyclical Scan Chain

In some compression methods such as Run-length and Golomb coding, a set of difference vectors is used in compression instead of the individual test vectors. As discussed earlier in this chapter, this technique achieves more compression because of long runs of 0s in the difference vectors. To convert the difference vectors, and obtain the original data, a CSC [21] is used. A CSC consists of several flip-flops and an XOR gate. A CSC should be the same size as the internal scan chain of the CUT, and its contents should not be overwritten when the system clock is applied to the CUT. Figure 10.22 shows the architecture of a CSC [21].

An m-bit vector shifting out from an m-bit CSC register over the next m clock cycles is the bit-by-bit XOR of the existing data in the register before the m clock cycles begin. The CSC register must be initialized to all 0s before serial difference data begin to come in.

CSC can be configured using the chip boundary scan, or using a scan chain in a different system clock domain [21]. In either case, its size must be the same as that of the scan chain it is preparing data for, and its contents should not be altered during its operation. Figure 10.23 shows configuring boundary scan of *Core 1* as CSC of CUT. If the length of the boundary scan chain used for CSC of a CUT is larger than the size of the CUT internal scan chain, the boundary scan chain can be broken up at the right position, and a feedback from the appropriate scan element is used for the CSC XOR input.

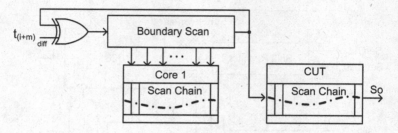

Fig. 10.23 Configuring boundary scan of *Core 1* as CSC of CUT

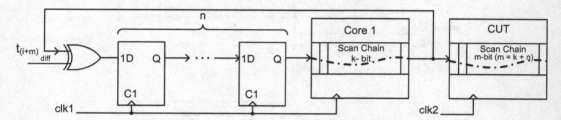

Fig. 10.24 Using internal scan chain of *Core 1* to build the CSC of the CUT

Using internal scan of a core for CSC of another core is also possible [21]. This is illustrated in Fig. 10.24. If the length of the internal register of a core used for CSC of a CUT is larger than the size of the CUT internal scan chain, the same can be done as was done for the boundary scan chain, above. However, if the length of the internal scan chain of core used (*Core 1*) is less than the length of CUT internal scan chain (which is the case in Fig. 10.24), several flip-flops are placed before the input of the internal scan chain of *Core 1* to form a CSC with the same length as the CUT internal scan chain. To make sure that the contents of the register used for CSC in this case is not altered, the CSC should be in a different clock domain than the CUT.

10.3.3   Code-based Decompression

Decompression architecture for decompressing data that is encoded using code-based schemes is generally similar to the architecture shown in Fig. 10.19. In this section, the decoder hardware is discussed for each of the code-based methods discussed in Sect. 10.2.1. For Huffman coding scheme, the architecture of *Synchronizer 2* is also briefly discussed.

10.3.3.1   Huffman

While Huffman code gives the optimum compression for a test set divided into a particular set of fixed length symbols, it generally requires a very large decoder [6]. A Huffman code with L_{sym}-bit block size requires a finite state machine (FSM) with up to $2^{L_{sym}}$ states. Therefore, the number of states grows exponentially as the block size increases. To reduce this complexity, several variations of Huffman code are used. One such variation is Selective Huffman coding that we discussed in Sect. 10.2.1.1, which only encodes the most frequent symbols. In this method, if n codewords are to be decoded, the FSM requires at most $n + L_{sym}$ states for decoding data. Several extra states are also needed for controlling the synchronization logic.

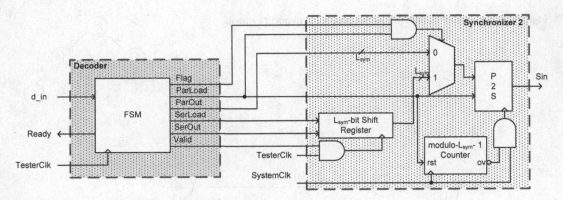

Fig. 10.25 Block diagram of Selective Huffman decoder and synchronization logic

In a Selective Huffman codeword, if the first bit is 0, it indicates that the following L_{sym} bits specify the symbol; otherwise, the following bits specify the encoded symbol. Because of this two-mode operation, the part of the hardware that is responsible for producing serial data for the scan register (*Synchronizer 2*) has to be able to take either serial data or parallel data from the decoder. Therefore, this part of the hardware becomes an overlay of the two hardware structures discussed for the synchronizer shown in Figs. 10.20 and 10.21. Figure 10.25 shows the block diagram of Selective Huffman decoder and its synchronizer.

Using the tester clock, the decoder receives 1-bit data, *d_in* after asserting *Ready* signal, and it controls the operation of *Synchronizer 2* by setting *SerLoad*, *ParLoad*, *Valid*, and *Flag* output signals appropriately. *Flag* is 1 if a symbol is decoded and parallel data are prepared, and it is 0, otherwise. After decoding the encoded data, *ParLoad* becomes 1, and *ParOut* is loaded into *P2S* and shifted into the CUT internal scan chain with the system clock. If no decoding is being done, *Flag* becomes 0, *SerLoad* output is asserted, and each bit of data is shifted into the shift register of *Synchronizer 2*.

The shift register clock is controlled by the *Valid* signal coming from *Decoder*. After receiving a complete symbol, *ParLoad* becomes 1, the shift register output is loaded into *P2S*, and the counter is reset. Decoding the data and the operation of the synchronizer are controlled by the FSM.

Finite state machine. An FSM is used for decoding codewords generated by Selective Huffman coding, and for the control of the synchronizer. This FSM uses the tester clock.

If the first bit of the codeword is 1, it indicates that the subsequent bits form a prefix-free variable length code. In this case, in each tester clock cycle, one bit of the codeword is considered, and the FSM goes to the corresponding state. After decoding the codeword, *Valid* and *ParLoad* output signals become 1, and L_{sym}-bit data block is loaded into *P2S*. When the decoder loads the data block into *P2S*, and for the next L_{sym} cycles the data block is shifted into the CUT scan chain.

If the first bit of the codeword is 0, the original data should be shifted into the CUT scan chain without being decoded. Therefore, in each tester clock cycle a 1-bit data is put on *SerOut* signal, and *Valid* signal is set to 1, which enables the clock of the synchronizer shift register. At the same time, the decoder also asserts *SerLoad*, causing *SerOut* to be shifted into the shift register of the synchronizer. After the completion of loading the original data into the shift register, FSM sets *ParLoad* to 1, to load the content of the shift register to *P2S* and reset the counter.

To illustrate the FSM, consider Example 2 presented earlier in this chapter. The Huffman tree of the example is shown in Fig. 10.26 for reference. As shown in the figure and explained in the example, only the most frequent symbols are encoded. According to Fig. 10.4b symbols 0010 (S_0), 0100 (S_1), and 0110 (S_2) are the most frequent symbols. Therefore, these symbols are encoded, and others are transmitted as usual, serially.

Fig. 10.26 Selective Huffman tree of the three most frequent symbols

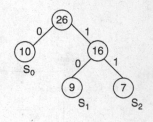

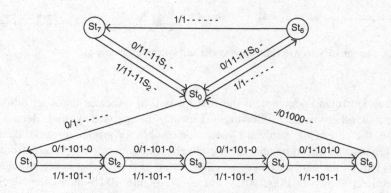

Fig. 10.27 FSM for decoding the Selective Huffman code of Example 2

Figure 10.27 shows the state diagram for decoding codewords and controlling the synchronizer. The input and output signals of the FSM are in the form of *d_in/Ready, ParLoad, SerLoad, Flag, Valid, ParOut, SerOut*. The input *d_in* is the serial data from the tester channel that is synchronized with *Synchronizer 1* (Fig. 10.19).

The FSM shown in Fig. 10.27 is constructed according to the Huffman tree of Fig. 10.26. State St_0 is the initial state. States St_1 to St_4 are entered if the first bit that appears on *d_in* is 0, and they shift out the original data when no decoding is needed. State St_6 is entered if the first bit that appears on *d_in* is 1 that indicates a symbol is coming on the rest of bits of *d_in*. States St_6 and St_7 decode the rest of the bits on *d_in* using the Selective Huffman tree that is explained below. State St_5 is used for synchronization with the output synchronizer. Signals issued to the synchronizer are shown on the right-hand side of the slash (/) on the edges coming out of the states.

Since there are three symbols, two bits are needed to decode them. St_6 is entered to start this decoding. If the first data bit received in this state is a 0, then we are decoding S_0 that is the left leaf in the tree of Fig. 10.26. In this case, St_6 produces S_0 and control returns to St_0. On the other hand, if the first bit received in St_6 is a 1, then S_1 or S_2 are being decoded (two right leafs in Fig. 10.26), and to distinguish between these two, control goes to state St_7. In this state, depending on *d_in* value of 0 or 1, S_1, or S_2 is decoded. This FSM can be realized with three flip-flops.

10.3.3.2 Dictionary-based

Figure 10.28 shows the architecture of the decoder of a Dictionary-based method. This decoder feeds the *Parallel Synchronizer* logic (Fig. 10.20) which will then generate synchronized data for the CUT scan chain.

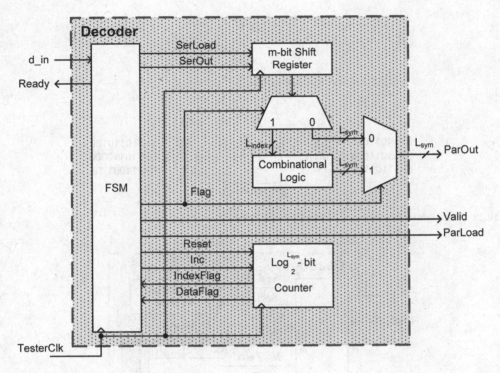

Fig. 10.28 Dictionary-based decoder

The decoder receives serial data from the tester channel. The first bit of the codeword is stored as *Flag*. Value 1 for this flag indicates that what follows is a dictionary word. In this case, the following L_{index} bits specify the position of the symbol in the dictionary. For the collection of the index, the FSM enables *SerLoad* to shift the next L_{index} bits. Using this index, the L_{sysm}-bit symbol is constructed using a combinational logic or memory lookup at the indexed address.

If the first bit of the codeword is 0, the word does not exist in the dictionary, and the following L_{sym} bits are the actual data. In this case, the incoming data is shifted into the shift register through *SerOut* output in L_{sym} tester clock cycles.

When the original data has been decoded and it is ready, *Valid* and *ParLoad* are issued. This causes *ParOut* to be loaded into *P2S* of the synchronizer (see Fig. 10.20), which makes it go into the scan register with the system clock.

Whether we are shifting data or index, the $log_2^{L_{sym}}$-bit counter of Fig. 10.28 is used for keeping track of the number of bits shifted. When index bits are shifted, *IndexFlag* is asserted, whereas shifting symbol causes *DataFlag* to be issued. The operations of the counter and the shift register are controlled by the FSM that is operation next.

Finite State Machine. The FSM shown in Fig. 10.29 is used for controlling operations of the counter and the shift register. The input and output signals of the FSM are arranged as: *d_in, IndexFlag, DataFlag/Ready, Reset, Inc, SerLoad, ParLoad, SerOut, Valid, Flag.*

The FSM receives new bit of data when it is ready (*Ready* = 1) and goes to St_1 to start receiving data from the tester channel. As shown in Fig. 10.29, in this state, the counter is set to 0 by putting a 1 on *Reset* (second output signal on the right hand side of the slash (/)). In this state, the decoder receives its first bit of the codeword. Depending on receiving a 1 or a 0, the FSM goes in state St_2 or St_3, respectively. The former state issues signals for collecting an index; and in the latter state the actual data is collected. Collection of data continues until the counters issue *IndexFlag* or *DataFlag*.

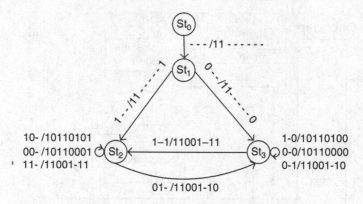

Fig. 10.29 State transition diagram of Dictionary-based FSM

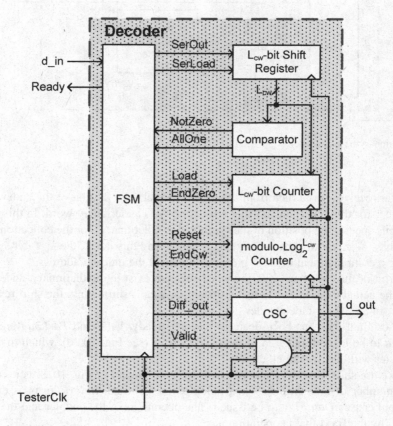

Fig. 10.30 Run-length decoder

10.3.3.3 Run-length

Figure 10.30 shows the block diagram of the Run-length decoder. Since the original data is generated by the Run-length decoder as a series of bits, *Serial Synchronizer* shown in Fig. 10.21 can be used to synchronize the generated data with the clock of the system. The synchronization logic is not discussed here, and signals related to it are not shown in this figure.

As discussed in Sect. 10.2.1.3, the size of codeword is fixed and is equal to L_{cw}. Therefore, each symbol is at most $2^{L_{cw}} - 1$ bits, and consists of runs of 0s followed by a 1, or runs of 0s equal to $2^{L_{cw}} - 1$.

Since a codeword specifies the number of consecutive 0s in a symbol, for decoding it, we need a counter of size L_{cw} bits, as shown in Fig. 10.30. Since codeword bits are received serially, a shift register is used to collect the codewords before assigning them to the counter. We also need a counter of size $log_2^{L_{cw}}$ bits for keeping track of the number of shifts. The FSM shown here resets this counter when it receives the first bit of data, and the counter counts up until it reaches $L_{cw} - 1$. At this time, *EndCw* is asserted and the FSM loads contents of the shift register into the L_{cw}-bit counter by asserting *Load*. This counter counts down until it reaches 0, and the decoder puts a 0 on *Diff_out* output signal while counting is being done. The L_{cw}-bit counter asserts *EndZero* output signal when it reaches 0. If all bits of codeword are not 1, it means that the symbol includes runs of 0s, followed by a 1, which will be generated on *Diff_out*. Otherwise, the symbol contains only 0s. After generating each bit of data, *Valid* is asserted (see lower part of Fig. 10.30), the clock of CSC is enabled, and *Diff_out* signal is shifted into CSC to obtain the actual data from the difference.

10.3.3.4 Golomb

Figure 10.31 shows the block diagram of Golomb decoder. The *Serial Synchronizer* shown in Fig. 10.21 can be used to synchronize the generated data with the clock of the system. The synchronization logic is not discussed here, and hence the required signals for synchronization are not shown in this figure.

As shown in Fig. 10.31, a log_2^m-bit counter is required to decode the codeword. A codeword to be decoded has a prefix and a tail. The data bits generated by the FSM are the bits of difference vectors that are shifted into the CSC of Fig. 10.31 to obtain the original test data. The *Ready* signal in Fig. 10.31 is used to get another bit of data when the decoder is ready. This signal is needed because the length of a codeword in Golomb compression technique is variable, and synchronization is required between the tester and the decoder. An FSM controls the operation of the counter.

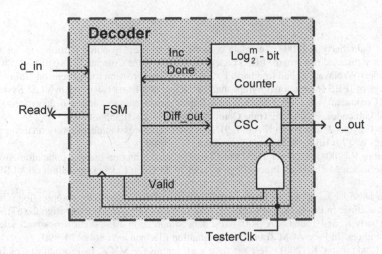

Fig. 10.31 Golomb decoder

10.3.4   Scan-Based Decompression

As mentioned, scan-based compressions compress data by sharing the same set of data among several scan chains. Decompression methods for these techniques are just the on-chip distribution logic structures that distribute test data to the scan chains that are the recipients of the shared data. Decompression hardware for the broadcast scan, Illinois scan, multiple-input broadcast scan, and reconfigurable broadcast scan are those shown in Figs. 10.14–10.17, respectively.

10.4   Summary

Reducing test time has been our biggest concern in all the chapters we have discussed so far. We directly or indirectly try to reduce test time. This chapter targeted this subject very directly, and by trying to reduce test data entering a chip. The first part of the chapter discussed compression techniques. We focused on compression techniques for which a decompression could easily and efficiently be made in hardware. Several methods of code-based compression techniques that perform well for test data were discussed. We then discussed scan-based compression techniques that can only be regarded as compression when dealing with test data and data for internal or boundary scan of a chip.

After a thorough discussion of these two compression techniques, we focused our attention on their corresponding decompressions. For the code-based decompressions, we showed hardware architectures for on-chip decoding. We discussed architectures for several code-based techniques and concluded that hardware realization of the techniques we presented were generally brief and did not create a large overhead on the chip area. For the scan-based, we showed that the decompression hardware, per se, was not really necessary. For the scan-based decompressions, data coming into the chip are decompressed as they arrive, and in most cases a few gates are all that is needed for decompressions. For the scan-based techniques, the burden is on the test generation that considers the scan structures. This chapter serves well as the last chapter for a series of chapters that all try to reduce test time one way or another.

References

1. Chandra A, Chakrabarty K (2003) Test data compression and test resource partitioning for system-on-a-chip using frequency-directed run-length (FDR) codes. Proc IEEE Trans Comput 52:(8):1076–1088
2. Karimi F, Meleis W, Navabi Z, and Lombardi F (2002) Data compression for System-on-Chip testing using ATE. In Proceedings of IEEE International Symposium on Defect and Fault Tolerance in VLSI Systems, 166, 2002
3. Chandra A, Chakrabarty K (2001) System-on-a-chip test-data compression and decompression architectures based on Golomb codes. Proc IEEE Trans Comput-Aided Des 20(3):355–368
4. Jas A, Ghosh-Dastidar J, and Touba N A (1999) Scan vector compression/decompression using statistical coding. In Proceedings of 17th IEEE VLSI Test Symposium, 114, 1999
5. Basu K, Mishra P (2008) A Novel Test-Data Compression Technique using Application-Aware Bitmask and Dictionary Selection Methods. In Proceedings of 18th ACM Great Lakes symposium on VLSI (GLSVLSI'08), 83–88, 2008
6. Jas A, Ghosh-Dastidar J, Ng M, and Touba N A (2003) An efficient test vector compression scheme using selective Huffman coding. In Proceedings of IEEE Transaction on Computer-Aided Design 22(67):97–806
7. Li L, Chakarbarty K, and Touba N A (2003) Test data compression using dictionaries with selective entries and fixed-length indices. In Proc ACM Trans Des Automation Electron Syst 8(4):470–490
8. Chandra A, Chakrabarty K (2001) Test resource partitioning for SOCs. In Proceedings of International Test Conference 18(5):80–91

9. Hamzaoglu I, Patel JH (1999) Reducing test application time for full scan embedded cores. In Proceedings of International Symposium on Fault-Tolerant Computing, 260–267, 1999

10. Reddy SM, Miyase K, Kajihara S, and Pomeranz I (2002) On test data volume reduction for multiple scan chain designs. In Proceedings of IEEE VLSI Test Symposium pp. 103–108

11. Lee KJ, Chen JJ, and Huang CH (1998) Using a single input to support multiple scan chains. In Proceedings of International Conference on Computer-Aided Design, 74–78, 1998

12. Shah MA, Patel JH (2004) Enhancement of the Illinois scan architecture for use with multiple scan inputs. In Proceedings of Annual Symposium on VLSI, 167–172, 2004

13. Samaranayake S, Gizdarski E, Sitchinava N, Neuveux F, Kapur R, and Williams TW (2003) A reconfigurable shared scan-in architecture. In Proceedings of IEEE VLSI Test Symosium, 9–14, 2003

14. Tang H, Reddy SM, and Pomeranz I (2003) On reducing test data volume and test application time for multiple scan chain designs. In Proceedings of International Test Conference, 1079–1088, 2003

15. Chandra A, Yan H, and Kapur R (2007) Multimode Illinois scan architecture for test application time and test data volume reduction. In Proceedings of IEEE VLSI Test Symposium, 84–92, 2007

16. Krishna CV, Jas A, and Touba NA (2002) Reducing test data volume using LFSR reseeding with seed compression. In Proceedings of IEEE International Test conference, 321–330, 2002

17. Wang LT, Wu CW, and Wen X (2006) VLSI Test Principles and Architectures: Design for Testability, Morgan Kaufmann, July 2006

18. Huffman DA (1952) A method for the construction of minimum redundancy codes. Proc IRE 40(9):1098–1101

19. Lee KJ, Chen JJ, and Huang CH (1999) Broadcasting test patterns to multiple circuits. in Proceedings of IEEE Transaction on Computer-Aided Design 18(12):1793–1802

20. Sitchinava N, Samaranayake S, Kapur R, Gizdarski E, Neuveux F, and Williams TW (2004) Changing the scan enable during shift. In Proceedings of IEEE VLSI Test Symposium, 73–78, 2004

21. Jas A, Touba NA (1998) Test vector compression via cyclical scan chains and its application to testing core-based designs. In Proceedings of IEEE International Test Conference, 458–464, 1998

Chapter 11
Memory Testing by Means of Memory BIST

This chapter is on memory testing, and our focus is on memory BIST (MBIST) structures. In today's technology, there is hardly any chip that does not contain some form of a memory. In addition, there is hardly any system that does not contain several dedicated memory chips. In the small scale, memories come as register files as part of digital system, and in the large scale, memories come as chips with memory cells that are reaching 1G cells in the next few years. Furthermore, memories come in many forms of volatile, nonvolatile, static, and dynamic, each of which has its own subcategories and structures. Regardless of the size, type, and hardware structures, memory cannot be tested in the same way as logic testing and it requires its own test methods.

The next section in this chapter discusses some of the peculiarities of memories and explains why memory testing is different than logic testing. The section after that discusses memory structures illustrating the hardware that is to be tested. Memory fault model, which is primarily a functional model, is discussed next. We then discuss some of the most commonly used memory testing techniques. This discussion brings us to the main topic that is treated in this chapter that is built-in self-test structures for memories.

11.1 Memory Testing

Memory categories include volatile and nonvolatile. Volatile memories are RAMs (random access memories), which can be static or dynamic (SRAM and DRAM). Static RAMs retain their stored value while being powered, whereas dynamic RAMs require refreshing. Nonvolatile memories are ROM (read only memory), PROM (programmable ROM), EPROM (erasable PROM), UVPROM (UV erasable PROM), EEPROM (electrically erasable PROM), and flash memories.

Memories are used as part of a system, they come as memory cores, they are used as part of programmable devices, or they come as standalone memory chips. Regardless of the type, size, and structure, they form sequential logic of many storage cells. In the earlier parts of this book, we showed how sequential circuits are turned into combinational ones to take advantage of combinational test methods. Unfortunately, because of hardware overhead, same cannot be done for memories. On the contrary, treating memories as sequential circuits with the number of cells reaching 10^9 in the next few years and wanting to test a circuit with 2^{1G} states is an absolute impossibility.

In spite of all these difficulties, memory testing must be done and is unavoidable. Since every system uses memory in one form or the other, system functionality heavily depends on its memory.

Z. Navabi, *Digital System Test and Testable Design: Using HDL Models and Architectures*,
DOI 10.1007/978-1-4419-7548-5_11, © Springer Science+Business Media, LLC 2011

This issue has become more critical in the recent years, especially due to the fact that more than 80% of the chip area in an SOC is occupied by various forms of memories.

11.2   Memory Structure

As mentioned earlier, memories come in many forms and sizes, and we cannot treat each separately. However, their common structures for reading and or writing addressed locations make their testing follow the same basic rules. This section discusses this structure focusing more on SRAMs than on other memory types.

Figure 11.1 shows the basic structure of a memory array that consists of $n \times m$ cells arranged in an n by m array. The hardware of the memory consists of the array of cells, decoders for decoding row and column addresses, read and write logic, and a logic block for handling input and output to and from the memory array. The exact arrangement of memory cells in the array depends on the space limitations on a chip, and not on memory word length and address space.

Other than the *IO Logic* block and the memory cells, hardware structures for the rest of a memory consist of basic logic structures like what we have been dealing with in all the preceeding chapters. The *IO Logic* block consists of analog sense amplifiers and some logic for selection of data coming from the memory array, or writing data into the array. The hardware for the individual cells very much depends on the type of memory. Figure 11.2 shows an SRAM and a DRAM cell.

Although all, but the *IO Logic* and the cells, can be tested by conventional test techniques presented in the earlier parts of this book, such is not done. Partly, this is because doing so would leave us with parts that are analog in nature (i.e., *IO Logic* and array of cells), and testing them would require analog techniques that are far more complex than our digital testing techniques. However, looking at the memory block as a whole, we can expect that with proper testing techniques and a proper fault model, faults in the surrounding logic as well as the array itself can be detected.

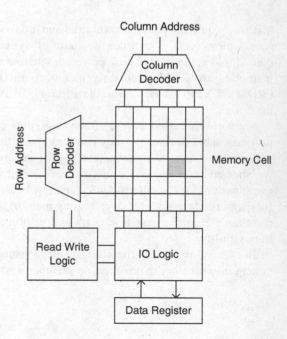

Fig. 11.1 Memory structure

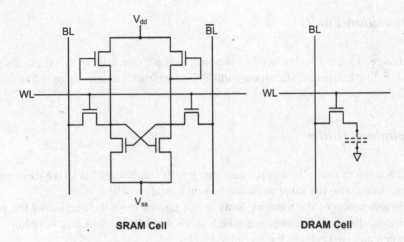

Fig. 11.2 Memory cells

11.3 Memory Fault Model

As mentioned earlier, a memory block consists of analog and digital parts. What makes memory testing simpler than having to test an array of $n \times m$ analog cells is the surrounding logic that digitizes our interface with the memory. Nevertheless, the majority of the hardware we are testing has properties that are not easily mapped to our logic stuck-at fault (SA) model.

Furthermore, a structural fault model for the memories, and trying to test for such faults, one by one, (as we did in logic testing) would require large resources because of the number of cells and possible structural faults that each cell may have.

The solution for memory testing is using a functional fault model instead of a structural one, and considering faults related to the surrounding logic and the memory array all at the same time.

Memory testing involves testing of the array for write and read operations, and thus is considered functional. Although this functional test primarily targets the array itself, it is expected that faults in the surrounding logic will also be detected by the same tests. Consider for example, a word line (*WL*) of an SRAM cell (Fig. 11.2) is stuck-at 0 due to faults in the *Row Decoder* logic (Fig. 11.1). Writing a value (say 0) into a cell addressed by this word line, and then trying to read it will result in reading back the previous value, and not the new value, 0. First, this is a functional test of the memory cell because we are testing it for the storage function that it is supposed to perform. In addition, failure of this test could be detecting *IO logic* functional fault in sensing or driving *BL* and $\overline{BL}$ lines coming from the cell. Memory fault models are described below.

11.3.1 Stuck-at Faults

The stuck-at memory functional fault is the collection of circumstances that make it look like the targeted cell has a 1 or a 0 that is permanently stored in it. To test for such a fault, 0 should be written and then read back. Then the same thing for 1. If in both cases, the same value is read back, then the cell stored value is said to be stuck-at that value.

11.3.2   Transition Faults

If a cell can change to v from $\bar{v}$, but not the other way around, then it looks like the cell's stored value cannot make a $\bar{v}$ to v transition. Transition faults (TF) are either 1-transition or 0-transition faults.

11.3.3   Coupling Faults

Because of adjacency of the cells, and because many cells share word lines (*WL*) and bit lines (*BL*), writing into one can cause the same value written into another cell.

Since the organization of the memory array is not known from the outside of the memory, we cannot tell which cells share bit lines and which share word lines, therefore coupling faults (CFs) can exist between any two cells in the memory.

To test for a CF, write 0s in all cells, then write a 1 in one cell. Read all other cells back to see if any, except the one that a 1 has been written into has a value of 1. Repeat this process for the opposite value.

Other CF types are *inversion*, *idempotent*, and *k*-cell coupling. The inversion coupling is when writing into a cell inverts the value of another. The *idempotent* coupling is when a specific transition (0 to 1 or 1 to 0) changes the coupled cell to a particular value (0 or 1). Lastly, *k*-cell coupling is when CFs occur between more than two cells, and k in this case is the number of participating cells.

11.3.4   Bridging and State CFs

Bridging faults (BFs) are caused by two or more bit lines (*BL*) that are shorted together. Depending on the logic behind the short, the resulting value of the short behaves as AND or OR logic of the two cells. Some BFs are detected by stuck-at faults. State coupling faults are similar to BFs, except that the resulting short only affects one cell, keeping the value of the other cell intact.

There are several other fault models that do not occur as often. Several books on testing have a through presentation of these materials [8].

11.4   Functional Test Procedures

The last section explained some possible functional effects memory cells can have on each other. From this explanation it can easily be understood that to completely and exhaustively test a memory array, every cell must be read and written into with 0 and 1, while reading all other cells to see how they are affected. Obviously, exhaustively executing this procedure can lead to long test times. Although there are memory test procedures that do variations of this exhaustive test, there are others that try to reduce time complexity of testing to $O(N)$, where N is the total number of memory bits, by making certain approximations.

11.4.1   March Test Algorithms

This section discusses March tests that are of $O(N)$ complexity. March tests are named so, because starting with the first memory location a 1 (or a 0) is written while locations previous to that keep their written 1 (or 0) values. So it appears like 1s (or 0s) are marching in from location 0 to the last location in the memory.

For the start, we explain the March C-testing that is at memory test algorithm of the general memory March testing category [5]. March C begins with 0s in every memory location (all N 1-bit locations). We then start reading from location 0 in ascending order and check if 0s are actually there. Every location that is checked it is then written with a 1. So 1s march in ascending order. This checking and writing 1s continues until we read the last bit of the memory. At this time all memory bits contain 1s. The process now reverses. We read 1s in descending order and switch each read location to 0. In the reverse process, 0s march in from location N - 1 in descending order. The same test is then repeated starting with all 1s in N locations.

The process for most March tests can be explained by paragraphs like the above for March C. However, for ease of explanation, a notation has been devised by Van de Goor [5] a subset of which is shown in Fig. 11.3. This notation unambiguously specifies the testing procedure, and the number of reads and writes are easily seen that determine the order of a test procedure.

11.4.2   March C- Algorithm

As an example of using the notations of Fig. 11.3 consider March C- memory test algorithm as described below:

$$March\ C- : \{\Updownarrow (w0); \Uparrow (r0, w1); \Uparrow (r1, w0); \Downarrow (r0, w1); \Downarrow (r1, w0); \Updownarrow (r0)\}$$

Steps listed below are the interpretation of this test. The list below or the above expression show that March C- uses 10 read/write operations, and it therefore of $10N$ order.

R	Memory Read operation
r0	Read a 0 from the memory
r1	Read a 1 from the memory
W	Memory Write Operation
w0	Write a 0 to the memory
w1	Write a 1 to the memory
$\Uparrow$	Increasing memory address ordering
$\Downarrow$	Decreasing memory address ordering
$\Updownarrow$	There is no difference between different addressing orders.

Fig. 11.3 A subset of March test notations

Steps in March C-Test:

1. Write 0s to all cells in any order ($\updownarrow$ ($w0$)).
2. Read from the lowest address (expected read value is 0), write a 1 at this address, and repeat until the highest address is reached ($\Uparrow$ ($r0, w1$)).
3. Read from the lowest address (expected read value is 1), write a 0 at this address and repeat until the highest address is reached ($\Uparrow$ ($r1, w0$)).
4. Read from the highest address (expected read value is 0), write a 1 at this address, and repeat until the lowest address is reached ($\Downarrow$ ($r0, w1$)).
5. Read from the highest address (expected read value is 1), write a 0 at this address, and repeat until the lowest address is reached ($\Downarrow$ ($r1, w0$)).
6. Read from all cells in any order (expected read value is 0) ($\updownarrow$ ($r0$)).

11.4.3   MATS+ Algorithm

As another example of a memory testing algorithm consider MATS+. This algorithm is less complex than March C– and detects fewer faults. With the notation described above, MATS+ can be represented as follows:

$$\text{MATS+} : \{ \updownarrow \ (w0); \Uparrow \ (r0, w1); \Downarrow \ (r1, w0) \}$$

Steps in MATS+:

1. Write 0s to all cells in any order ($\updownarrow$ ($w0$)).
2. Read from the lowest address (expected read value is 0), write a 1 at this address, and repeat until the highest address is reached ($\Uparrow$ ($r0, w1$)).
3. Read from the highest address (expected read value is 1), write a 0 at this address, and repeat until the lowest address is reached ($\Downarrow$ ($r1, w0$)).

11.4.4   Other March Tests

Using the notation of Fig. 11.3, other March tests are described in Fig. 11.4. These algorithms are different in the fault types (those discussed in Sect. 11.3) that they detect. MBIST being the focus of

Algorithm	Descrption	Ref.
MATS	{$\updownarrow$ (w0):$\updownarrow$ (r0, w1);$\updownarrow$ (r1)}	[1,2]
MATS+	{$\updownarrow$ (w0):$\Uparrow$ (r0, w1);$\Downarrow$ (r1, w0)}	[3,4]
MATS++	{$\updownarrow$ (w0):$\Uparrow$ (r0, w1);$\Downarrow$ (r1, w0, r0)}	[5]
MARCH X	{$\updownarrow$ (w0):$\Uparrow$ (r0, w1);$\Downarrow$ (r1, w0); $\updownarrow$ (r0)}	[5]
MARCH C-	{$\updownarrow$ (w0):$\Uparrow$ (r0, w1); $\Uparrow$ (r1, w0); $\Downarrow$ (r0, w1); $\Downarrow$ (r1, w0); $\updownarrow$ (r0)}	[6]
MARCH A	{$\updownarrow$ (w0):$\Uparrow$ (r0, w1, w0, w1); $\Uparrow$(r1, w0, w1); $\Downarrow$ (r1, w0, w1, w0); $\Downarrow$ (r0, w1, w0)}	[7]
MARCH Y	{$\updownarrow$ (w0):$\Uparrow$ (r0, w1, r1); $\Downarrow$ (r1, w0, r0); $\updownarrow$ (r0)}	[5]
MARCH B	{$\updownarrow$ (w0):$\Uparrow$ (r0, w1, r1, w0, r0, w1); $\Uparrow$(r1, w0, w1); $\Downarrow$ (r1, w0, w1, w0); $\Downarrow$ (r0, w1, w0)}	[7]

Fig. 11.4 March test algorithms

Fig. 11.5 March test algorithm complexities

Algorithm	Complexity
MATS	4n
MATS+	5n
MATS++	6n
MARCH X	6n
MARCH C–	10n
MARCH A	15n
MARCH Y	8n
MARCH B	17n

this chapter, we leave comparison of these algorithms for the interested reader to several good books and articles on the topic. Figure 11.5 shows order of complexity of the algorithms in Fig. 11.4.

11.5 MBIST Methods

There are several reasons for doing memory testing by a dedicated hardware component that is incorporated into the memory structure. First, memory test patterns are very regular and can easily be created by simple counters and shift registers. Second, memory testing involves several iterations of writing and reading data to and from memory that can easily be kept track of with counters counting in sequential order.

Finally, because memories play an important role in reliability of systems they are used in, they should be tested regularly without having to remove them from the system. This section presents several MBIST architectures that implement algorithms described in the previous section or variations of them.

11.5.1 Simple March MBIST

As our first example, we implement Simple March memory test algorithm using a MBIST. The architecture, controller, and corresponding Verilog codes will be discussed here.

11.5.1.1 Simple March MBIST Architecture

Figure 11.6 shows the MBIST architecture. The memory to be tested is shown in gray, and solid line blocks show the test circuitry. Test data that are to be applied are generated by this MBIST circuitry and applied to the memory. As data are being read from the memory, they are compared with the reproduction of the same data that was written into specific memory locations. After writing and reading all locations, we expect all data read from the memory to be the same as those that were written into it.

Counter. Input data, address, and switching between reading and writing the memory are provided by a counter. The least significant bits of the counter provide addressing for all locations of the memory. The counter bit to the left of the address group of bits toggles between write and read operations. The three most significant bits of the counter are decoded to generate eight test vectors for testing memory words.

Figure 11.7 shows the Verilog code of this counter. The counter carry-out (*cout*) becomes "1" when the count reaches its maximum.

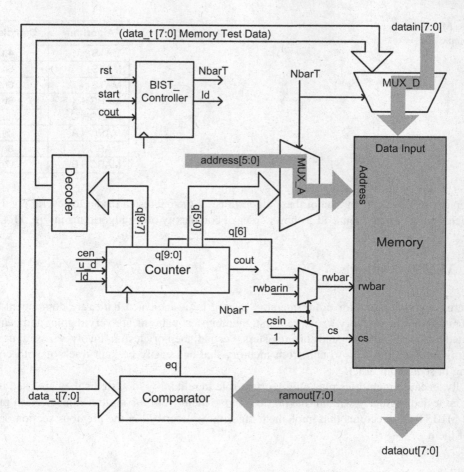

Fig. 11.6 Memory BIST for simple March algorithm

```
module counter
    #(parameter length = 10) (d_in, clk, ld, u_d, cen, q, cout);

    input [length-1:0] d_in;
    input clk, ld, u_d, cen;
    output [length-1:0] q;
    output cout;

    reg [length:0] cnt_reg;

    always @ (posedge clk) begin
        if ( cen ) begin
            if ( ld )
                cnt_reg <= {1'b0, d_in};
            else if ( u_d )
                cnt_reg <= cnt_reg + 1;
            else
                cnt_reg <= cnt_reg - 1;
        end
    end

    assign q = cnt_reg[length-1:0];
    assign cout = cnt_reg[length];

endmodule
```

Fig. 11.7 Simple March MBIST counter

Fig. 11.8 Simple March
BIST decoder: test pattern
generation

Test Pattern	Input	Decoder Output
0	000	00000000
1	001	00001111
2	010	00110011
3	011	01010101
4	100	11111111
5	101	11110000
6	110	11001100
7	111	10101010

```
module decoder (input [2:0] in, output [7:0] out);
  wire [7:0] out_temp;

  assign out_temp = (in[1:0] == 2'b 11) ? 8'b 01010101 :
                    (in[1:0] == 2'b 10) ? 8'b 00110011 :
                    (in[1:0] == 2'b 01) ? 8'b 00001111 :
                    (in[1:0] == 2'b 00) ? 8'b 00000000 :
                    8'b zzzzzzzz;

  assign out = (in[2] == 1'b 0) ? out_temp : ~ out_temp;
endmodule
```

Fig. 11.9 Simple March BIST MBIST decoder

Decoder. The test data decoder uses a 3-bit input vector to lookup memory test patterns shown in Fig. 11.8. The Verilog code of the decoder is shown in Fig. 11.9.

Multiplexers. The multiplexers of the BIST architecture of Fig. 11.6 select between normal memory inputs and BIST provided inputs. When *NbarT* is "0," the memory is working in the normal mode and when this input becomes "1" it operates in test mode.

Comparator. Initially, the same test pattern is written into all memory locations, and then these data are read out from all locations. As data are being written and read, the decoder input and thus test patterns remain unchanged. A comparator checks memory data with decoder output. When memory is being tested and data are being read, the comparator should have same data on both its inputs.

11.5.1.2 Test Session

A test session begins when the counter is all 0s and ends when the counter reaches all 1s. Starting with all 0s, test pattern 0 is written in location 0. As the counter is incremented, this same pattern is written into all memory locations. When all memory locations are written into, the counter increments, causing the least significant part of it (address bits) to roll over to all 0s, and the *rwbar* bit becomes 1. When this happens, the same data will start being read from all memory locations. When this is done, the *rwbar* bit becomes 0, address starts back at 0, and the next test pattern starts being written into all locations. This process continues for all eight test patterns. When done, all test patterns have been written into and read from all memory locations. While this is happening, the comparator checks for a mismatch and issues an error if it finds one.

11.5.1.3　Simple March BIST Controller

The BIST controller starts the counter when it receives the *start* signal and waits for the carry-out (*cout*) of the counter. The Verilog code of this controller is shown in Fig. 11.10.

11.5.1.4　Simple March BIST Structure

The Verilog code of Fig. 11.11 shows the complete BIST structure including the RAM that is being tested. Components instantiated in this description are according to the diagram of Fig. 11.6. In addition to the components instantiated, this code has an **always** statement that issues the *fail* flag if the comparator finds a mismatch.

11.5.1.5　BIST Tester

The memory and its BIST are tested in the *BIST_tester* testbench. This testbench is shown in Fig. 11.12. The testbench initially loads external file data into the memory at time 5 ns when *operate* becomes 1. Then at some arbitrary times, data are written into and read from the memory. The BIST test session begins when *start* becomes 1 at time 50 ns. Testing continues until all RAM locations have been tested. While the memory is being tested, external read and write operations are ignored.

```verilog
module BIST_controller (input start, rst, clk, cout, output NbarT,ld);

    reg current = reset;

    parameter reset = 1'b 0, test = 1'b 1;

    always @ (posedge clk) begin
        if (rst)
            current <= reset;
        else
            case(current)
                reset: if (start)
                            current <= test;
                       else
                            current <= reset;
                test: if (cout)
                            current <= reset;
                      else
                            current <= test;

                default:
                            current <= reset;
            endcase
    end

    assign NbarT = (current == test) ? 1'b 1 : 1'b 0;
    assign ld = (current == reset) ? 1'b 1 : 1'b 0;

endmodule
```

Fig. 11.10　Simple March MBIST controller

```
module BIST #(parameter size = 6, length =8)
              (start,rst,clk,csin,rwbarin,opr,address,
               datain,dataout,fail);
   input start, rst, clk, csin, rwbarin, opr;
   input [size-1: 0] address;
   input [length-1: 0] datain;
   output [length-1: 0] dataout;
   output fail;
   reg fail;
   reg [9:0] zero;

   wire cout, ld, NbarT, cs, rwbar, gt, eq, lt;
   wire [9:0] q;
   wire [7:0] data_t;
   wire [length-1:0] ramin, ramout, bit_array;
   wire [size-1:0] ramaddr;
   reg [length-1:0] ram_testvalue;
   reg conv_enable;
   integer i, index, power, mult;
   integer faulty_adr, faulty_bit ;
   initial zero = 10'b 0000000000;


   BIST_controller CNTRL (start, rst, clk, cout, NbarT, ld);
   counter CNT (zero, clk, ld, 1'b1, 1'b1, q, cout);
   decoder DEC (q[9:7], data_t);
   multiplexer #(8) MUX_D (datain, data_t, NbarT, ramin);
   multiplexer #(6) MUX_A (address, q[5:0], NbarT, ramaddr);

   assign rwbar = (~NbarT) ? rwbarin : q[6];
   assign cs = (~NbarT) ? csin : 1'b 1;

   RAM MEM (ramaddr, ramin, cs, rwbar, opr, ramout);
   comparator CMP (data_t, ramout, gt, eq, lt);

   always @ (posedge clk) begin
      if (NbarT && rwbar && opr)
         if (~eq) begin
            fail <= 1'b1;
         end else begin
            fail <= 1'b0;
         end
   end

   assign dataout = ramout;
endmodule
```

Fig. 11.11 Simple March BIST structure

11.5.2 March C- MBIST

March C- algorithm was discussed in Sect. 11.4. With minor changes to the architecture shown in Fig. 11.6, architecture for the implementation of March C- algorithm is obtained and is shown in Fig. 11.13. Notice the extra logic for the control of the memory *rwbar* in this architecture. The counter in simple March is replaced by a counter-sequencer that handles the ten phases of March C- test. Here, we are handling a memory that has a word-length of eight bits and we are testing the memory in 1-bit vertical slices. This word-level implementation is not an exact implementation of March C- test.

```
module BIST_tester ();
    reg [7:0] ramin;
    reg [5:0] addr;
    reg cs, rwbar, start;
    reg rst, clk;
    reg operate;
    wire [7:0] ramout;
    wire fail;

    initial begin
        cs = 0;
        rwbar = 1;
        start = 0;
        rst = 0;
        operate = 0;
        clk = 0;
    end

    BIST UUT (start,rst,clk,cs,rwbar,operate,addr,ramin,ramout,fail);

    always #5 clk = ~clk;

    initial begin
        #5 operate = 1'b1;
        #5 ramin = 8'b11110001;
        #5 cs = 1'b1;
        #5 addr = 6'b101100;
        #10 ramin = 8'b00101100;
        #10 addr = 6'b101110;
        #10 start = 1'b1;
        #140 rwbar = 1'b1;
        #147 cs = 1'b0;
        #463 operate =1'b0;
    end
endmodule
```

Fig. 11.12 Simple March memory BIST testbench

11.5.2.1 March C- BIST Counter-sequencer

Address, direction of address generation (from highest to lowest or visa versa), test data, and switching between reading and writing the memory are handled a 13-bit counter. The six least significant bits of the counter-sequencer provide addressing for all locations of the memory. The next four bits to the left of the address group of bits specify the ten consecutive read/write operations of March C-. Also these four bits along with the three most significant bits of the counter-sequencer are decoded to generate the appropriate test vectors for testing the memory words.

According to March C- algorithm, in Steps 1 and 6, the address increments (or decrements) and, for the specific address, only read or write operation will take place. In all other steps, both read and write operations must be done before going to the next address. To implement this, when the counter-sequencer is in Steps 1 or 6, it behaves normally and increments or decrements by 1 with every clock cycle. In all other steps, the address of memory (the six least significant bits ($cnt\_reg[5:0]$) do not change for two consecutive cycles, but $cnt\_reg[9:6]$ increments and then decrements by 1. The Verilog code of this counter is shown in Fig. 11.14.

11.5.2.2 Decoder

The test data decoder uses a 7-bit input vector to generate memory test patterns and the value that is to be compared with the value read from the memory. For every bit of each memory word, all six

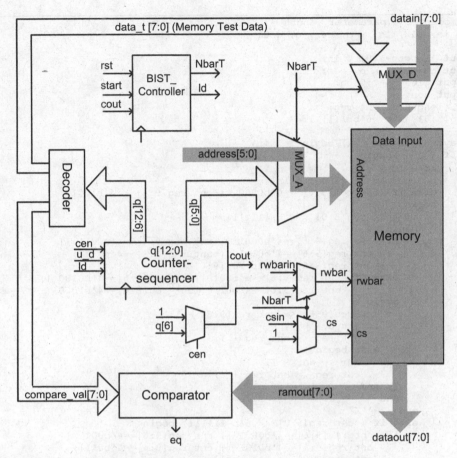

Fig. 11.13 Memory BIST for March C- algorithm

steps of March C- algorithm (that contains ten operations) are exercised. So using the three most significant input bits, the bit within the word that the above ten operations must be performed on is specified. The decoder specifies which test value must be written into the memory word. The Verilog code of the decoder is shown in Fig. 11.15.

The remaining parts of the March C- MBIST hardware remain the same as those discussed for the Simple March tester.

11.5.3 Disturb MBIST

Disturb testing is for studying the robustness of the data storage of the flash or DRAM cells when the state of a neighboring cell is changing. In disturb testing, a checkerboard pattern is written into the entire array in the "disturb write state." After the data have had time to settle, it is read back in the "disturb read state." This is a quick check for gross data retention faults. For a memory size of N bits, the complexity of disturb test algorithm is $O(N)$.

An MBIST for this method has the same basic architecture as that of Fig. 11.6, with a difference in the way test data are generated. In Fig. 11.6, the decoder uses a 3-bit input vector to look up the appropriate memory test pattern, while in this method, a walking-0 circuit is used to generate the new test pattern. In disturb testing MBIST, an initial value will be written to all memory cells, and after a delay, when data are being read from the memory, it is compared with the written data.

```verilog
module counter #(parameter length = 13) // counter-sequencer module
   ( d_in, clk, ld, u_d, cen, q, cout );

   input [length-1:0] d_in;
   input clk, ld, u_d, cen;
   output [length-1:0] q;
   output cout;
   reg [length:0] cnt_reg;
   reg [5:0] adr = 6'b 111111;

   always @(posedge clk) begin
      if ( cen )begin
         if ( ld )
            cnt_reg <= {1'b0, d_in};
         else if (cnt_reg[9:6] > 4'b0000 && cnt_reg[9:6] < 4'b1001)
         begin
            if (cnt_reg[5:0] < 6'b 111111)
            begin
               if (cnt_reg[9:6]==4'b0001 || cnt_reg[9:6]==4'b0011 ||
                   cnt_reg[9:6]==4'b0101 || cnt_reg[9:6]==4'b0111)
                  cnt_reg[9:6] <= cnt_reg[9:6] + 1;
               else if (cnt_reg[9:6]==4'b0010||cnt_reg[9:6]==4'b0100 ||
                       cnt_reg[9:6]==4'b0110||cnt_reg[9:6]==4'b1000)
               begin
                  cnt_reg[9:6] <= cnt_reg[9:6] - 1;
                  if ( u_d )
                     cnt_reg <= cnt_reg + 1;
                  else begin
                     adr <= adr - 1;
                     cnt_reg <= cnt_reg + 1;
                  end
               end
            end
            else if ( cnt_reg[5:0] == 6'b 111111) begin
               if (cnt_reg[9:6]==4'b0001 || cnt_reg[9:6]==4'b0011 ||
                   cnt_reg[9:6]==4'b0101 || cnt_reg[9:6]==4'b0111)
                  cnt_reg[9:6] <= cnt_reg[9:6] + 1;
               else if (cnt_reg[9:6]==4'b0010||cnt_reg[9:6]==4'b0100||
                       cnt_reg[9:6]==4'b0110||cnt_reg[9:6]==4'b1000)
               begin
                  if ( u_d )
                     cnt_reg <= cnt_reg + 1;
                  else begin
                     adr <= adr - 1;
                     cnt_reg <= cnt_reg + 1;
                  end
               end
            end
         end else if ( u_d ) begin
            cnt_reg <= cnt_reg + 1;
         end else begin
            adr <= adr - 1;
            cnt_reg <= cnt_reg + 1;
         end
         if (cnt_reg[9:6] == 4'b 1010) begin
            cnt_reg[9:6] <= 4'b 0000;
            cnt_reg[length:10] <= cnt_reg[length:10] + 1;
         end
      end
   end

   assign q = u_d ? cnt_reg[length-1:0] : {cnt_reg[length-1:6], adr};
   assign cout = cnt_reg[length];

endmodule
```

Fig. 11.14 March C- MBIST counter-sequencer

```verilog
module decoder (in, cen, out, lastValue);
   input [6:0] in;
   input cen;
   output reg [7:0] out;
   output reg [7:0] lastValue;
   reg [7:0] tempShift, zero;
   integer shiftAmount, i;

   initial begin
      tempShift = 8'b 10000000;
      zero = 8'b 00000000;
      shiftAmount = 0;
   end

   always @(in) begin
      if (cen) begin
         shiftAmount = 0;
         for (i=6; i>3; i=i-1)
            if (in[i] == 1'b1)
               shiftAmount = shiftAmount*2 + 1;
            else
               shiftAmount = shiftAmount*2;
         if (in[3:0]==4'b0000||in[3:0]==4'b0100||in[3:0]==4'b1000)
         begin
            out = 8'b 00000000;
            if (in[3:0] == 4'b 0000)
               lastValue = 8'b 00000000;
            else
               lastValue = tempShift >> shiftAmount;
         end else if (in[3:0]==4'b0010||in[3:0]==4'b 0110)
         begin
            out = tempShift >> shiftAmount;
            lastValue = 8'b 00000000;
         end
      end
      else out = 8'b zzzzzzzz;
   end

endmodule
```

Fig. 11.15 March C- MBIST decoder

Next, the initial data are rotated to the left to form the new test pattern. This process continues until the initial value is rotated by the number of bits in each memory word (m). This results in checking both transitions 0 to 1 and 1 to 0 for each memory cell. After writing and reading all locations, we expect all data read from the memory to be the same as the values they were initialized to.

11.5.3.1 Disturb BIST Walking-0

The walker test pattern generator uses a 3-bit input vector to determine the number of left rotations that must be applied to the initial test value. The memory test patterns are shown in Fig. 11.16. The Verilog code of the walker test generator is shown in Fig. 11.17.

11.5.3.2 Disturb BIST Structure

The Verilog code of Fig. 11.18 is a partial code of disturb test MBIST that includes the memory being tested. Components instantiated in this description are according to the diagram of Fig. 11.6,

Fig. 11.16 Walking-0; test
pattern generation

Test Pattern	Input	Walker Output
0	000	01111111
1	001	11111110
2	010	11111101
3	011	11111011
4	100	11110111
5	101	11101111
6	110	11011111
7	111	10111111

```verilog
module walk (input [7:0] in, input [2:0] iteration, output [7:0] out);
   assign out = (iteration == 3'b 000) ? in :
                (iteration == 3'b 001) ? {in[6:0], in[7]} :
                (iteration == 3'b 010) ? {in[5:0], in[7:6]} :
                (iteration == 3'b 011) ? {in[4:0], in[7:5]} :
                (iteration == 3'b 100) ? {in[3:0], in[7:4]} :
                (iteration == 3'b 101) ? {in[2:0], in[7:3]} :
                (iteration == 3'b 110) ? {in[1:0], in[7:2]} :
                (iteration == 3'b 111) ? {in[0], in[7:1]} :
                8'b zzzzzzzz;
endmodule
```

Fig. 11.17 Memory BIST pattern generator walking-0

```verilog
module BIST #(parameter size = 6, length =8)
   (start,rst,clk,csin,rwbarin,opr,address,datain,dataout,fail);
   . . .

   BIST_controller CNTRL (start, rst, clk, cout, NbarT, ld);
   counter CNT (zero, clk, ld, 1'b1, 1'b1, q, cout);
   walk WALK (ram_testvalue, q[9:7], data_t);
   multiplexer #(8) MUX_D (datain, data_t, NbarT, ramin);
   multiplexer #(6) MUX_A (address, q[5:0], NbarT, ramaddr);

   assign rwbar = (~NbarT) ? rwbarin : q[6];
   assign cs = (~NbarT) ? csin : 1'b 1;

   RAM MEM (ramaddr, ramin, cs, rwbar, opr, ramout);
   comparator CMP (data_t, ramout, gt, eq, lt);

   always @ (posedge clk) begin
      if (NbarT && rwbar && opr)
         if (~eq) begin
            fail <= 1'b1;
            faulty_adr <= ramaddr;
         end else begin
            fail <= 1'b0;
         end
   end

   . . .
   // calculate faulty bit number to report
   . . .
   assign dataout = ramout;
endmodule
```

Fig. 11.18 Disturb memory BIST structure

with this difference that a decoder is substituted with a walking-0 circuit. In this code, the first **always** block is used to issue the *fail* flag if the comparator finds a mismatch. It also specifies the faulty address the faulty bit. The code of this MBIST is a synthesizable Verilog whose hardware correspondence can easily be understood by inspection.

11.6 Summary

In this chapter, we covered memory testing while emphasizing on MBIST hardware structures. For test algorithms, we only presented the basic algorithms and avoided the theories that have led to the development of such algorithms. In presenting MBIST hardware structures, we showed that other than a few minor differences in the way MBIST counter and decoder work, most MBISTs follow the same template for their hardware structures. The architecture we presented applies to large memory chips, as well as short on-chip register files.

References

1. Knaizuk[AU3] J Jr, Hartmann CRP (1977) An optimal algorithm for testing stuck-at faults in Random Access Memories. IEEE Trans Comput C-26(11):1141–1144
2. Nair R (1979) Comments on an optimal algorithm for testing stuck-at faults in Random-Access Memories. IEEE Trans Comput C-28(3):258–261
3. Abadir MS, Reghbati JK (1983) Functional testing of Semiconductor Random Access Memories. ACM Comput Surv 15(3):175–198
4. Winegarden S, Pannell D (1981) Paragons for memory test. In: Proceedings of the International Test Conference, Oct. 1981, pp. 44–48
5. van de Goor AJ (1991) Testing semiconductor memories: theory and practice. Wiley, Chichester, UK.
6. Marinescu M (1982) Simple and efficient algorithms for functional RAM Testing. In: Proceedings of the International Test Conference, Nov. 1982, pp. 236–239
7. Suk DS, Reddy SM (1981) A March test for functional faults in Semiconductor Random-Access Memories. IEEE Trans Comput C-30(12):982–985
8. Bushnell ML, Agrawal VD (2000) Essentials of electronic testing for digital, memory & mixed-signal VLSI circuits. Kluwer, Norwell, MA.

Appendix A
Using HDLs for Protocol Aware ATE[1]

This Appendix describes the issues involved in transitioning from a design simulation environment to a physical test environment, especially in the case of complex (multicore) SOC devices. The article also discusses the advantages of using HDL code directly for programming of the tester hardware, as opposed to the traditional, bit-level language used in the past. In this case, the ability to be able to directly interpret and execute the HDL commands on the tester hardware is referred to as having "Protocol Aware" (PA) [1] capability. As the name suggests, this implies that the tester hardware has the ability, via reconfigurable firmware, to understand and execute high-level commands on physical hardware in the DUT's native "Protocol" language. "Protocol Aware" is a larger umbrella that not only allows the tests to be constructed in an HDL format, but is also able to translate that into physical bits on test hardware.

A.1   Motivation

Modern semiconductor devices often behave in a nondeterministic manner not only in their end application but during test execution on ATE as well. This is the result of design methodologies that allow the assembly of the device from a library of IP blocks. These IP blocks often support specific industry standard protocols such as JTAG, DDR memory buses, PCI Express, etc. While the operation of any individual block may be predictable, the relationship between the timing of protocols is often not. Today's SOC ATE does not deal well with ambiguity. Any deviation from expected device behavior will cause that device to fail the ATE test, both during engineering development or production. Functionally testing devices that exhibit nondeterministic behavior is extremely difficult on current generation ATE.

A.2   Protocol Aware ATE

To deal with DUT nondeterministic behaviors, as part of the next round of UltraFLEX digital instruments Teradyne is developing a new ATE architecture – Protocol Aware ATE (PA). This project will require new software, hardware, and firmware. The intelligence required to handle protocols is contained in a FPGA on the ATE Pin Electronics instrument that can be reprogrammed based on the particular protocols required by any individual device program.

---

[1] This text is taken from a paper by Eric Larson, Teradyne 2008 with his permission. The original paper can be downloaded from the publisher's site for this book, or by contacting Teradyne for this and other related materials.

Z. Navabi, *Digital System Test and Testable Design: Using HDL Models and Architectures*,
DOI 10.1007/978-1-4419-7548-5, © Springer Science+Business Media, LLC 2011

The list of potential protocols to support is endless and clearly they cannot be supported at once. Some are so low in volume that it may not be worth the effort. Others may be too complex to implement in a practical manner. The hardware and software implementation of PA must be flexible enough to provide a solution for many different protocols. Some of these protocols have similar characteristics and can be thought of as a Protocol Family. Below is a partial list of popular protocols and potential groupings.

Low speed serial and parallel:

- JTAG
- MDIO
- SRAM
- Flash

DRAM:

- DDR, DDR2, DDR3
- LPDDR, LPDDR2
- GDDR3, GDDR4, GDDR5

High speed serial:

- PCI Express
- SATA
- DigRF
- Serial RapidIO

A.3 Protocol Aware ATE Implementation

While Protocol Aware ATE requires a new architecture and cannot be simply dropped into to existing instruments it does offer the potential to increase the quality and reduce the cost of test for complex SOC devices.

A possible architecture for the implementation of a PA ATE involves the addition of a FPGA to standard ATE Digital Instruments. The purpose of the FPGA is to emulate the operation of selected DUT protocols. This requires that the ATE software and hardware support reprogramming of the FPGA to act properly depending on the protocol required. Some protocols, JTAG for example, are slow speed and serial in nature and require only a few connections to the device. Others such as DDR2 and DDR3 are much higher in speed and parallel in nature, requiring dozens of ATE channels to work closely together to interpret and respond to command and data information from the DUT. This "Protocol Engine" architecture allows handshaking between the DUT and the ATE instrument with the ATE interpreting instructions from the selected Protocol and responding accordingly. Response time will naturally be determined by the latency between the DUT launching information to the ATE, the ATE instrument interpreting the information and sending the response to the DUT. Keeping this latency as short as possible is a key design parameter for any Protocol Aware instrument (Fig. A.1).

In addition to emulating the desired Protocol, the instrument must also support classic Digital ATE test functionality such as Scan, DFT, functional test, and characterization. The user must be able to select between "normal" and Protocol Aware operation during both engineering and production test.

One key requirement is the ability to read and write internal DUT registers in a simple and straightforward manner, similar to the high level language used in simulation and bench instruments. A properly implemented Protocol Aware solution will allow the user to enter a read or write command along with the associated address and payload data and have the DUT immediately respond.

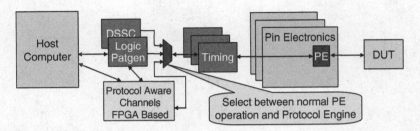

Fig. A.1 Protocol Aware digital instrument architecture

This can be achieved by use of present HDLs. With this, recreating sets of transactions from simulation or bench instrument on ATE will no longer require translation to the low level language of ATE patterns. Instead of appearing as a pseudo-random group of 1s and 0s the DUT interaction will be at a high level of abstraction, like Verilog and VHDL.

A.4 Limitations

Limitations come with every project and Protocol Aware ATE is no exception. The most obvious issue is the huge and growing number of protocols. It is clear that not all protocols are created equal, either in ease of implementation or popularity. Initial solutions will cover a set of popular protocols with an expanded list available over time.

The speed of ATE PA engines is limited by a couple of bus characteristics and ATE attributes. If the bus requires I/O handshaking the round-trip delay of the pin electronics along with processing time in the FPGA may limit speed to that of low speed protocols. Buses that do not require handshaking can generally be supported up to much higher speeds, limited by the fundamental operating frequency of the FPGA.

A.5 Conclusions

Protocol Aware ATE is a new architecture and all indications are that as a concept it is very appealing to a broad set of ATE users, both existing and potential. Implemented properly PA ATE can provide immediate payback by improving test development time and reducing customer time-to-market. In the long run additional benefits around better fault coverage will also become apparent.

This concept signals a fundamental shift in SOC ATE architecture. Future digital instruments will be designed to be Protocol Aware. While starting with digital, PA capability applies to analog and mixed signal instruments as well.

References

1. Molavi S, Evans A, Clancy R (2008) Protocol Aware test methodologies using today's ATE. Proceedings of 17th IEEE Asian test symposium, Sapporo, Japan, November 2008
2. Evans C (2007) The New ATE:Protocol Aware, Proceedings of 2007 IEEE International Test Conference, Santa Clara, CA, October 2007

Appendix B
Gate Components for PLI Test Applications

A set of PLI functions has been developed for gate level fault simulation and other test applications. These test applications are based on gate level descriptions of a circuit being processed. The headers of the gates used for this purpose are shown here. The PLI functions work only if these gates are used in the netlist.

```verilog
//Buffer:
module bufg #(parameter tphl = 1, tplh = 1)
   (out,in);
input in;
output out;

//Not:
module notg #(parameter tphl = 1, tplh = 1)
   (out,in);
input in;
output out;

//And:
module and_n #(parameter n = 2, tphl = 1, tplh = 1)
   (out,in);
input [n-1:0] in;
output out;

//Or:
module or_n #(parameter n = 2, tphl = 1, tplh = 1)
   (out,in);
input [n-1:0] in;
output out;

//Nand:
module nand_n #(parameter n = 2, tphl = 1, tplh = 1)
   (out,in);
input [n-1:0] in;
output out;
```

```
//Nor:
module nor_n #(parameter n = 2, tphl = 1, tplh = 1)
   (out,in);
input [n-1:0] in;
output out;


//Xor:
module xor_n #(parameter n = 2, tphl = 1, tplh = 1)
   (out,in);
input [n-1:0] in;
output out;


//Xnor:
module xnor_n #(parameter n = 2, tphl = 1, tplh = 1)
   (out,in);
input [n-1:0] in;
output out;


//Fan_Out:
module fanout_n #(parameter n = 2,tphl = 3, tplh = 5)
   (in, out);
input in;
output [n-1:0] out;


//Primary Input:
module pin #(parameter n = 1)
   (in, out);
input [n-1:0] in;
output [n-1:0] out;


//Primary Output:
module pout #(parameter n = 1)
   (in, out);
input [n-1:0] in;
output [n-1:0] out;


//D Flip Flop:
module dff #(parameter tphl = 0, tplh = 0)
(Q, D, C, CLR, PRE, CE, NbarT, Si, global_reset);
input D, C, CLR, PRE, CE, NbarT, Si, global_reset;
output reg Q;
```

Appendix C
Programming Language Interface Test Utilities

A set of utilities for performing test application in Verilog testbenches have been developed[1]. Using these utilities, we are able to use a Verilog testbench as a programming platform, a virtual test, or for evaluating testability of our DFT methods. In the testbench, the Verilog model of our circuit is instantiated and the utilities discussed below facilitate access to the circuit's internal lines and gates for performing various test applications. The utilities are developed in Verilog programming language interface (PLI), and can be invoked as tasks in Verilog testbenches.

C.1 Stuck-at Fault Injection

This PLI utility takes the full name of the site of fault (wire) and the fault value and performs the stuck-at fault injection.

Function call	$InjectFault(wire, FaultValue);
Example	$ InjectFault(FA_inst.sum, 1'b1);

C.2 Fault Removal

The $RemoveFault task takes the full name of the site of fault (wire) and removes the injected stuck-at fault from this wire.

Function call	$RemoveFault(wire);
Example	$RemoveFault(FA_inst.sum);

C.3 Transient Fault

The PLI transient fault injection takes the full name of site of fault (wire), the fault value and duration of existence of fault; then performs the transient fault injection. This fault injection does not need a fault removal function, because the injected fault will be removed after the defined fault duration.

Function call	$TransientFault(wireName, FaultValue, faultDuration)
Example	$TransientFault(FA_inst.sum, 1'b1, 2)

---

[1]Managing the development and developing the PLI Test package has been done by Nastaran Nemati.

C.4   Bridging Fault

In order to perform bridging fault injection, the full name for two wires that are bridged and the mode of bridging ("and" or "or") must be specified. Very similar to the other fault removal functions, for bridging fault removal, the site of fault must be specified.

Function call	$BridgingFault(wire1, wire2, BridgingMode);
Example	$BridgingFault(FA_inst.sum, FA_inst.cout, "and");

Function call	$RemoveBridgingFault(wire1, wire2);
Example	$RemoveBridgingFault(FA_inst.sum, FA_inst.cout);

C.5   Coupling Fault

Injection and removal of coupling fault is possible by passing the full name of two coupled wires to the related PLI function.

Function call	$CouplingFault(wire1,wire2);
Example	$CouplingFault(FA_inst.sum, FA_inst.cout));

Function call	$RemoveCouplingFault(wire1, wire2);
Example	$RemoveCouplingFault(FA_inst.sum, FA_inst.cout);

C.6   Parallel Fault

Using PLI facilities, parallel faults can be injected and removed in the design under test. For this purpose, the number of parallel faults or the parallel factor, the list of sites of faults and their related stuck-at values must be defined.

Function call	$ParInjectFault (parallelFactor, wireList, stuckAtList);
Example	$ParInjectFault (3, FA_inst.sum, FA_inst.cout, FA_inst.cin, 1'b1, 1'b0, 1'b0);

In order for parallel fault removal, only the parallel factor and wire list are required.

Function call	$ParInjectFault (parallelFactor, wireList);
Example	$ParRemoveFault (3, FA_inst.sum, FA_inst.cout, FA_inst.cin);

C.7   Fault Collapsing

The fault collapsing function provided by PLI routines is based on line-oriented fault collapsing and requires the name of the design under test and the output file to store the fault list.

Function call	$FaultCollapsing (DUT, outFile);
Example	$FaultCollapsing (FA_inst, "FA.flt");

C.8 Forward Walker

PLI is capable of providing utilities to find one or all of the paths from one wire or input of the design under test to the outputs. Possible modes for this task are "ONE" or "ALL", which specify if one or all of the forward paths found from the internal node to the primary outputs must be recorded.

Function call	$forward_walker(startWire, mode);
Example	$forward_walker(FA_inst.A, "ONE");

C.9 Backward Walker

Using PLI, you can find one or all of the paths from one wire or output back to the inputs of the design under test. Possible modes for this task are "ONE" or "ALL", which specify if one or all of the backward paths found from the internal node to the primary outputs must be recorded.

Function call	$backward_walker(startWire, mode);
Example	$backward_walker(FA_inst.Sum, "ALL");

C.10 Find Cone

The cone that is driven by a certain wire can be found and recorded.

Function call	$find_cone(startWire);
Example	$find_cone(FA_inst.Cin);

C.11 Loader_Driver Finder

The information regarding the gate or module driving a wire (driver), or being driven by one (load), is provided by the following PLI functions.

Function call	$load (wire);
Example	$load (FA_inst.Sum);

Function call	$driver (wire);
Example	$driver (FA_inst.Cin);

C.12 X-Path Check

During test applications such as deterministic test generation, it may be necessary to check for the existence of an x-path from an internal node to the primary outputs. In this case by specifying the name of the wire and the mode of finding x-path, the related PLI function can be used. Possible modes for x-path checking are "ONE" or "ALL" which specify if one or all of the x-paths found from the internal node to the primary outputs must be recorded.

Function call	$x_path_check(wire, mode);
Example	$x_path_check(FA.Cin, "ALL");

C.13   SCOAP Parameters

To calculate SCOAP testability parameters for combinational circuits, the name of the design under test and the output file must be passed to the related function.

Function call	$SCOAP(DUT, outFile);
Example	$SCOAP(testbench_FA, "FA.scp");

C.14   Signal Activity

The number of times that an event occurs on a certain wire or on all of the wires in the design can be observed and recorded. The mode of finding signal activity specifies if the activity of one particular wire or all wires must be calculated, and if the result must be recorded in an output file or must be printed in the simulator console. Based on the selected mode, the other required arguments (i.e., DUT, wire, and output file) must also be specified. As shown in the examples, in either case, the unnecessary arguments can be easily ignored.

Function call	$SignalActivities (mode, DUT, wire, outFile);
Example	$SignalActivities ("ONE_Print", testbench_FA.sum);
Example	$SignalActivities ("ALL_File", testbench_FA,"FA.sga");

C.15   Enable Disable

For some test applications having the capability of enabling some modules and disabling the others is useful. In that case the full name of the considered component and the mode ("enable"/"disable") must be specified.

Function call	$enableDisable(DUT, DUT.component, mode);
Example	$enableDisable(4bitAdder_inst. FA_inst1,0); //disable
Example	$enableDisable(4bitAdder_inst. FA_inst2,1); //enable

Appendix D
IEEE Std. 1149.1 Boundary Scan Verilog Description

The complete Verilog description for the Standard IEEE 1149.1 is included in this appendix. This code was described in Chap. 8 and used in examples of this chapter.

```
///////////////////////////////////////////////////////////////////
//      Multiplexer 2 - 1
///////////////////////////////////////////////////////////////////
module MUX2_1 (i1, i2, sel, out);
   input i1;
   input i2;
   input sel;
   output out;
   assign out = sel ? i2 : i1;
endmodule


///////////////////////////////////////////////////////////////////
//      Multiplexer 4 - 1
///////////////////////////////////////////////////////////////////
module MUX4_1 (i1, i2, i3, i4, sel, out);
   input i1;
   input i2;
   input i3;
   input i4;
   input [1:0] sel;
   output out;
   assign out = (sel == 2'b00) ? i1 :
                (sel == 2'b01) ? i2 :
                (sel == 2'b10) ? i3 :i4;
endmodule


///////////////////////////////////////////////////////////////////
//      D FlipFlop
///////////////////////////////////////////////////////////////////
module D_FF( D, CLK, RstBar, Q);
   input D, CLK, RstBar;
   output Q;
   reg Q;
   always @(posedge CLK or negedge RstBar) begin
      if(RstBar == 1'b0)
           Q = 1'b0;
        else if (CLK == 1'b1)    Q = D;
   end
endmodule
```

```
/////////////////////////////////////////////////////////////////
//      Boundary scan register cell description
/////////////////////////////////////////////////////////////////
module BSRegister1bit(DIN, SIN, ShiftBR, UpdateBR, ClockBR, RstBar,
                      ModeControl, SOUT, DOUT);
    input DIN, SIN, ShiftBR, UpdateBR;
    input ClockBR, RstBar, ModeControl;
    output SOUT, DOUT;
    wire D_DF1;
    reg Q_DF1, Q_DF2;
    assign D_DF1 = ShiftBR ? SIN: DIN;
    always @(posedge ClockBR, negedge RstBar)
    if(!RstBar) Q_DF1 <= 0; else Q_DF1 <= D_DF1;
    always @(posedge UpdateBR, negedge RstBar)
    if(!RstBar) Q_DF2 <= 0; else Q_DF2 <= Q_DF1;
    assign DOUT = ModeControl? Q_DF2: DIN;
    assign SOUT = Q_DF1;
endmodule

/////////////////////////////////////////////////////////////////
//      Instruction register cell description
/////////////////////////////////////////////////////////////////
module InstructionRegister1bit (DIN, SIN, ShiftIR, UpdateIR, ClockIR,
                      RstBar, SOUT, DOUT);
    input DIN, SIN, ShiftIR, UpdateIR;
    input ClockIR, RstBar;
    output SOUT;
    output reg DOUT;
    wire D_DF1;
    reg Q_DF1;
    assign D_DF1 = ShiftIR ? SIN: DIN;
    always @(posedge ClockIR, negedge RstBar)
    if(!RstBar) Q_DF1 <= 0; else Q_DF1 <= D_DF1;
    always @(posedge UpdateIR, negedge RstBar)
    if(!RstBar) DOUT <= 0; else DOUT <= Q_DF1;
    assign SOUT = Q_DF1;
endmodule

/////////////////////////////////////////////////////////////////
//            ByPass Register description
/////////////////////////////////////////////////////////////////

module BS_BYR (DIN, SIN, ShiftBY, ClockBY, RstBar, SOUT);
    input DIN, SIN;
    input ShiftBY, ClockBY, RstBar;
    output reg SOUT;
    wire D_DF;
    assign D_DF = ShiftBY ? SIN: DIN;
    always @(posedge ClockBY, negedge RstBar)
    if(!RstBar) SOUT <= 0; else SOUT <= D_DF;
endmodule

/////////////////////////////////////////////////////////////////
//            Tri-State
/////////////////////////////////////////////////////////////////
module tristate( in,   enable, out);
    input in , enable;
    output out;
    assign out = (enable == 1'b1) ? in : 1'bZ;
endmodule
```

```
/////////////////////////////////////////////////////////////////
//   Instruction register description
/////////////////////////////////////////////////////////////////
module BS_IR  #(parameter Length = 3)(DIN, SIN, ShiftIR, ClockIR, UpdateIR,
                RstBar, SOUT, DOUT);
    input [Length -1 : 0] DIN;
    input SIN , UpdateIR, ClockIR, ShiftIR, RstBar;
    output SOUT;
    output [Length -1 : 0] DOUT;
    wire [Length -1 : 1] SIN_im;
    genvar i;
    generate for (i=0; i<Length; i=i+1) begin
        if(i == 0) begin
            InstructionRegister1bit l1 (.DIN(DIN[i]),.SIN(SIN),
                                        .ShiftIR(ShiftIR),.ClockIR(ClockIR),
                                        .UpdateIR(UpdateIR),.SOUT(SIN_im[1]),
                                            .DOUT(DOUT[i]),.RstBar(RstBar));
        end
        else if (i > 0 & i < Length - 1) begin
            InstructionRegister1bit l2_n(.DIN(DIN[i]),.SIN(SIN_im[i]),
                                        .ShiftIR(ShiftIR),.ClockIR(ClockIR),
                                        .UpdateIR(UpdateIR),.SOUT(SIN_im[i+1]),
                                        .DOUT(DOUT[i]), .RstBar(RstBar));
        end
        else if (i == Length - 1) begin
            InstructionRegister1bit ln(.DIN(DIN[i]),.SIN(SIN_im[i]),
                                       .ShiftIR(ShiftIR),.ClockIR(ClockIR),
                                       .UpdateIR(UpdateIR), .SOUT(SOUT),
                                           .DOUT(DOUT[i]),.RstBar(RstBar));
        end
    end
    endgenerate
endmodule
```

```
/////////////////////////////////////////////////////////////////
//   Boundary scan register description
/////////////////////////////////////////////////////////////////
module BS_BSR  #(parameter Length = 8)(DIN, SIN, UpdateBR, ClockBR, ShiftBR, ModeControl, RstBar,
SOUT, DOUT);
    input [Length -1 : 0] DIN;
    input SIN , UpdateBR, ClockBR, ShiftBR, ModeControl, RstBar;
    output SOUT;
    output [Length -1 : 0] DOUT;
    wire [Length -1 : 1] SIN_im;
    genvar i;
    generate for (i=0; i<Length; i=i+1) begin
        if(i == 0) begin
            BSRegister1bit l1 (.DIN(DIN[i]), .SIN(SIN), .ShiftBR(ShiftBR),
                               .UpdateBR(UpdateBR), .ClockBR(ClockBR),
                               .RstBar(RstBar),.ModeControl(ModeControl),
                               .SOUT(SIN_im[1]), .DOUT(DOUT[i]));
        end
        else if (i > 0 & i < Length - 1) begin
            BSRegister1bit l2_n (.DIN(DIN[i]), .SIN(SIN_im[i]),
                                 .ShiftBR(ShiftBR),.ModeControl(ModeControl),
                                 .ClockBR(ClockBR),.UpdateBR(UpdateBR),
                                 .SOUT(SIN_im[i + 1]),.DOUT(DOUT[i]),
                                 .RstBar(RstBar));
        end
        else if (i == Length - 1)begin
            BSRegister1bit ln (.DIN(DIN[i]), .SIN(SIN_im[i]), .ShiftBR(ShiftBR),
                               .ModeControl(ModeControl), .ClockBR(ClockBR),
                               .UpdateBR(UpdateBR), .SOUT(SOUT),.DOUT(DOUT[i]),
                               .RstBar(RstBar));
        end
    end
    endgenerate
endmodule
```

```
//////////////////////////////////////////////////////////////
//    TAP controller description
//////////////////////////////////////////////////////////////
module TAPController (TMS, TCLK, RstBar, sel, Enable,ShiftIR, ClockIR,
                      UpdateIR, ShiftDR, ClockDR, UpdateDR);
    input TMS;
    input TCLK;
    output reg RstBar = 1'b0;
    output sel;
    output reg Enable;
    output reg ShiftIR, ClockIR, UpdateIR;
    output reg ShiftDR, ClockDR, UpdateDR;
    parameter [3:0]Test_Logic_Reset =4'b1111;
    parameter [3:0]Run_Test_Idle =4'b1100;
    parameter [3:0]Select_DR_Scan =4'b0111;
    parameter [3:0]Capture_DR =4'b0110;
    parameter [3:0]Shift_DR =4'b0010;
    parameter [3:0]Exit1_DR =4'b0001;
    parameter [3:0]Pause_DR =4'b0011;
    parameter [3:0]Exit2_DR =4'b0000;
    parameter [3:0]Update_DR =4'b0101;
    parameter [3:0]Select_IR_Scan =4'b0100;
    parameter [3:0]Capture_IR =4'b1110;
    parameter [3:0]Shift_IR =4'b1010;
    parameter [3:0]Exit1_IR =4'b1001;
    parameter [3:0]Pause_IR =4'b1011;
    parameter [3:0]Exit2_IR =4'b1000;
    parameter [3:0]Update_IR =4'b1101;

    reg [3:0]TAP_STATE = Test_Logic_Reset;
    always @(posedge TCLK) begin
        case (TAP_STATE)
            Test_Logic_Reset :
                if (TMS == 1'b0) TAP_STATE = Run_Test_Idle;
                 else if(TMS == 1'b1) TAP_STATE = Test_Logic_Reset;
            Run_Test_Idle:
              if (TMS == 1'b1) TAP_STATE = Select_DR_Scan;
             else if (TMS == 1'b0) TAP_STATE = Run_Test_Idle;
            Select_DR_Scan :
                if (TMS == 1'b0) TAP_STATE = Capture_DR;
                else if (TMS == 1'b1) TAP_STATE = Select_IR_Scan;
            Capture_DR:
                    if (TMS == 1'b0) TAP_STATE = Shift_DR;
                else if (TMS == 1'b1)TAP_STATE = Exit1_DR;
            Shift_DR:
                    if (TMS == 1'b1) TAP_STATE = Exit1_DR;
                else if (TMS == 1'b0) TAP_STATE = Shift_DR;
            Exit1_DR:
                if (TMS == 1'b0) TAP_STATE = Pause_DR;
                  else if (TMS == 1'b1) TAP_STATE = Update_DR;
            Pause_DR:
                if (TMS == 1'b1) TAP_STATE = Exit2_DR;
                else if (TMS == 1'b0) TAP_STATE = Pause_DR;
            Exit2_DR:
              if (TMS == 1'b1) TAP_STATE = Update_DR;
                  else if (TMS == 1'b0) TAP_STATE = Shift_DR;
            Update_DR:
                if (TMS == 1'b0) TAP_STATE = Run_Test_Idle;
                    else if (TMS == 1'b1) TAP_STATE = Select_DR_Scan;
            Select_IR_Scan:
                if (TMS == 1'b0) TAP_STATE = Capture_IR;
              else if (TMS == 1'b1)TAP_STATE = Test_Logic_Reset;
            Capture_IR:
                    if (TMS == 1'b0) TAP_STATE = Shift_IR;
                else if (TMS == 1'b1) TAP_STATE = Exit1_IR;
```

```
                    Shift_IR:
                        if (TMS == 1'b1) TAP_STATE = Exit1_IR;
                          else if (TMS == 1'b0) TAP_STATE = Shift_IR;
                    Exit1_IR:
                        if (TMS == 1'b0) TAP_STATE = Pause_IR;
                          else if (TMS == 1'b1) TAP_STATE = Update_IR;
                    Pause_IR:
                        if (TMS == 1'b1) TAP_STATE = Exit2_IR;
                          else if (TMS == 1'b0) TAP_STATE = Pause_IR;
                    Exit2_IR:
                        if (TMS == 1'b1) TAP_STATE = Update_IR;
                          else if (TMS == 1'b0) TAP_STATE = Shift_IR;
                    Update_IR:
                        if (TMS == 1'b0) TAP_STATE = Run_Test_Idle;
                          else if (TMS == 1'b1) TAP_STATE = Select_DR_Scan;
             endcase
        end // END ALWAYS
        always @(negedge TCLK) begin
                    RstBar   = 1'b1;
                    Enable   = 1'b0;
                    ShiftIR  = 1'b0;
                    ShiftDR  = 1'b0;
                    ClockIR  = 1'b1;
                    UpdateIR = 1'b0;
                    ClockDR  = 1'b1;
                    UpdateDR = 1'b0;
             case (TAP_STATE)
                Test_Logic_Reset:
                    RstBar = 1'b0;
                Shift_IR: begin
                    Enable   = 1'b1;
                    ShiftIR  = 1'b1;
                    ClockIR  = 1'b0; end
                Shift_DR: begin
                    Enable   = 1'b1;
                    ShiftDR  = 1'b1;
                    ClockDR  = 1'b0; end
                Capture_IR:
                    ClockIR  = 1'b0;
                Update_IR:
                   UpdateIR = 1'b1;
                Capture_DR:
                    ClockDR  = 1'b0;
                Update_DR:
                    UpdateDR = 1'b1;
             endcase
        end //end always
    assign sel = TAP_STATE[3];
endmodule

/////////////////////////////////////////////////////////////
//    Decoder Unit description
/////////////////////////////////////////////////////////////
module Decoder (Instruction, TCLK, ShiftIR, UpdateIR, ClockIR, ShiftDR,
                UpdateDR, ClockDR,DRInShiftBR, DRInUpdateBR, DRInClockBR,
                DRInTMS, DROutShiftBR, DROutUpdateBR,DROutClockBR, DROutTMS,
                IRShiftIR, IRUpdateIR,IRClockIR, ShiftBY, ClockBY,
                Select_DR);
    input  [2:0] Instruction;
    input  TCLK, ShiftIR, UpdateIR, ClockIR, ShiftDR, UpdateDR, ClockDR;
    output reg DRInShiftBR, DRInUpdateBR, DRInClockBR, DRInTMS;
    output reg DROutShiftBR, DROutUpdateBR, DROutClockBR, DROutTMS;
    output reg IRShiftIR, IRUpdateIR, IRClockIR;
    output reg ShiftBY, ClockBY;
    output reg [1:0] Select_DR;
    parameter [2:0]bypass_instruction  = 3'b111;
    parameter [2:0]intest_instruction  = 3'b011;
```

```verilog
parameter [2:0]sample_instruction  = 3'b010;
parameter [2:0]preload_instruction = 3'b001;
parameter [2:0]extest_instruction  = 3'b000;
parameter [1:0]sel_br   = 2'b11;
parameter [1:0]sel_dr   = 2'b10;
parameter [1:0]sel_ir   = 2'b01;
parameter [1:0]sel_none = 2'b00;
always @(Instruction, ShiftIR, ShiftDR, ClockIR, ClockDR, UpdateIR,
        UpdateDR, TCLK) begin

   // set ALL output signals to their inactive states
   DRInShiftBR   = 1'b0;
   DRInUpdateBR= 1'b0;
   DRInClockBR = 1'b0;
   DRInTMS = 1'b0;
  DROutShiftBR = 1'b0;
   DROutUpdateBR = 1'b0;
   DROutClockBR = 1'b0;
   DROutTMS = 1'b0;
   IRShiftIR = 1'b0;
   IRUpdateIR = 1'b0;
   IRClockIR = 1'b0;
   ShiftBY = 1'b0;
   ClockBY = 1'b0;
   Select_DR = sel_none;
   // compute the value OF signals
   case (Instruction)
     bypass_instruction: begin
         if (ShiftDR == 1'b1)
             Select_DR = sel_br;
         else
            Select_DR = sel_ir;
        ClockBY = (~ClockDR & TCLK);
        ShiftBY = ShiftDR;
        IRClockIR = (~ClockIR & TCLK);
     IRShiftIR = ShiftIR;
        IRUpdateIR    = (UpdateIR & TCLK);
     end
        sample_instruction : begin
          if (ShiftDR == 1'b1)
              Select_DR = sel_dr;
            else
              Select_DR = sel_ir;
           DRInClockBR = ~ClockDR & TCLK;
           DRInUpdateBR = UpdateDR;
           DRInShiftBR = ShiftDR;
           DROutClockBR = ~ ClockDR & TCLK;
           DROutUpdateBR = UpdateDR;
           DROutShiftBR = ShiftDR;
        IRClockIR = ~ ClockIR & TCLK;
           IRShiftIR = ShiftIR;
           IRUpdateIR = UpdateIR & TCLK;
        end
        preload_instruction : begin
           if (ShiftDR == 1'b1)
            Select_DR = sel_dr;
           else
               Select_DR = sel_ir;
         DRInClockBR   = ~ ClockDR & TCLK;
           DRInUpdateBR= UpdateDR;
           DRInShiftBR = ShiftDR;
           DROutClockBR          = ~ ClockDR & TCLK;
           DROutUpdateBR     = UpdateDR;
           DROutShiftBR      = ShiftDR;
        IRClockIR      = ~ ClockIR & TCLK;
           IRShiftIR   = ShiftIR;
           IRUpdateIR  = UpdateIR & TCLK;
        end
```

```verilog
                    extest_instruction: begin
                      if(ShiftDR == 1'b1)
                          Select_DR = sel_dr;
                        else
                    Select_DR = sel_ir;
                        DRInShiftBR = ShiftDR;
                        DRInClockBR = ~ ClockDR & TCLK;
                        DRInUpdateBR= UpdateDR;
                        DROutTMS = 1'b1;
                        DROutShiftBR = ShiftDR;
                        DROutClockBR = ~ClockDR & TCLK;
                        DROutUpdateBR        = UpdateDR;
                        IRClockIR    = ~ClockIR & TCLK;
                        IRShiftIR    = ShiftIR;
                        IRUpdateIR   = UpdateIR & TCLK;
                    end
                    intest_instruction: begin
                        if(ShiftDR == 1'b1)
                           Select_DR = sel_dr;
                        else
                           Select_DR = sel_ir;
                        DRInShiftBR = ShiftDR;
                        DRInClockBR = ~ ClockDR & TCLK;
                        DRInUpdateBR= UpdateDR;
                        DRInTMS = 1'b1;
                        DROutShiftBR = ShiftDR;
                        DROutClockBR = ~ClockDR & TCLK;
                        DROutUpdateBR        = UpdateDR;
                        IRClockIR = ~ClockIR & TCLK;
                        IRShiftIR = ShiftIR;
                        IRUpdateIR = UpdateIR & TCLK;
                    end
                    default: begin
                    if (ShiftDR == 1'b1)
                        Select_DR = sel_br;
                    else
                        Select_DR = sel_ir;
                    ClockBY = (~ClockDR & TCLK);
                    ShiftBY = ShiftDR;
                    IRClockIR = (~ClockIR & TCLK);
                    IRShiftIR = ShiftIR;
                    IRUpdateIR = (UpdateIR & TCLK);
              end
          endcase
      end
endmodule
```

Appendix E
Boundary Scan IEEE std. 1149.1 Virtual Tester

What follows is the Verilog code of the virtual tester discussed in Chap. 8. This is brought here for reference when studying the details of events taking place in the example of Chap. 8. Furthermore, this test bench can be used as a virtual tester template for other circuits with the Standard IEEE std. 1149.1 boundary scan. In which case, only the size of test data and response will have to be set according to those of CUT.

```
/////////////////////////////////////////////////////////////
// Virtual tester Description
/////////////////////////////////////////////////////////////
module Virtual_Tester(input start , output finish);
   parameter testVectorName1 = "CPUin.tvf";
   parameter testVectorName2 = "CPUout.tvf";
   parameter memLength1 = 3;
   parameter memLength2 = 5;
   parameter intestingoutput = "intestingresponse.out";
   parameter extestingoutput = "extestingresponse.out";
   reg TCLK =1'b0;
   reg clk =1'b0;
   wire intesting , extesting;
   wire lastdata;
   wire  [0:8]inputDatai;
   wire  [0:15]inputDatao;
   reg g_reset= 1'b0;
   wire endIn, endOut, ready, outready;
   wire [0:15]outputData ;
   wire clkenable;
   wire stop;

   BS_Driver  #(9, 16, 1, 3, 5) BS_D (.rst(g_reset),.clk (clk),
              .TCLK(TCLK),.intesting(intesting),
              .extesting(extesting), .lastdata(lastdata),
              .inputDatai(inputDatai),.inputDatao(inputDatao),
              .endIn(endIn), .endOut(endOut), .ready(ready),
              .outready(outready),.outputData(outputData),
              .clkenable(clkenable));

   IO_Driver IO(.start(start), .clk(clk), .TCLK(TCLK), .endIn(endIn),
               .endOut(endOut), .ready(ready),.outready(outready),
               .outputData(outputData),.g_reset(g_reset), .stop(stop),
               .intesting(intesting),.extesting(extesting),
               .lastdata(lastdata),.inputDatai(inputDatai),
               .inputDatao(inputDatao));

   assign finish = stop;
   always #1 TCLK = ~TCLK;
   always  #1 begin
      if (clkenable ==1'b1)
         #1 clk = ~clk;
      else
         clk =1'b0;
   end

   initial begin
      g_reset = 1'b1;
      #1;
      g_reset = 1'b0;
   end
endmodule
```

```
/////////////////////////////////////////////////////////////////
// Boundary Scan Driver of Virtual Tester
/////////////////////////////////////////////////////////////////

module BS_Driver #(parameter inputLength = 9,parameter outputLength = 16,
                   parameter outputLength2=1,parameter instruction_length=3,
                   parameter numberOfClk = 5)
                   (input rst,input clk,input TCLK,input intesting,input
                   extesting,input lastdata,input[0:inputLength-1]inputDatai,
                   input [0:outputLength-1]inputDatao,output reg endIn,
                   output reg endOut, output reg ready, output reg outready,
                   output[0:outputLength-1]outputData, output reg clkenable);
    parameter [2:0]bypass_instruction   = 3'b111;
    parameter [2:0]intest_instruction    = 3'b011;
    parameter [2:0]sample_instruction    = 3'b010;
    parameter [2:0]preload_instruction   = 3'b001;
    parameter [2:0]extest_instruction    = 3'b000;
    reg [0:inputLength-1] in_signali;
    reg [0:outputLength-1] in_signalo;
    reg [0:2*instruction_length-1]instruction =
        {bypass_instruction,bypass_instruction};
    reg [0:outputLength-1] out_signal;
    reg TDI = 1'b0;
    reg TMS = 1'b0;
    reg Ckin     = 1'b0;
    //reg rst  = 1'b0;
    reg [0:8] Pin        ;
    wire TDOF;
    wire PoutF;
    integer i, j = 0;
    integer k =0;
    system #(9,1) FUT( .TDI(TDI), .TMS(TMS), .TCLK(TCLK), .Ckin(clk),
                    .g_reset(rst), .Pin(Pin), .TDO(TDOF), .Pout(PoutF));
    always @( intesting | extesting ) begin

/*******************************************************/
/*                    intesting                        */
/*******************************************************/
    if (intesting == 1'b1) begin
        endIn = 1'b0;
         endOut = 1'b0;
         ready = 1'b0;
         clkenable =1'b0;
        for (i =0; i < 5; i = i+1) begin
            @ (negedge TCLK)
            TMS = 1'b1 ;
        end // 5 clock with TMS = 1 resets the state machine!
        // loading preload instruction //
        instruction = {bypass_instruction,preload_instruction};
         @ (negedge TCLK)
        TMS   = 1'b0 ;
        // go to Run_Test_Idle
         @ (negedge TCLK)
        TMS   = 1'b1 ;
        // go to Select_Dr_Scan
        @ (negedge TCLK)
        TMS       = 1'b1 ;
        // go to Select_IR_scan
         @ (negedge TCLK)
        TMS   = 1'b0 ;
        // go to Capture_IR
         @ (negedge TCLK)
        TMS   = 1'b0 ;
        // go to Shift_IR
```

```verilog
for ( i = 0; i < 2*instruction_length-1; i = i+1) begin
   @(negedge TCLK) begin
      TDI    = instruction[i];
        TMS  = 1'b0 ;
        // stay in Shift_IR, shift n-1 first bit of instruction
   end
end
@(negedge TCLK) begin
   TDI = instruction[2*instruction_length -1] ;
   TMS = 1'b1 ;
end
// go to Exit1_IR and load preload instruction in IR
@(negedge TCLK)
TMS  = 1'b1 ;
// go to Update_IR
@(negedge TCLK)
TMS  = 1'b1 ;
// go to Select_DR scan
@(negedge TCLK)
TMS  = 1'b0 ;
// go to Capture_DR
@(negedge TCLK) begin
   TMS = 1'b0 ;
   ready = 1'b1;
end
@(posedge TCLK) begin
   in_signali = inputDatai;
   ready = 1'b0;
end
// go to Shift_DR;
for ( i = 0; i < inputLength-1; i = i+1) begin
   @(negedge TCLK) begin
    TDI     = in_signali[i] ;
       TMS  = 1'b0 ;
       //stay in Shift_DR, shift the n-1 firstbit ;
   end
end
@(negedge TCLK) begin
   TDI = in_signali[inputLength-1];
   TMS = 1'b1 ;
   // go to Exit1_DR and shift last bit
end
@(negedge TCLK)
TMS  = 1'b1 ;
// go to Update_DR
@(negedge TCLK)
TMS  = 1'b1 ;
// go to Select_DR_Scan
@(negedge TCLK)
TMS  = 1'b1 ;
// go to Select_IR_Scan
@(negedge TCLK)
TMS  = 1'b0 ;
// got to Capture_IR
@(negedge TCLK)
TMS  = 1'b0 ;
// go to Shift_IR
instruction = {bypass_instruction,intest_instruction};
for ( i = 0; i < 2*instruction_length -1; i = i+1) begin
   @(negedge TCLK) begin
      TDI    = instruction[i];
        TMS  = 1'b0 ;
   end
   // stay in Shift_IR
end
```

```verilog
@(negedge TCLK) begin
    TDI = instruction[2*instruction_length -1] ;
    TMS = 1'b1 ;
    // go to Exit_IR and load intest instruction
  end
@(negedge TCLK)
  TMS   = 1'b1 ;
  // go to Update_IR
@(negedge TCLK)
  TMS   = 1'b0 ;
  // go to Run_Test_Idle
  @(negedge TCLK) begin
    TMS = 1'b0 ;
    clkenable =1'b1;
  end
  for ( i =0; i < numberOfClk; i = i+1) begin
    @(negedge TCLK) begin
        TMS     = 1'b0 ;
    end
  end// stay in Run_Test_Idle
  clkenable = 1'b0;
        ///////////////////////////////////////////////////////////////
/*                  loop until last data shift in                  */
  ///////////////////////////////////////////////////////////////
  while (lastdata != 1'b1) begin
    @(negedge TCLK) begin
        ready =1'b1;
        TMS   = 1'b1 ;
    end// go to Select_DR_Scan
    @(negedge TCLK) begin
        in_signali = inputDatai;
        ready=1'b0;
        outready =1'b0;
        TMS   = 1'b0 ;
        // go to Capture_DR
    end
    @(negedge TCLK)
    TMS = 1'b0 ;
    // go to Shift_DR
    for (i =0; i < 2; i=i+1) begin
        @(negedge TCLK) begin
            TDI = 1'b0 ;
            TMS = 1'b0 ;
            // stay in Shift_DR
        end
    end
    for (i =0; i < outputLength; i=i+1) begin
        @(negedge TCLK) begin
            TDI = 1'b0 ;
            TMS = 1'b0 ;
            out_signal[i] = TDOF;
            // stay in Shift_DR
        end
    end
    for (i =0; i< inputLength-1; i=i+1) begin
        @(negedge TCLK) begin
            TDI = in_signali[i] ;
            TMS = 1'b0 ;
            // stay in Shift_DR
        end
    end
    @(negedge TCLK) begin
        TDI = in_signali[inputLength-1] ;
        TMS   = 1'b1 ;
      out_signal[outputLength+1]     = TDOF ;
        // go to Exit1_DR
    end
```

```verilog
    @(negedge TCLK) begin
        outready = 1'b1;
         TMS   = 1'b1 ;
         // go to Update_DR
    end
    @(negedge TCLK) begin
       TMS    = 1'b0 ;
         outready =1'b0;
         // go to Run_Test_Idle
    end
    @(negedge TCLK) begin
       TMS    = 1'b0 ;
         clkenable = 1'b1;
    end
    for ( i =0; i < numberOfClk; i = i+1) begin
        @(negedge TCLK) begin
            TMS = 1'b0 ;
        end
    end // stay in Run_Test_Idle
    clkenable = 1'b0;
  end // last input data
  @(negedge TCLK)
  TMS   = 1'b1 ;
  // go to Select_DR_Scan
  @(negedge TCLK)
  TMS   = 1'b0 ;
  // go to Capture_DR_Scan
  @(negedge TCLK)
  TMS   = 1'b0 ;
  // go to Shift_DR
  for (i =0; i < 2; i=i+1) begin
      @(negedge TCLK) begin
         TDI = 1'b0 ;
          TMS   = 1'b0 ;
          // stay in Shift_DR
      end
  end
  for (i =0; i < outputLength-1; i=i+1) begin
      @(negedge TCLK) begin
         TDI = 1'b0 ;
          TMS   = 1'b0 ;
          out_signal[i] = TDOF;
          // stay in Shift_DR
      end
  end
  @(negedge TCLK) begin
    TDI = 1'b0 ;
     TMS = 1'b1 ;
     out_signal[outputLength-1] = TDOF;
     // go to Exit1_DR
  end
  @(negedge TCLK) begin
     outready = 1'b1;
     TMS = 1'b1 ;
     // go to Update_DR
  end
@(negedge TCLK) begin
     outready =1'b0;
     TMS = 1'b1 ;
     // go to Select_DR_Scan
  end
  @(negedge TCLK)
  TMS   = 1'b1 ;
// go to Select_IR_Scan
  @(negedge TCLK) begin
     TMS = 1'b1 ;
```

```
                 // go to Test_Logic_reset;
          end
          @(posedge TCLK) begin
              endIn = 1'b1;
              endOut = 1'b0;
          end
      end // for end intesting

  /*******************************************************************************/
  /*                              Extest testing                              */
  /*******************************************************************************/
      else if (Extesting == 1'b1) begin
          endOut = 1'b0;
          endIn = 1'b0;
          outready = 1'b0;
          ready =1'b0;
          for (i =0; i < 5; i = i+1) begin
              @(negedge TCLK)
              TMS = 1'b1 ;
          end // 5 clock with TMS = 1 resets the state machine!
          // loading preload instruction //
          instruction = {bypass_instruction, preload_instruction};
          @(negedge TCLK)
          TMS   = 1'b0 ;
          // go to Run_Test_Idle
          @(negedge TCLK)
            TMS   = 1'b1 ;
            // go to select_DR_Sscan
          @(negedge TCLK)
            TMS   = 1'b1 ;
            // go to Select_IR_Sscan
            @(negedge TCLK)
            TMS   = 1'b0 ;
            // go to Capture_IR
            @(negedge TCLK)
            TMS   = 1'b0 ;
            // go to Shift_IR
            for ( i = 0; i < 2*instruction_length-1; i = i+1) begin
                @(negedge TCLK) begin
                    TDI     = instruction[i];
                    TMS   = 1'b0 ;
                      // stay in Shift_IR, shift the first bit 1'b1 ;
                end
            end
            @(negedge TCLK) begin
              TDI = instruction[2*instruction_length -1] ;
              TMS = 1'b1 ;
              // go to Exit1_IR and load preload instruction in IR
            end
            @(negedge TCLK)
            TMS   = 1'b1 ;
            // go to Update_IR
            @(negedge TCLK) begin
              ready =1'b1;
              TMS = 1'b1 ;
              // go to Select_DR_Scan
            end
            @(negedge TCLK) begin
              in_signalo = inputDatao;
              ready = 1'b0;
              TMS = 1'b0 ;
              // go to Capture_DR
            end
            @(negedge TCLK)
            TMS   = 1'b0 ;
            // go to Shift_DR, shift the first bit ;
```

```
  for ( i = 0; i < outputLength; i = i+1) begin
      @(negedge TCLK) begin
          TDI =   in_signalo[i] ;
          TMS  = 1'b0 ;
            // stay in Shift_DR, shift the second bit ;
      end
  end
  for ( i = 0; i < inputLength-1; i = i+1) begin
      @(negedge TCLK) begin
          TDI    = 1'b0;
          TMS  = 1'b0 ;
            // stay in Shift_DR, shift the second bit ;
      end
  end
  @(negedge TCLK) begin
      TDI = 1'b0;
      TMS = 1'b1 ;
      // go to Exit1_DR
  end
  @(negedge TCLK)
  TMS  = 1'b1 ;
  // go to Update_DR
  @(negedge TCLK)
  TMS  = 1'b1 ;
  // go to Select_DR_Scan
  @(negedge TCLK)
  TMS  = 1'b1 ;
  // go to Select_IR_Scan
 @(negedge TCLK)
  TMS  = 1'b0 ;
  // go to Capture_IR
 @(negedge TCLK)
 TMS  = 1'b0 ;
 // go to shift_IR, first bit
 instruction = {sample_instruction, extest_instruction};
 for ( i = 0; i < 2*instruction_length -1; i = i+1) begin
     @(negedge TCLK) begin
         TDI    = instruction[i];
         TMS  = 1'b0 ;
     end
end
  @(negedge TCLK) begin
     TDI = instruction[2*instruction_length -1] ;
     TMS = 1'b1 ;
     // go to Exit1_IR and load intest instruction
  end
  @(negedge TCLK)
  TMS  = 1'b1 ;
  // go to Update_IR
  while(lastdata != 1'b1) begin
      @(negedge TCLK)
      TMS = 1'b1 ;
      // go to Select_DR_Scan
      @(negedge TCLK)
      TMS = 1'b0 ;
      // go to Capture_DR
      @(negedge TCLK) begin
        ready = 1'b1;
          TMS  = 1'b0 ;
      end
        @(posedge TCLK) begin
           in_signalo = inputDatao;
           ready = 1'b0;
        end
        // go to Shift_DR
        for ( i = 0; i < outputLength2; i = i+1) begin
```

```verilog
            @(negedge TCLK) begin
               TDI = 1'b0 ;
                 TMS = 1'b0 ;
                 // stay in Shift_DR
            end
         end
         for ( i = 0; i < outputLength; i = i+1) begin
            @(negedge TCLK) begin
               TDI = in_signalo[i];
                 TMS = 1'b0 ;
                 out_signal[i] = TDOF;
            end
         end
         for( i = 0; i < inputLength-1; i=i+1) begin
            @(negedge TCLK) begin
                 TDI = 1'b0 ;
                 TMS = 1'b0 ;
            end
         end
         @(negedge TCLK) begin
           TDI = 1'b0 ;
             TMS = 1'b1 ;
             // go to Exit1_DR
         end
         @(negedge TCLK) begin
             outready = 1'b1;
             TMS = 1'b1 ;
         end
         @(posedge TCLK)
         outready = 1'b0;
         // go to Update_DR
   end // last data
   @(negedge TCLK)
   TMS = 1'b1 ;
   // go to Select_DR_Scan
   @(negedge TCLK)
   TMS = 1'b0 ;
   // go to Capture_DR
   @(negedge TCLK)
   TMS = 1'b0 ;
   // go to Shift_DR, shift out first TDO
   for ( i = 0; i < outputLength2; i = i+1) begin
       @(negedge TCLK) begin
         TDI = 1'b0 ;
         TMS = 1'b0 ;
       end
   end
end
for ( i = 0; i < outputLength; i = i+1) begin
   @(negedge TCLK) begin
       TDI = 1'b0;
       TMS    = 1'b0 ;
       out_signal[i] = TDOF;
   end
end
@(negedge TCLK) begin
   TDI = 1'b0 ;
   TMS = 1'b1 ;
   // go to Exit1_DR
end
@(negedge TCLK) begin
   outready =1'b1;
   TMS = 1'b1 ;
end
@(posedge TCLK)
outready =1'b0;
// go to Update_DR
```

```verilog
        @(negedge TCLK)
        TMS = 1'b1 ;
        // go to Select_DR_Scan
        @(negedge TCLK)
        TMS = 1'b1 ;
        // go to Select_IR_Scan
        @(negedge TCLK)
      TMS = 1'b1 ;
        @(posedge TCLK) begin
            endOut = 1'b1;
            endIn = 1'b0;
        end
    end// end extesting
  end // always
  assign outputData = out_signal;
endmodule
```

```verilog
////////////////////////////////////////////////////////////
// IO driver Module of Virtual Tester
////////////////////////////////////////////////////////////
module IO_Driver #(parameter testVectorName1 = "CPUin.tvf", parameter
                   testVectorName2 = "CPUout.tvf", parameter memLength1 = 3,
                   parameter memLength2 = 5,
                   parameter intestingoutput = "intestingresponse.out",
                   parameter extestingoutput = "extestingresponse.out")
                  (start, clk, TCLK, endIn, endOut, ready, outready,
                   outputData,g_reset, stop, intesting, extesting,
                   lastdata, inputDatai, inputDatao);
    input start;
    input TCLK ;
    input clk;
    output reg intesting , extesting;
    output reg lastdata;
    output reg [0:8]inputDatai;
    output reg [0:15]inputDatao;
    input g_reset;
    input endIn, endOut, ready, outready;
    input [0:15]outputData ;
    output reg stop;
    reg [0:8]TESTin   [0: memLength1-1];
    reg [0:15]TESTout[0: memLength2-1];
    integer f1, f2, f3, f4;
    integer i , numOfvector, status;

    initial begin
       f3 = $FOPEN( intestingoutput, "w");
       f4 = $FOPEN( extestingoutput, "w");
       // memory loading
     $DISPLAY("\nstart initializing testin memory...");
       f1 = $FOPEN( "CPUin.txt", "r");
       numOfvector = 0;
       while( !$FEOF(f1) ) begin
          $FSCANF(f1,"%b ",TESTin[numOfvector]);
          numOfvector = numOfvector + 1;
       end //while
       $DISPLAY("end initializing testin memory...\n");
       $DISPLAY("\nstart initializing testout memory...");
       f2 = $FOPEN( "CPUout.txt", "r");
       numOfvector = 0;
       while( !$FEOF(f2) ) begin
          $FSCANF(f2,"%b ",TESTout[numOfvector]);
          numOfvector = numOfvector + 1;
       end //while
       $DISPLAY("end initializing testout memory...\n");
       intesting =1'b0;
```

```
         extesting = 1'b0;
         lastdata = 1'b0;
         stop = 1'b0;
         i =0;
end
always @ (start, ready, outready, endIn, endOut) begin
    if (start == 1'b1) begin
         intesting = 1'b1;
         i =0;
         lastdata = 1'b0;
    end
    else if (ready == 1'b1 & intesting == 1'b1) begin
         inputDatai = TESTin[i];
         i= i+1;
         if(i == memLength1) begin
             lastdata = 1'b1;
         end
    end
    else if (ready == 1'b1 & extesting == 1'b1) begin
         inputDatao = TESTout[i];
         i= i+1;
         if(i == memLength2) begin
             lastdata = 1'b1;
         end
    end
    else if(outready == 1'b1 & intesting == 1'b1) begin
         $DISPLAY ("%b\n" , outputData);
         $FWRITE(f3,"%b\n",outputData);
    end
    else if(outready == 1'b1 & extesting == 1'b1) begin
          $DISPLAY ("%b\n" , outputData);
          $FWRITE(f4,"%b\n",outputData);
    end
    else if (endIn == 1'b1) begin
         intesting = 1'b0;
         #1;
         extesting = 1'b1;
         i =0;
         lastdata = 1'b0;
    end
    else if (endOut == 1'b1) begin
         extesting = 1'b0;
         stop = 1'b1;
         i=0;
         lastdata = 1'b0;
    end
    end
endmodule
```

```
///////////////////////////////////////////////////////////
// Testbench for virtual tester
///////////////////////////////////////////////////////////
module External();
    reg start;
    wire stop;
    Virtual_Tester uu( .start(start) , .finish  (finish));

    initial begin
         start = 1'b0;
         #2 ;
         start = 1'b1;
         #1;
         start = 1'b0;
    end
    always @ (finish) begin
         if (finish == 1'b1) begin
           $DISPLAY ("end of test");
           $STOP;
         end
    end
endmodule
```

Appendix F
Generating Netlist by Register Transfer Level Synthesis (*NetlistGen*)

In order to perform test applications such as fault collapsing and fault simulation that have been done in many instances in this book, a netlist of primitive logic gates is needed. To be able to obtain this netlist from register transfer level (RTL) descriptions, a synthesis program is required. The synthesis program must be able to generate its hardware using our gate primitives (Appendix B) in order for our PLI functions (Appendix C) to be able to perform their test related tasks. For this purpose, *NetlistGen*[1] that is a netlist generation program has been developed. In the background, *NetlistGen* uses the Web Version of Xilinx ISE (free version) for RT level synthesis, and it translates its intermediate NGC output of the synthesized design into the netlist required by test PLI functions. The end result is a synthesis program going from synthesizable Verilog code to a netlist of primitives. This appendix shows the procedures for installing and using this program.

F.1   Installing and Configuring NetlistGen

For using the netlist generator you should first install Xilinx ISE web pack edition. The installation process is listed below:

- Download Windows based Xilinx ISE web pack edition from: http://www.xilinx.com.
- Click on "setup.exe" and follow the setup process.

  - Choose Web Pack edition for installation.
  - Leave the default options of checked boxes unchanged.
  - Click on the install button to start installation.

F.2   Synthesis and Netlist Generation

After installing Xilinx ISE, for generating a Verilog netlist from a behavioral Verilog code, the *NetlistGen.exe* file should be copied in the location as the behavioral description of your Verilog file. A partial Verilog code showing the required input/output format and a part of the RT level behavioral code is shown in Fig. F.1.

---

[1]This program has been developed by Hashem Haghbayan.

```
module Controller ( reset, clk, op_code, rd_mem, wr_mem, ir_on_adr, pc_on_adr,  . . .
   . . .
   input clk;
   input[1:0]op_code ;
   output rd_mem;
   output wr_mem;
   output ir_on_adr;
   output pc_on_adr;
   . . .
   reg [1:0] present_state, next_state;

   always @( posedge clk )
      if( reset ) present_state <= `Reset;
      else present_state <= next_state;
   always @(present_state) begin
      . . .
      case( present_state )
         . . .
         `Fetch : begin
               next_state = `WaitState;
               pc_on_adr=1'b1; rd_mem=1'b1; data_on_dbus=1'b1;
               ld_ir=1'b1; inc_pc=1;
         end
         . . .
      endcase
   end
endmodule
```

Fig. F.1 Port declaration format and sample RT level code

As shown, port declaration format for behavioral top module should list input output declarations after the module header. *NetlistGen* cannot translate input or output in the port identifier list. Same applies to port reg declarations.

When the RT level input is prepared as above, the following steps will result in the generation of a netlist.

- Run *NetlistGen*.
- When prompted, choose a name for your project. This name will become the project name of the synthesis process of ISE.
- When prompted for the name of the module, enter the name of the top level module of your project. The name of the Verilog file of the top level module should be the same as the name of the module. After that, *NetlistGen* reads the behavioral Verilog file. If you want to add any other Verilog files to your design, you can add them in the next step. Added files should be available in the root directory.
- Now you should synthesize your design with Xilinx ISE. For that, choose "y" to synthesize your design. For netlist generation, the *NetlistGen_V2* netlist generator uses the intermediate EDIF format that is obtained from Xilinx Synthesis Technology (*xse.exe*) NGC output.
- When running *NetlistGen_V2*, ISE installation path must be known. For this, ISE version 12.2 path is assumed. Enter the proper installation path if you are using a different ISE version.
- After selecting the location of ISE installation, the synthesis process will start. The process will be completed successfully if there are no compilation and synthesis errors.
- *NetlistGen* will generate the netlist in the *netlist_MODULENAME_V1.v* file. Part of what is automatically generated for example of Fig. F.1 is shown in Fig. F.2.

```
module Controller_net(global_reset, reset,
clk,
op_code,
rd_mem,
wr_mem,
. . .
   input clk;
   input[1:0]op_code ;
   . . .
   output rd_mem;
   output wr_mem;
wire_2_4,
wire_2_5;

pin #(2) pin_0 ({op_code[0], op_code[1]}, {op_code_0, op_code_1});

fanout_n #(2, 0, 0) FANOUT_8 (wire_8, {wire_8_0, wire_8_1});
fanout_n #(2, 0, 0) FANOUT_9 (wire_9, {wire_9_0, wire_9_1});
fanout_n #(8, 0, 0) FANOUT_10 (wire_3, {wire_3_0, wire_3_1, wire_3_2, wire_3_3, wire_3_4,
. . .
and_n #(2, 0, 0) AND_11 (pc_on_adr, {wire_2_4, wire_4_4});
notg #(0, 0) NOT_6 (wire_11, reset);
and_n #(2, 0, 0) AND_12 (wire_10, {wire_2_5, wire_11_0});
and_n #(2, 0, 0) AND_13 (wire_13, {wire_11_1, wire_4_5});
or_n #(2, 0, 0) OR_4 (rd_mem, {wire_8_1, wire_7_2});
and_n #(4, 0, 0) AND_15 (wr_mem, {op_code_0_4, wire_1_8, wire_3_7, wire_6_3});
dff INS_1 (wire_3, wire_10, clk, 1'b0, 1'b0, 1'b1, NbarT, Si, global_reset);
dff INS_2 (wire_2, wire_12, clk, 1'b0, 1'b0, 1'b1, NbarT, Si, global_reset);
endmodule
```

Fig. F.2 Netlist generated by *NetlistGen*

Index

A

AC, 8, 17–18, 43, 47, 66, 67, 253, 255, 330
Access routine, 56–58
AC test, 8, 17
Activation, 47, 83, 84, 112, 127, 131–133, 135, 178, 180–182, 186, 188, 190–193, 195, 200, 214, 219, 222, 271, 276, 302
Adding BIST components, 342
Adding BIST hardware, 332, 333
Ad Hoc testability, 12
Adjustable expected coverage-per-test, 167–170
Aliasing, 117, 119, 122, 297, 299, 312, 314, 318, 320, 322, 340
Analog converter, 18
AND-bridging faults, 74
Appearance fault, 67, 88, 111, 112, 135
Arbitrary waveform generator, 17
Architecture design, 298
Asynchronous reset, 214, 234, 264, 302
Asynchronous set, 214
ATE. *See* Automatic test equipment
ATE architecture and instrumentation, 17–19
ATPG. *See* Automatic test pattern generation
At-speed testing, 8
Automatic test equipment (ATE), 7, 8, 13–20, 53, 223, 224, 238, 257, 261, 281, 285, 292, 294–299, 325, 329, 345
Automatic test pattern generation (ATPG), 116, 143, 201, 204, 209, 211, 235, 347, 357, 359
AWG. *See* Arbitrary waveform generator

B

Backtrace, 194, 197
Backtracing, 195–197
Backtrack, 187, 196
Bandwidth, 345
Basic CPT implementation, 137–138
Basic PODEM, 191–193
Basic TG procedure, 177–180
Bed-of-nails probing, 261
Behavioral description, 3, 29, 104, 105, 107, 108, 154, 229, 234, 236, 253, 335
Behavioral level, 29, 30
Behavioral level simulation, 15

Behavioral model, 5, 24, 236
Behavioral testbench, 55
BEST. *See* Built-in evaluation and self-test
BF. *See* Bridging fault
Bidirectional pin, 266
Bidirectional signal, 18
Bidirectional signal power, 18
Bidirectional switches, 77, 78
BILBO. *See* Built-in logic block observer
BILBO test process, 329
Binary counter, 300, 301, 303
Binding, 343
BIST. *See* Built-in self-test
BISTed CUT model, 335, 342
Bit line (BL), 258, 377, 378
Black box, 213
Block coverage, 54, 56
Board level, 12, 261277–281, 320–321, 326
Bonding, 14
Boolean difference, 72, 143–145
Boolean equation, 157
Boolean expression, 29, 30, 32, 36
Boolean function, 29, 30, 71, 83, 94, 145
Boundary register, 265, 266, 268, 270–273, 275, 276, 281, 283, 287, 290, 291, 324, 325
Boundary scan
    standard, 261–264, 275, 290, 292
Boundary scan description language (BSDL), 261, 290–292
Branch coverage, 54
Bridging effect, 73, 310
Bridging fault, 73–76, 84–87, 378
Broadcast scan, 357–361, 372
BSDL. *See* Boundary scan description language
Buffer memory, 270
Built-in evaluation and self-test (BEST), 321–322
Built-in logic block observer (BILBO), 320, 328–329
Built-in self-test (BIST)
    architectures, 16, 295–298, 312, 317, 319–330, 343, 383
    controller, 295, 297–300, 312–314, 317–321, 323, 325, 327, 330, 332–335, 337–339, 342, 343, 384
    procedure, 299, 300, 312, 320, 343
    tester, 384
Bus conflict, 262

Printed in the United States
By Bookmasters